# Topics in Current Genetics 22

Series Editor: *Stefan Hohmann*



For further volumes:
http://www.springer.com/series/4735

José Pérez-Martín • Antonio Di Pietro

# Morphogenesis and Pathogenicity in Fungi

Springer

*Editors*
José Pérez-Martín
Centro Nacional de Biotecnologia
CSIC
Department of Microbial
Biotechnology
Campus de Cantoblanco
Madrid
Spain
jperez@cnb.csic.es

Antonio Di Pietro
University of Cordoba
Department of Genetics
Campus Rabanales, Edificio
Gregor Mendel
Cordoba
Spain
ge2dipia@uco.es

ISBN 978-3-642-42834-0 ISBN 978-3-642-22916-9 (eBook)
DOI 10.1007/978-3-642-22916-9
Springer Heidelberg Dordrecht London New York



Printed on acid-free paper

Springer is part of Springer Science+Business Media (www.springer.com)

*Topics in Current Genetics* publishes review articles of wide interest in volumes that center around a specific topic in genetics, genomics as well as cell, molecular and developmental biology. Particular emphasis is placed on the comparison of several model organisms. Volume editors are invited by the series editor for special topics, but further suggestions for volume topics are highly welcomed.

Each volume is edited by one or several acknowledged leaders in the field, who ensure the highest standard of content and presentation. All contributions are peer-reviewed.

All volumes of Topics in Current Genetics are part of the Springer eBook Collection. The collection includes online access to more than 3,500 newly released books, book series volumes and reference works each year. In addition to the traditional print version, this new, state-of-the-art format of book publications gives every book a global readership and a better visibility.

*Editorial office*:
Topics in Current Genetics
Series Editor: Stefan Hohmann
Cell and Molecular Biology
Göteborg University
Box 462
40530 Göteborg, Sweden
FAX: +46 31 7862599
E-mail: editor@topics-current-genetics.se

# Preface

Fungal diseases cause human suffering and enormous economic losses. New approaches for antifungal control are instrumental in meeting the threat imposed by these infectious agents. Such conceptual advances are only possible through the identification of novel biochemical and molecular targets in the fungal cell. The great diversity that exists among pathogenic fungi in terms of lifestyles, infection strategies, and disease symptoms represents a major challenge in the search for common antifungal targets, because it is likely that different attributes will be important for different fungi to cause disease. However, all pathogens share a common need for making the appropriate developmental decisions during induction of the pathogenic program. Such developmental switches are often associated with differentiation processes that require a reset of the cell cycle and the induction of a new morphogenetic program. The ability of pathogenic fungi to modulate their cell cycle and morphogenesis is emerging as a key determinant for successful infection. In the past, experimental approaches to understand regulation of the pathogenic developmental program have focused on the study of signal transduction pathways and transcriptional changes. Today, the study of how genetic virulence programs control morphogenesis and cell cycle offers exciting opportunities to explore the molecular basis of fungal pathogenicity under new angles, which at the same time are complementary to previous approaches in the field.

In the last years there has been a tremendous increase in the amount of research and relevant data on the relationships between morphogenesis, cell cycle, and regulation of virulence programs in pathogenic fungi. This progress has been the subject of a number of excellent reviews over the years. So far, however, the field has lacked a monographic collection where many of these important contributions and views are joined together. In this book we attempted to cover a broad spectrum of taxonomically and biologically diverse fungal pathogens, as well as a number of key topics related to morphogenesis and pathogenesis. The double aim was to provide introductory material for the nonspecialist and also to offer the most recent and updated views on these crucial cell biological processes.

The book is organized in 13 chapters written by qualified researchers on each topic. Each chapter contains an introduction followed by an in-depth and up-to-date analysis of the current state of knowledge on the subject. The chapters were conceived and written autonomously, so that they can be read independently, even though this may have resulted in some repetitiveness of basic concepts.

Finally, we want to thank each and every author for his/her excellent contribution. We are also indebted to the anonymous reviewers who read and made important suggestions to improve the chapters. We also thank Stefan Hohmann for his constant encouragement during the editing process and the staff at Springer Verlag for their continuous help and support to make this book possible.

Madrid/Córdoba
José Pérez-Martín and Antonio Di Pietro

# Contents

# Contributors

**Andrew J. Alspaugh** Departments of Medicine and Molecular Genetics/Microbiology, Duke University School of Medicine, DUMC 3355, 1543 Duke Hospital, South Durham, NC 27710, USA, andrew.alspaugh@duke.edu

**Elizabeth Ballou** Departments of Medicine and Molecular Genetics/Microbiology, Duke University School of Medicine, DUMC 3355, 1543 Duke Hospital, South Durham, NC 27710, USA, elizabeth.ballou@duke.edu

**Martine Bassilana** Centre National de la Recherche Scientifique and Université de Nice-Sophia Antipolis, Institute of Developmental Biology and Cancer, CNRS-UMR6543 Faculté des Sciences, Parc Valrose, 06108 Nice, France, mbassila@unice.fr

**Matthias Becker** Institute of Molecular Biosciences, Massey University, Private Bag 11 222, Palmerston North 4442, New Zealand, M.Becker@massey.ac.nz

**Yvonne Becker** Institute of Molecular Biosciences, Massey University, Private Bag 11 222, Palmerston North 4442, New Zealand, Y.E.Rolke@massey.ac.nz

**Alexandra Brand** Institute of Medical Sciences, University of Aberdeen, Aberdeen AB25 2ZD, UK, a.brand@abdn.ac.uk

**Gemma Cartwright** Institute of Molecular Biosciences, Massey University, Private Bag 11 222, Palmerston North 4442, New Zealand, G.M.Cartwright@massey.ac.nz

**Jaime Correa-Bordes** Departamento de Microbiología, Facultad de Ciencias, Universidad de Extremadura, Avda Elvas SN, 06071, Badajoz, Spain, jcorrea@unex.es

**Maurizio Del Poeta** Department of Biochemistry and Molecular Biology, Medical University of South Carolina, 173 Ashley Avenue, BSB 512A, Charleston, SC29425, USA, delpoeta@musc.edu

**Antonio Di Pietro** Department of Genetics, University of Cordoba, Campus Rabanales, Edificio Gregor Mendel, 14071, Cordoba, Spain, ge2dipia@uco.es

**André Fleissner** Institut für Genetik, Technische Universität Braunschweig, 38106, Braunschweig, Germany, A.Fleissner@tu-braunschweig.de

**Peter Follette** Faculté des Sciences, Université de Nice-Sophia Antipolis, Institute of Developmental Biology and Cancer, Parc Valrose, 06108, Nice, France, Peter.Follette@unice.fr

**Gustavo Henrique Goldman** Centro de Ciência e Tecnologia do Bioetanol and Faculdade de Ciências Farmacêuticas, Ribeirão Preto Universidade, São Paulo, Ribeirão Preto, 14040-903, Brazil, ggoldman@usp.br

**Neil A.R. Gow** School of Medical Sciences, Institute of Medical Sciences, University of Aberdeen, Foresterhill, Aberdeen, Scotland, UK, n.gow@abdn.ac.uk

**Steve D. Harris** Department of Plant Pathology, Center for Plant Science Innovation, University of Nebraska, Lincoln, NE, USA, sharri1@unlnotes.unl.edu

**Yasuyuki Kubo** Laboratory of Plant Pathology, Graduate School of Life and Environmental Sciences, Kyoto Prefectural University, Kyoto, 606-8522, Japan, y_kubo@kpu.ac.jp

**Iran Malavazi** Departamento de Genética e Evolução, Centro de Ciências Biológicas e da Saúde, Universidade Federal de São Carlos, São Carlos, Brazil, imalavazi@ufscar.br

**Connie B. Nichols** Departments of Medicine and Molecular Genetics/Microbiology, Duke University Medical Center, DUMC 3355, 1543 Duke Hospital South, Durham, 27710, NCUSA, connie.nichols@duke.edu

**José Pérez-Martín** Department of Microbial Biotechnology, Centro Nacional de Biotecnologia CSIC, Campus de Cantoblanco-UAM, 28049, Madrid, Spain, jperez@cnb.csic.es

**Elena Pérez-Nadales** Department of Genetics, University of Cordoba, Campus Rabanales, Edificio Gregor Mendel, 14071, Cordoba, Spain, ge2penae@uco.es

**Asfia Qureshi** Department of Biochemistry and Molecular Biology, Medical University of South Carolina, Charleston, SC29425, USA, qureshi@musc.edu

**Julia Schumacher** Institut für Biologie und Biotechnologie der Pflanzen, Westfälische Wilhelms Universität Münster, Schloßgarten 3, 48149, Münster, Germany, jschumac@uni-muenster.de

**Barry Scott** Institute of Molecular Biosciences, Massey University, Private Bag 11 222, Palmerston North 4442, New Zealand, d.b.scott@massey.ac.nz

**Paul Tudzynski** Institut für Biologie und Biotechnologie der Pflanzen, Westfälische Wilhelms Universität Münster, Schloßgarten 3, 48149, Münster, Germany, tudzyns@uni-muenster.de

**Carlos R. Vázquez de Aldana** Instituto de Biología Funcional y Genómica, CSIC-Universidad de Salamanca, Salamanca, Spain, cvazquez@usal.es

# Chapter 1
# Molecular Basis of Morphogenesis in Fungi

**Steven D. Harris**

**Abstract** In fungi, cellular morphogenesis is driven by localized membrane expansion and cell wall deposition. Variation in the geometry of fungal cells likely arises through the precise temporal and spatial regulation of these processes. Nevertheless, these modes of regulation are not well understood in filamentous fungi. This review focuses on three key aspects of fungal cellular morphogenesis: symmetry breaking, polarity maintenance, and septum formation. The mechanisms underlying cellular morphogenesis are summarized, with an emphasis on comparison to the model yeasts. In addition, mechanisms that coordinate morphogenesis with the yeast cell division cycle are briefly outlined. It is proposed that to some extent, analogous mechanisms function during fungal development to alter cell shape and size.

## Abbreviations

| | |
|---|---|
| CDK | Cyclin-dependent kinase |
| GAP | GTPase-activating protein |
| GEF | Guanine nucleotide exchange factor |
| MEN | Mitotic exit network |
| NETO | New end take-off |
| Nox | NADPH oxidase |
| PAK | p21-associated kinase |
| ROS | Reactive oxygen species |
| SIN | Septation initiation network |
| SPB | Spindle pole body |

S.D. Harris (✉)
Department of Plant Pathology, Center for Plant Science Innovation, University of Nebraska, Lincoln, NE, USA
e-mail: sharri1@unlnotes.unl.edu

J. Pérez-Martín and A. Di Pietro (eds.), *Morphogenesis and Pathogenicity in Fungi*,
Topics in Current Genetics 22, DOI 10.1007/978-3-642-22916-9_1,

| | |
|---|---|
| SRD | Sterol-rich domain |
| v-SNARE | Vesicle-soluble, NSF attachment receptor |
| Ts | Temperature sensitive |

## 1.1 Introduction

Fungal cells exhibit a diverse array of shapes and sizes. The basic unit of fungal growth is the hypha, in which membrane expansion and cell wall deposition are solely confined to the tip, resulting in the formation of an elongated tubular structure (Fig. 1.1). Results from elegant modeling experiments demonstrate that regulating the extent to which growth is directed to the tip can generate different cell shapes, such as the ovoid shape characteristic of budding yeast (Bartnicki-Garcia et al. 1989). Although the inventory of gene products involved in hyphal morphogenesis continues to expand (Harris 2010), the regulatory mechanisms that

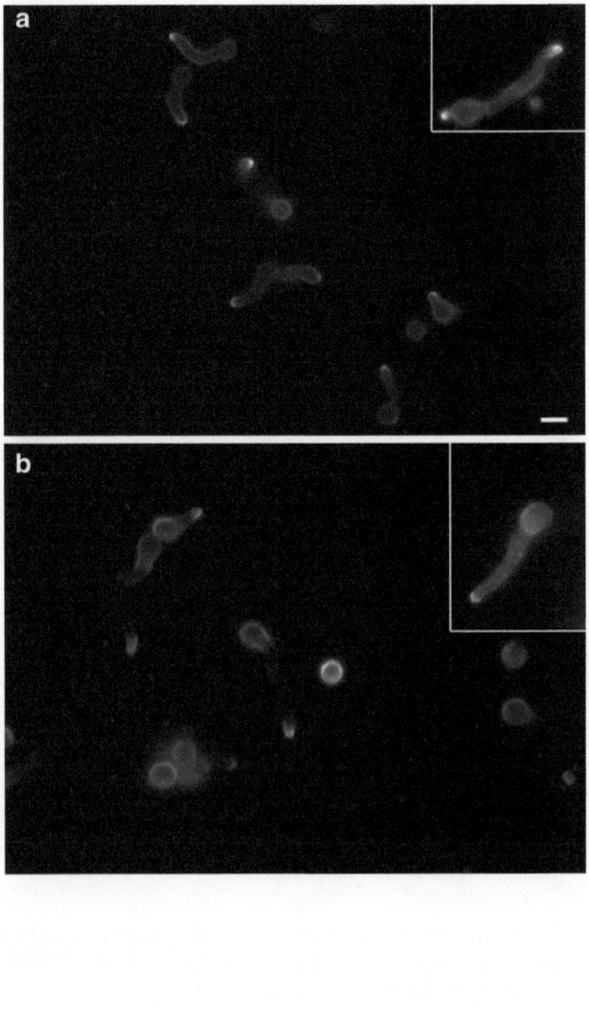


**Fig. 1.1** Cell surface expansion and cell wall deposition are localized to the hyphal tip. *A. nidulans* hyphae (strain FGSC28) were imaged after 8–10 h of growth in rich media. (**a**) Hyphae were stained with filipin to detect membrane sterols. (**b**) Hyphae were stained with FITC-conjugated wheat germ agglutinin or FITC-conjugated concanavalin A (inset) to detect sites of active cell wall deposition

determine where and when morphogenesis occurs remain poorly characterized. However, the need to understand these mechanisms has never been greater. Many human and plant pathogenic fungi adopt different cell shapes at different stages in their life cycles, and it has long been evident that the ability to do so determines the degree to which they cause disease (e.g., Lo et al. 1997; Hamer et al. 1989). It seems likely that novel control strategies will emerge when the mechanisms that regulate cellular morphogenesis are better understood.

The objective of this review is twofold. First, the gene products and mechanisms that have been implicated in hyphal morphogenesis will be summarized. This summary will focus on three key stages: the establishment of hyphal polarity (or symmetry breaking), the maintenance of a stable polarity axis, and septum formation. Comparative genomics approaches will be used to highlight important similarities, and differences, in the way that these stages are executed across the fungal kingdom. Second, potential strategies for the coordination of hyphal morphogenesis with the cell division cycle will be outlined. Because only very little is known about this in filamentous fungi, well-established examples of this coordination in the model yeasts *Saccharomyces cerevisiae* and *Schizosaccharomyces pombe* will be presented as useful guides.

## 1.2 Breaking Symmetry: Establishment of a Polarity Axis

When undergoing vegetative growth, filamentous fungi typically establish a polarity axis in two different contexts: during the germination of a spore, or upon formation of a branch from a preexisting hypha. In both cases, a more-or-less symmetrically shaped cell (i.e., a spore or an intercalary hyphal compartment) is altered such that all growth is directed to a specific cortical site, leading to the emergence of a new hypha. Three general processes are required to break symmetry and establish a new polarity axis (Nelson 2003). First, a polarity axis has to be specified. Second, the resulting positional information has to be transmitted to the morphogenetic machinery (i.e., the cytoskeletal and vesicle trafficking functions that directly contribute to the building of a new membrane and cell wall). Finally, the morphogenetic machinery must respond by aligning itself along the polarity axis.

### *1.2.1 Axis Specification*

In principle, the specification of a polarity axis entails marking a cortical site such that it is distinct from all other sites on the cell surface. How this is accomplished in filamentous fungi has largely been a mystery. On the contrary, the mechanisms that generate positional information in the model yeasts *S. cerevisiae* and *S. pombe* are reasonably well characterized (Chang and Peter 2003). Comparative genomic approaches have revealed that these mechanisms are conserved to some degree in filamentous fungi, and their analysis has revealed that they might have important roles in axis specification (Harris and Momany 2004; Fischer et al. 2008). In remains entirely possible, however, that polarity axes may arise via spontaneous mechanisms that do not rely upon a preset marker.

In *S. cerevisiae*, a new bud is generated with each passage through the cell division cycle. The site of bud emergence is specified by a group of proteins that comprise the bud-site selection module (Chant 1999; Park and Bi 2007). There are two distinct budding patterns, each specified by its own module. Key elements of the module that specifies the axial budding pattern include Axl1, Axl2, Bud3, and Bud4, whereas key elements of the module that specifies the bipolar budding pattern include Bud8, Bud9, Rax1, and Rax2. Annotation of multiple fungal genomes has revealed that the axial module is conserved (Harris and Momany 2004), although functional studies in *Neurospora crassa* and *Aspergillus nidulans* suggest that it regulates the formation of actin rings at septation sites rather than polarity establishment (Justa-Schuch et al. 2010; Si et al. 2010). The bipolar module is less well conserved (i.e., Bud8 and Bud9 homologs are found only within the Saccharomycotina), and the only characterized component, Rax1 (=RgaB in *A. nidulans*), regulates sexual development (Han et al. 2004). Accordingly, it seems likely that homologs of the yeast bud-site selection markers do not specify axes for the establishment of hyphal polarity in filamentous fungi.

Following cytokinesis, *S. pombe* cells initiate polarized growth from the old cell ends. Once they pass a specific point early in the subsequent cell cycle known as "new end take-off," or NETO, they also initiate growth from the new cell end (i.e., generated by the prior septation event; Martin and Chang 2005). This morphogenetic pattern is enforced by a regulatory system that employs microtubules to deliver marker proteins to cell ends, where they coordinate the organization of the actin cytoskeleton (Martin 2009). Key components of this regulatory system include (1) Tea1, the microtubule plus end-associated marker protein that is delivered to cell ends; (2) Mod5, a membrane-anchored protein located at cell ends that serves as a "receptor" for Tea1; and (3) Tea2, a kinesin motor protein that transports Tea1 along microtubules to the plus end. Unlike the bud-site selection module of *S. cerevisiae*, this regulatory system does appear to be functionally conserved in filamentous fungi. In particular, *A. nidulans* possess homologs of Tea1 (TeaA), Mod5 (TeaR), and Tea2 (KipA) that act in an analogous manner to position the Spitzenkorper at hyphal tips (Konzack et al. 2005; Takeshita et al. 2008), thereby ensuring directed hyphal growth. It is not yet clear whether this system is also conserved in other filamentous fungi, such as *N. crassa*.

Although the specification of a polarity axis has generally been thought to require some sort of cortical marker, results from studies using *S. cerevisiae* suggest that this is not necessarily so. In the absence of all known cortical markers (i.e., bud-site selection landmarks, and occupied pheromone receptors), yeast cells are still able to specify a polarity axis via a process referred to as spontaneous polarization (Slaughter et al. 2009). Among the important requirements for this process are two positive feedback loops. The first couples actin-mediated vesicle exocytosis with endocytosis to concentrate the monomeric GTPase Cdc42 (the key determinant of cell polarization in yeast; see below) on the plasma membrane. This loop makes specific assumptions about the association of Cdc42 with vesicles, and its existence remains controversial (Layton et al. 2011). The second feedback loop incorporates reaction-diffusion mechanisms that locally concentrate Cdc42 on the plasma membrane independent of the cytoskeleton (Kozubowski et al. 2008). It is entirely conceivable that a similar spontaneous polarization process could underlie axis specification during spore germination and/or hyphal branching in filamentous fungi. To date, there is no evidence that either of these symmetry-breaking events follows a predictable pattern akin to axial or bipolar budding in yeast (note that the emergence of the *second* germ tube from germinating spores does appear to follow a bipolar pattern; Harris 1999), which suggests that cortical markers may not exist. Moreover, the mechanisms that drive both positive feedback loops in yeast (i.e., localized exocytosis coupled to global endocytosis, and reaction-diffusion mechanisms) have each been implicated in the regulation of polarized hyphal growth (Regalado et al. 1996; Taheri-Talesh et al. 2008). Finally, the use of cortical markers is likely a necessity in yeast cells, which must reestablish a polarity axis during each cell cycle. Because polarity axes are sustained in hyphal cells, there presumably would be little need to "remember" the locations of polarization sites.

### *1.2.2 Transmission of Positional Information*

Whether generated by cortical markers or through spontaneous mechanisms, positional information must be relayed to the morphogenetic machinery for symmetry breaking to occur. In both *S. cerevisiae* and *S. pombe*, the Cdc42 GTPase module serves as the critical link that transmits positional information to the morphogenetic machinery (Park and Bi 2007; Perez and Rincon 2010). For example, during budding in *S. cerevisiae* (Park and Bi 2007), the presence of cortical markers triggers local activation of the Ras-like GTPase Rsr1/Bud1, which in turn activates Cdc42 via its guanine nucleotide exchange factor (GEF) Cdc24. Active GTP-bound Cdc42 subsequently functions through several effectors to organize the morphogenetic machinery. Among these effectors are the p21-associated kinases (PAKs) Ste20 and Cla4, the formin Bni1, Gic1/Gic2, and components of the exocyst. The collective activity of these effectors results in the local assembly of actin cables and septin structures that direct vesicle exocytosis to the incipient bud site.

In filamentous fungi, the mechanisms that transmit positional information to the morphogenetic machinery during hyphal growth are not well understood. Unlike *S. cerevisiae* or *S. pombe*, many filamentous fungi possess homologs of Cdc42 and the closely related GTPase Rac1 (Harris 2011). Although there appears to be considerable variation in how different fungi deploy these GTPases to regulate morphogenesis, an emerging theme is that at least one of them must be active to enable symmetry breaking. Furthermore, specific Cdc42 effectors appear to be functionally conserved (e.g., PAKs and formins) in filamentous fungi, whereas others are absent (e.g., Gic1/Gic2), and still others represent novel functions not found in yeast (e.g., NADPH oxidases) (Harris et al. 2009). Nevertheless, it remains unclear whether Cdc42 and/or Rac1 transmits positional information to the morphogenetic machinery during hyphal growth as per the yeast paradigm. This lack of understanding can be attributed in part to the fact that the regulators of Cdc42 and Rac1 [i.e., the GEFs and GTPase-activating proteins (GAPs)] are not as well characterized in filamentous fungi as in yeast. More importantly, as long as the source of positional information (if any) is unknown, it will be difficult to determine how Cdc42 and/or Rac1 might relay this information to the morphogenetic machinery.

### 1.2.3 Response to Positional Signals

In response to positional information, the morphogenetic machinery is locally reorganized such that cell surface expansion and cell wall deposition are directed to the specified site. The general cytological features of one such site, the hyphal tip, are depicted in Fig. 1.2 (Steinberg 2007; Taheri-Talesh et al. 2008; Harris 2010). Cytoplasmic microtubules mediate long-range vesicle transport to and from the tip region. Actin filaments provide "tracks" for localized myosin-mediated delivery of exocytic vesicles to sites of cell wall deposition at the hyphal tip. The assembly of these filaments is likely regulated in part by components of the polarisome. Actin patches, which associate with and drive internalization of endocytic vesicles, accumulate in a subapical "belt." Septin-based structures also localize just behind the tip (e.g., Lindsey et al. 2010), where among multiple functions they likely serve as diffusion barriers that help to maintain the polarity axis. Finally, the exocyst localizes to a discrete site at the immediate apex of the hyphal tip, where it facilitates docking of incoming exocytic vesicles. In general, these localization

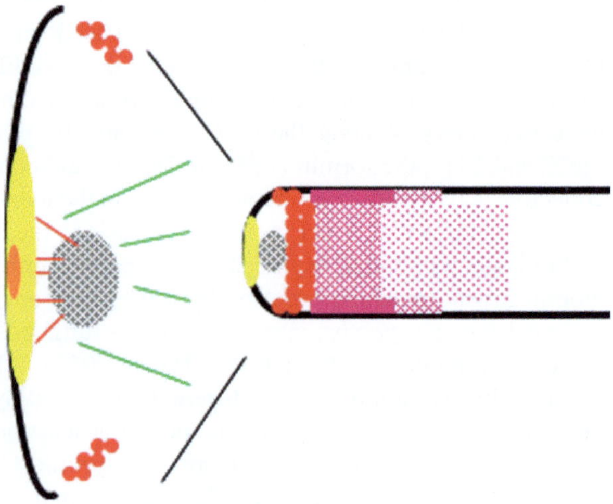

**Fig. 1.2** Schematic representation of the morphogenetic machinery at the hyphal tip. An expanded view is shown to the left. See text for details. *Red* = the actin cytoskeleton (*lines* = filaments; *spots* = patches), *green* = microtubules, *purple* = the septins, *stippled black* = the Spitzenkorper, *yellow* = the polarisome, and *orange* = the exocyst

patterns resemble those observed at bud sites, cell ends, and the tips of mating projections in yeast cells. Thus, it is perhaps not too surprising that the composition of the morphogenetic machinery is extensively conserved when filamentous fungi are compared to *S. cerevisiae* and *S. pombe*.

Despite the similarities with yeast, there are distinctive features that characterize the morphogenetic machinery of filamentous fungi. The most notable is the Spitzenkorper [i.e., "apical body"; a dynamic cluster of vesicles that typically resides in the subapical region just behind the hyphal tip (Fig. 1.2; Girbardt 1957; Harris et al. 2005; Steinberg 2007)]. The Spitzenkorper is thought to serve as a transit station that receives exocytic vesicles transported from the hyphal interior along microtubules, and dispenses these vesicles onward to the hyphal apex along actin filaments. Classical modeling experiments show that it is possible to generate a hypha-like cell shape simply by fixing the position of the Spitzenkorper behind the extending hyphal tip (Bartnicki-Garcia et al. 1989, 1995), which highlights the importance of the Spitzenkorper for hyphal growth. Consistent with this view, the Spitzenkorper appears to be a ubiquitous feature of fungal hyphae (Lopez-Franco and Bracker 1996) [e.g., a recent report describes a similar structure at the tips of *S. cerevisiae* mating projections (Chapa-y-Lazo et al. 2011)]. Nevertheless, a study of Spitzenkorper ontogeny in *N. crassa* found that spore germination and early stages of germling growth occur in the absence of a detectable Spitzenkorper (Araujo-Palomares et al. 2007). During this time, growth is erratic and occurs only in a fixed direction once a Spitzenkorper forms (i.e., in germling >150 mm in length). According to these results, the Spitzenkorper might be dispensable for axis specification, but is required for the subsequent stabilization of the axis.

## 1.3 Polarity Maintenance: Stabilization of Polarity Axes

Sustained polar growth is arguably the distinguishing feature of filamentous fungi. Once polarity is established, it is maintained for the entire lifetime of a hypha, typically being lost only in response to environmental insults or developmental signals. Increasing amounts of cytoplasmic volume and nuclei are accommodated by the generation of secondary polarity axes, which result in the formation of hyphal branches. In contrast to this growth pattern, yeast cells typically undergo repeated cycles of polarized and isotropic growth as they divide to produce new daughter cells. Accordingly, there is considerable interest in understanding the mechanisms that contribute to the maintenance of hyphal polarity, with a view to the idea that this might help to define what is unique about hyphae relative to other cell types. An essential feature of polarity maintenance is the ability to stabilize a polarity axis such that the flow of exocytic vesicles remains targeted to the hyphal tip. As noted above, the role of the Spitzenkorper in the control of vesicle trafficking to the tip is well established, and more recent studies also highlight the importance of endocytosis and plasma membrane organization in maintaining hyphal polarity. Another critical aspect of polarity maintenance is to ensure that no additional spurious polarity axes form in the vicinity of the hyphal tip (i.e., apical dominance). Localized production of reactive oxygen species (ROS) has recently been implicated in this function.

### 1.3.1 Endocytosis

In *S. cerevisiae*, endocytosis plays a critical role in the recycling of cell surface proteins and membranes (Toret and Drubin 2006). Well-characterized endocytic cargo proteins include mating pheromone receptors and vesicle-soluble NSF attachment protein receptors (v-SNAREs) (Kubler and Riezman 1993; Valdez-Taubus and Pelham 2003). For both these examples, endocytosis is required to maintain a polarized distribution of these proteins on the cell surface. In other cases (e.g., Cdc42), the role of endocytosis in concentrating proteins on the cell surface remains controversial. A recent study suggests that integral membrane proteins that diffuse slowly and are actively recruited into both exocytic and endocytic vesicles are the most likely candidates to be effectively polarized by endocytosis (Layton et al. 2011).

The mechanisms underlying endocytosis in *S. cerevisiae* have been characterized in considerable detail (Toret and Drubin 2006). A key step in the process is the assembly of actin patches that drive the internalization of endocytic vesicles. In both budding and fission yeast, actin patches are highly dynamic and typically accumulate in areas of active cell wall deposition (Doyle and Botstein 1996; Pelham and Chang 2001). Localization of actin and known patch components such as fimbrin has revealed similar dynamic behavior in the filamentous fungi *A. nidulans* and *N. crassa* (Araujo-Bazan et al. 2008; Taheri-Talesh et al. 2008; Upadhyay and Shaw 2008; Delgado-Alvarez et al. 2010). In actively growing hyphae, patches accumulate near the hyphal tip, but are depleted from the immediate apex of the hypha. This zone of depletion contains components of the morphogenetic machinery involved in vesicle exocytosis, including actin filaments, the exocyst, and v-SNAREs. Furthermore, characterization of fimbrin mutants shows that although they appear capable of specifying a polarity axis and initiating polar growth, they subsequently fail to maintain polarity (Upadhyay and Shaw 2008). Because these mutants are also defective for endocytosis, these observations provide compelling evidence that endocytosis is essential for the maintenance of hyphal polarity.

Results from the analysis of endocytosis in filamentous fungi support a model in which cell surface expansion and cell wall deposition occur at the immediate hyphal apex, while components involved in these processes are recycled by endocytosis in the subapical regions that flank the apex (Taheri-Talesh et al. 2008; Upadhyay and Shaw 2008; Steinberg 2007; Fig. 1.2). An attractive feature of this model is that it provides a potential mechanism by which one or more cortical markers could be concentrated at the apex, though these markers would presumably need to be integral membrane proteins that exhibit the characteristics noted above for yeast proteins likely to be polarized by endocytosis.

### 1.3.2 Membrane Organization

It is increasingly apparent that the plasma membranes of yeast and polarized animal cells are not homogeneous, but consists of a dynamic patchwork of domains that differ in lipid composition (Malinsky et al. 2010; Simons and Gerl 2010). These domains are referred to as lipid rafts or sterol-rich domains (SRDs), and are typically enriched for sterols and sphingolipids when compared to the rest of the plasma membrane. In *S. cerevisiae*, specific plasma membrane proteins preferentially localize to SRDs (Malinska et al. 2003; Malinsky et al. 2010), which distinguishes them from the general population of plasma membrane proteins. Critical functions attributed to SRDs include signal transduction and membrane trafficking (Simons and Gerl 2010). The role of SRDs in maintaining the polarity of budding yeast cells, particularly during the formation of mating projections, remains controversial (Valdez-Taubus and Pelham 2003). However, in polarized neurons, there is considerable support for the idea that SRDs play a pivotal role in axon growth and guidance (Guirland et al. 2004). In particular, they are thought to serve as platforms for signaling proteins that mediate responses to spatial and temporal cues that direct polarized growth.

There is growing evidence that highlights the importance of SRDs in the maintenance of hyphal polarity. In several filamentous fungi (e.g., *A. nidulans*, *Fusarium graminearum*, *Candida albicans*, and *Ustilago maydis*), the use of filipin to localize sterols has revealed that SRDs concentrate at hyphal tips (Martin and Konopka 2004; Pearson et al. 2004; Rittenour and Harris 2008; Canovas and Perez-Martin 2009). Functional studies show that disruption of SRD formation leads to loss of hyphal polarity. For example, perturbation of sphingolipid synthesis in *U. maydis* (i.e., using compound aureobasidin A) dramatically affects localization of the morphogenetic machinery at hyphal tips (Canovas and Perez-Martin 2009). Furthermore, in *A. nidulans*, there is growing evidence for the existence of two distinct ceramide pools that regulate polarized growth (Li et al. 2006). Notably, one pool appears to consist of glucosylceramides that are synthesized by a novel ceramide synthase unique to filamentous fungi (Rittenour et al. 2011). This result suggests the existence at hyphal tips of multiple SRDs with distinct sphingolipid and/or sterol compositions. Conceivably, each SRD could represent a spatially segregated signaling platform that contributes to polarity maintenance by responding to a specific set of internal or external cues. In such a scenario, the septins might play a key role as membrane barriers that demarcate distinct domains.

### 1.3.3 ROS and Apical Dominance

In polarized plant cells, localized accumulation of ROS serves as a marker for active growth sites. For example, in *Arabidopsis* root hairs, NADPH oxidase (Nox)-mediated ROS accumulation triggers localized calcium influx to establish a positive feedback loop that maintains polarized growth (Takeda et al. 2008). Nox-mediated ROS also regulates polarized growth in filamentous fungi (Takemoto et al. 2007), though what part calcium might play in its effects, if any, remains a mystery. In fungi such as *A. nidulans*, *Magnaporthe oryzae*, and the endophyte *Epichloe festucae*, the accumulation of ROS at hyphal tips appears to be important for enforcing apical dominance (Takemoto et al. 2006; Egan et al. 2007; Semighini and Harris 2008). For example, the failure to accumulate ROS in *A. nidulans* leads to an increase in abnormal apical branching. Additional studies are needed to further understand how the effects of ROS on polarity maintenance are mediated. Besides calcium, other potential targets of ROS could include actin, which is known to be very sensitive to cellular redox state (e.g., Farah and Amberg 2007).

In animal cells, Nox associates with multiple regulatory factors that tightly control its activity (Lambeth 2004). These include the GTPase Rac1, as well as the regulators $p67^{phox}$, $p47^{phox}$, and $p40^{phox}$. The conservation of Rac1 and $p67^{phox}$ (=NoxR) function in Nox-mediated ROS generation in filamentous fungi is well established (Cano-Dominguez et al. 2008; Semighini and Harris 2008; Tanaka et al. 2008). The recent identification of Bem1 and Cdc24 as fungal analogs of $p47^{phox}$ and $p40^{phox}$ further completes the picture of how Nox activity is controlled in fungi (Takemoto et al. 2011). However, it also suggests an intriguing link between Nox and well-known functions involved in the establishment and maintenance of cellular polarity. For example, by simultaneously regulating Nox and Cdc42, Bem1 and Cdc24 could conceivably spatially couple localized organization of the morphogenetic machinery with the production of ROS.

## 1.4 Septum Formation

Septa are cross-walls that partition hyphae into individual cells or compartments. Septum formation is not a strict requirement for polarized hyphal growth, as lower fungi and various mutant forms of higher fungi (e.g., *A. nidulans sep* mutants) are quite capable of rapid hyphal extension in the absence of septa. Nevertheless, for the filamentous fungi that do form septa, the process by which it does so provides some insight into the temporal and spatial coordination of morphogenetic events in hyphae. One example is the temporal coordination of septation with nuclear division. In hyphae, septum formation follows mitosis and is presumably dependent upon nuclear signals that are relayed by a conserved protein kinase cascade known as the septation initiation network (SIN). A second example is the spatial regulation of septum formation. At least in the Ascomycotina, hyphal compartments are typically multinucleate and exhibit a narrow size distribution. This implies that each mitotic event cannot be followed by the formation of a septum (i.e., otherwise compartments would all possess a single nucleus) and that there must be some way to measure the distance between adjacent septa. Finally, septation is driven by the assembly of an actin ring whose constriction is coincident with deposition of the cell wall material that forms the septum. Accordingly, it seems likely that constriction of the ring must be coordinated in a manner that might bear some similarity to events at the hyphal tip.

### 1.4.1 Temporal Coordination of Septum Formation

Past studies using *A. nidulans* firmly established that septum formation requires persistent signals that emanate from mitotic nuclei (Momany and Hamer 1997). By analogy to animal cells, it seems likely that these signals originate from the spindle itself or from its associated spindle pole bodies (SPBs). Although the nature and source of these signals still remain unknown, they are most likely transmitted to septation sites via the conserved SIN pathway. As characterized in *S. pombe* and *S. cerevisiae*, where it is known as the mitotic exit network (MEN), the SIN is essentially a GTPase-activated cascade of protein kinases (Simanis 2003). In *S. pombe*, activation of the SIN coordinates constriction of the actin ring with the completion of mitosis and septum assembly (Krapp and Simanis 2008). Upon destruction of the mitotic cyclin-dependent kinase (CDK) Cdc2–Cdc13, components of the SIN associate with the newly formed SPB before subsequent relocalization to the division site. In particular, the relocalization of the protein kinase Sid2 (along with its cofactor Mob1) to the division site is thought to serve as the trigger for septum formation. Besides CDK destruction, activation of the SIN also depends on the polo kinase Plo1. At this time, the mechanisms by which the SIN regulates actin ring dynamics and septum assembly in *S. pombe* are not fully understood. However, key effectors include the Cdc14-like phosphatase Clp1 and the F-BAR protein Cdc15 (Chen et al. 2008; Hachet and Simanis 2008), both of which are intimately involved in assembly and constriction of the ring.

Annotation of the *A. nidulans* genome sequence revealed that whereas the core GTPase and protein kinase components of the SIN are conserved in filamentous fungi, novel (or weakly conserved) scaffold proteins appear to mediate association of the SIN with SPBs (Kim et al. 2009). Characterization of the SIN in *A. nidulans* has uncovered other notable differences compared to that in *S. pombe*. First, the SIN is required for assembly of the actin ring in *A. nidulans*, not just for its constriction (Bruno et al. 2001). Second, localization of SIN components to SPBs is not required for SIN function (Kim et al. 2009). Third, SIN components appear to remain associated with the actin ring as it constricts (Kim et al. 2009). Finally, the *A. nidulans* homolog of polo kinase (PloA) has no detectable role in septation (Bachewich et al. 2005), which implies that it does not activate the SIN as in *S. pombe*. Collectively, these results suggest that the SIN plays a critical role in the coordination of actin ring assembly with nuclear division in filamentous fungi. As in *S. pombe*, the relevant SIN targets remain unknown. However, recent studies in *N. crassa* and *A. nidulans* support a model whereby the SIN acts via the Rho4 GTPase module (particularly the GEF Bud3) to recruit the formin SepA/Bni1 locally to initiate the assembly of actin rings at septation sites (Justa-Schuch et al. 2010; Si et al. 2010; Fig. 1.3).

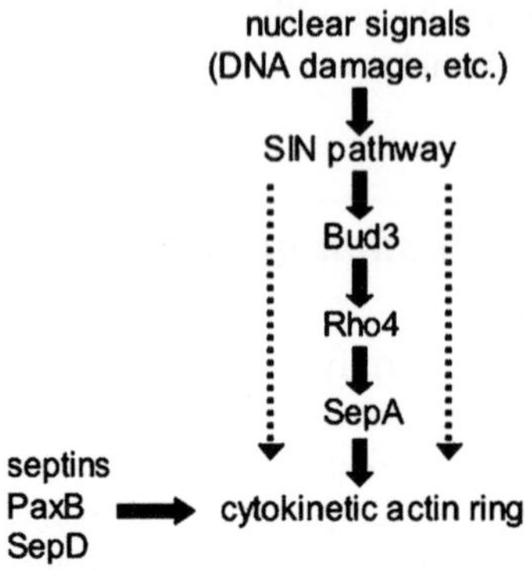


**Fig. 1.3** A model signaling pathway for the temporal regulation of septum formation (largely based on results from experiments using *A. nidulans*). The SIN pathway responds to nuclear signals to promote activation of the GTPase Rho4 via its GEF Bud3. Activated Rho4 in turn recruits the formin SepA to trigger the formation of the actin ring. *Dashed arrows* indicate that the SIN likely acts through multiple mechanisms to regulate ring formation. Several additional proteins (septins, paxilllin, and SepD) act at later stages to control constriction and stability of the ring

## *1.4.2 Spatial Coordination of Septum Formation*

The relatively consistent spacing of septa within a typical fungal hypha implies the existence of a mechanism that spatially regulates septum formation. Past studies with *A. nidulans* demonstrated that the primary determinants of septal position are nuclei (Wolkow et al. 1996). In particular, mutants defective in nuclear distribution were exploited to show that septa only formed in and around clusters of nuclei, whereas they were absent from hyphal regions that were devoid of nuclei. Nevertheless, there are two examples where the presence of nuclei is not sufficient to trigger septum formation. First, despite the presence of actively dividing nuclei, septa do not form in the vicinity of the hyphal tip (i.e., usually in the proximal half of the hyphal tip cell). It seems reasonable to speculate that the mechanisms that enforce apical dominance, such as localized ROS accumulation, might also prevent septation near the tip. Second, septa are not formed adjacent to each other. That is, if a given mitosis triggers septum formation, it is extremely unlikely that immediately adjacent mitotic events will also do so. A plausible explanation for this is that newly forming septa generate a repressive signal that blocks the formation of additional septa at nearby sites. In support of this notion, the observation that forced expression of the developmental regulator BrlA in growing *A. nidulans* hyphae results in compartments that generally possess a single nucleus (i.e., all mitotic events were presumably followed by septum formation) (S. Harris, unpublished observations) suggests that many cortical sites that are competent for septation are not used. A candidate target for the proposed repressive signal could be the SIN, which is subject to negative regulation in *S. pombe* (e.g., Johnson and Gould 2011). On the contrary, the nature of the potential signals itself remains a complete mystery.

### 1.4.3 *Coordination of Actin Ring Constriction with Cell Wall Deposition*

The process of septation in fungal hyphae can be broken down into the following discrete steps (Sharpless and Harris 2002; Justa-Schuch et al. 2010; A. Virag and S. Harris, manuscript in preparation; Fig. 1.4): (1) an actin ring is assembled at the septation site, (2) a thin septum appears while the actin ring still spans the width of the hypha, (3) the septum becomes thicker as the actin ring constricts, and (4) as deposition of the septum is completed, a small ring remains that likely encircles the septal pore. Studies using *S. pombe* suggest that actin ring constriction is coordinated with deposition of the septum (Liu et al. 1999) and that the ring itself might serve as a positional marker that guides the delivery of cell wall material (Liu et al. 2002; Viestica et al. 2008). A similar relationship might also hold true in filamentous fungi. For example, *Ashbya gossypii bud3* mutants form improperly oriented actin rings that lead to abnormal patterns of cell wall deposition at septation sites (Wendland 2003). Additional results from the characterization of *A. nidulans* septation mutants provide some initial insight into the potential coordination of ring constriction with septum deposition. In particular, both the temperature-sensitive *sepD5* mutation (the identity of the affected gene remains unknown) and the Δ*paxB* deletion (PaxB is the *A. nidulans* homolog of the *S. pombe* Pxl1 paxillin; Ge and Balasubramanian 2008) cause a severe delay at step (2) of the sequence described above (Fig. 1.4). This observation suggests that the step where a thin septum appears while the actin ring still spans the width of the hypha might represent a critical transition point that precedes the commitment to ring constriction and deposition of the full septum. Moreover, it is tempting to speculate that a crucial event that might occur at this transition is a switch from one Rho GTPase to another. According to this model, Rho4 would direct the assembly of the actin ring and the initial stages of septum deposition before Rho1 "stepped in" to complete the

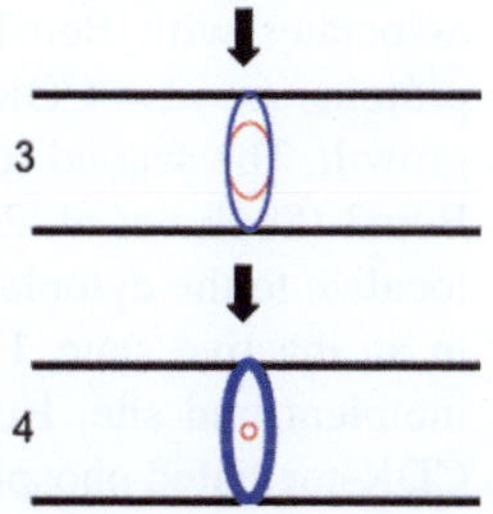


**Fig. 1.4** Four stages of septum formation. See text for details. *Red* denotes the cytokinetic actin ring, whereas *blue* represents the septal wall material

deposition of the full septum. Further studies are clearly needed to test this model and to more deeply explore the relationship between actin ring constriction and septum deposition in fungal hyphae.

## 1.5 Coordination of Morphogenesis to Regulate Cell Shape

For yeast cells, the precise coordination of morphogenesis with nuclear division and cellular growth plays a crucial role in the determination of cell shape.

This relationship is perhaps best exemplified in *S. cerevisiae*, where morphogenesis is coordinated with nuclear division at multiple points in the cell cycle (Lew and Reed 1995). At these transition points (i.e., START in late G1, G2/M), distinct CDK complexes trigger the appropriate set of events required for faithful nuclear division. Notably, these same complexes also activate specific morphogenetic responses. For example, at START, the same CDK complex that promotes the initiation of DNA replication also directs the organization of the bud site to enable polarity establishment. Later, distinct CDK complexes that promote events underlying mitosis also govern the switch from apical to isotropic bud growth. Finally, destruction of the latter complex following mitosis also causes relocalization of the morphogenetic machinery to the mother-bud neck in preparation for cytokinesis. Although these coregulatory steps are usually sufficient to ensure the coordination of bud morphogenesis with nuclear division, they sometimes fail as a result of environmental or stochastic perturbations. In such cases, morphogenetic defects (e.g., alterations in the geometry of the mother-bud neck) activate a checkpoint that delays the activation of mitotic CDK complexes (Keaton and Lew 2006), thus prohibiting nuclear division until normal bud morphology has been restored. Ultimately, the collective effect of these regulatory mechanisms is to ensure that a bud of proper size and shape is ready to receive the newly formed daughter nucleus.

Recent studies have provided important insight into the molecular mechanisms by which *S. cerevisiae* CDK complexes impinge upon the regulatory pathways that control symmetry breaking. It is not too surprising that these mechanisms focus on regulating the timing and location of Cdc42 activation. One set of mechanisms acts via the Cdc42 GEF Cdc24. Prior to START, Cdc24 is sequestered in the nucleus through association with Far1 (Shimada et al. 2000). However, CDK-mediated phosphorylation of Far1 results in its destruction, thereby releasing Cdc24 and permitting its relocalization to the incipient bud site. At the bud site, Cdc24 also associates with Bem1, Boi1, and Boi2 (McCusker et al. 2007). The latter two proteins are also CDK substrates whose phosphorylation is required for polarized growth. The second set of mechanisms acts via the Cdc42 GAPs Rga2, Bem2, and Bem3 (Sopko et al. 2007; Knaus et al. 2007). Prior to START, Bem2 and Bem3 localize to the cytoplasm and cell surface, where they presumably maintain Cdc42 in an inactive state. Following START, along with Rga2, they accumulate at the incipient bud site. However, Rga2 and Bem3 are each inactivated as a result of CDK-mediated phosphorylation. Collectively, these two sets of mechanisms ensure that active GTP-bound Cdc42 is present only at the incipient bud site coincident with passage through START. Later in the cell cycle, additional mechanisms further coordinate morphogenesis with nuclear division. For example, CDK complexes likely work in conjunction with the polo kinase to activate the Rho1 GTPase locally at the mother-bud neck to enable the assembly of the cytokinetic actin ring (Yoshida et al. 2006).

Despite differences in cell shape compared to *S. cerevisiae*, the coordination of morphogenesis with nuclear division is also a fundamental feature of the *S. pombe* cell cycle. The best example of this is the transition from monopolar to bipolar growth (i.e., NETO), which occurs early in the G2 phase of the cell cycle (Martin and Chang 2005). Like budding yeast, Cdc42 is essential for polarity establishment in fission yeast (Miller and Johnson 1994). Accordingly, it is strongly suspected that the ability of CDK complexes to phosphorylate Cdc42 GEFs and GAPs is conserved in *S. pombe*, and would provide an effective mechanism that coordinates NETO with cell cycle progression (Moseley and Nurse 2009). Later in the *S. pombe* cell cycle, signals that emanate from the growing tips determine the timing of mitotic entry by regulating the activation of the mitotic CDK complex (Moseley et al. 2009; Martin and Berthelot-Grosjean 2009). Essentially, the protein kinase Pom1 forms a gradient that is highest at the cell ends and lowest in the middle. As the cell ends grow apart, Pom1 is depleted from the cell middle, which then triggers a regulatory pathway that promotes activation of the mitotic CDK complex. This mechanism, as well as the morphogenetic checkpoint of *S. cerevisiae*, represents two examples where morphogenetic inputs determine the timing of nuclear division and further emphasize that the coordination of morphogenesis with nuclear division is a "two-way street."

### 1.5.1 Coordination of Morphogenesis in Hyphal Cells

In filamentous fungi such as *A. nidulans*, the term duplication cycle is used to describe events during hyphal growth that are analogous to those observed in the cell cycle of uninucleate cells (i.e., *S. cerevisiae* and *S. pombe*) (Fiddy and Trinci 1976). These events include the doubling of cell mass, and nuclear and cell division. However, there are notable differences relative to uninucleate cells. First, cell division is not accompanied by cell separation, and second, in most cases cells are multinucleate. Because of these differences, it is not immediately apparent that cellular morphogenesis needs to be tightly coordinated with nuclear division. Indeed, the analysis of conditional lethal mutations that affect nuclear division or polarity establishment suggests that the two processes are completely independent. For example, temperature sensitive (Ts) *A. nidulans* mutants defective in the establishment of hyphal polarity (e.g., *podD* and *swoA*) accumulate large numbers of nuclei (i.e., >20; Harris et al. 1999; Momany et al. 1999), whereas Ts mutants defective in nuclear division (e.g., *nimA* and *nimT*) form long hyphae despite possessing only a single nucleus (Morris 1975). These observations imply that there are no checkpoints that delay nuclear division when morphogenesis is compromised (or vice versa). Nevertheless, there are key morphological transitions where it is reasonable to expect some degree of coordination with nuclear division. Good examples might include the establishment of hyphal polarity during spore germination, and the formation of lateral branches from an existing hypha. Future studies will be needed to determine the validity of this assumption and whether the relevant mechanisms are similar to those that coordinate morphogenesis in yeast cells.

### *1.5.2 Developmental Regulation of Morphogenesis*

Fungal development typically encompasses the formation of asexual or sexual reproductive structures that bear spores. For many plant pathogenic fungi, the formation of infection structures such as appressoria also represents a type of development. In either case, development often involves the cessation of hyphal extension and a transition from the formation of multinucleate hyphal cells to cells that are uninucleate (or possess a very small number of nuclei). Accordingly, these transitions are presumably accompanied by much tighter coordination of cellular morphogenesis with nuclear division than what is normally seen in hyphae. For example, the development of asexual conidiophores in *A. nidulans* is initiated from a multinucleate foot cell that generates an airborne stalk (Timberlake 1990). Once it reaches a certain height, the stalk depolarizes and forms a multinucleate vesicle. Next, synchronized budding from the vesicle produces a layer of cells termed metulae. Most importantly, this division represents a transition to uninucleate growth, as metulae and all subsequent cell types, including the conidiogenous phialides and the conidiospores themselves, possess only a single nucleus. Although it is likely that cell cycle dynamics are altered upon the switch to uninucleate growth, this has yet to be experimentally verified. However, it has been shown that CDK activity increases ~tenfold during asexual development in *A. nidulans*, which could account for a more rapid cell cycle (Ye et al. 1999). More intriguingly, greater Cdk activity may also provide a mechanism to tighten the coordination of cellular morphogenesis with nuclear division, perhaps through the modulation of Cdc42 and other Rho-related GTPases. This could conceivably lead to the termination of polarized growth in a manner similar to that observed during bud morphogenesis in *S. cerevisiae*. In this context, a comparative analysis of the latter steps of conidiation to yeast budding would likely be instructive in revealing mechanisms that coordinate morphogenesis with nuclear division during fungal development.

## References

Araujo-Bazan L, Penalva MA, Espeso EA (2008) Preferential localization of the endocytic internalization machinery to hyphal tips underlies polarization of the actin cytoskeleton in *Aspergillus nidulans*. Mol Microbiol 67:891–905

Araujo-Palomares CL, Castro-Longoria E, Riquelme M (2007) Ontogeny of the Spitzenkorper in germlings of *Neurospora crassa*. Fungal Genet Biol 44:492–503

Bachewich C, Masker K, Osmani S (2005) The polo-like kinase PLKA is required for initiation and progression through mitosis in the filamentous fungus *Aspergillus nidulans*. Mol Microbiol 55:572–587

Bartnicki-Garcia S, Hergert F, Gierz G (1989) Computer simulations of fungal morphogenesis and the mathematical basis for hyphal tip growth. Protoplasma 153:46–57

Bartnicki-Garcia S, Bartnicki DD, Gierz G, Lopez-Franco R, Bracker CE (1995) Evidence that Spitzenkorper behavior determines the shape of a fungal hypha: a test of the hyphoid model. Exp Mycol 19:153–159

Bruno K, Morrell JL, Hamer JE, Staiger CJ (2001) SEPH, a Cdc7p orthologue from *Aspergillus nidulans*, functions upstream of actin ring formation during cytokinesis. Mol Microbiol 42:3–12

Cano-Dominguez N, Alvarez-Delfin K, Hansberg W, Aguirre J (2008) NADPH oxidases NOX-1 and NOX-2 require the regulatory subunit NOR-1 to control cell differentiation and growth in *Neurospora crassa*. Eukaryot Cell 7:1352–1361

Canovas D, Perez-Martin J (2009) Sphingolipid biosynthesis is required for polar growth in the dimorphic pathogen *Ustilago maydis*. Fungal Genet Biol 46:190–200

Chang F, Peter M (2003) Yeasts make their mark. Nat Cell Biol 5:294–299

Chant J (1999) Cell polarity in yeast. Annu Rev Cell Dev Biol 15:365–391

Chapa-y-Lazo B, Lee S, Regan H, Sudbery P (2011) The mating projections of Saccharomyces cerevisiae and Candida albicans show key characteristics of hyphal growth. Fungal Biol 115(6):547–556

Chen CT, Feoktistova A, Chen JS, Shim YS, Clifford DM et al (2008) The SIN kinase Sid2 regulates cytoplasmic retention of the *S. pombe* Cdc14-like phosphatase Clp1. Curr Biol 18:1594–1599

Delgado-Alvarez DL, Callejas-Negrete OA, Gomez N, Freitag M, Roberson RW et al (2010) Visualization of F-actin localization and dynamics with live cell markers in *Neurospora crassa*. Fungal Genet Biol 47:573–586

Doyle T, Botstein D (1996) Movement of yeast cortical actin cytoskeleton visualized in vivo. Proc Natl Acad Sci USA 93:3886–3891

Egan M, Wang ZY, Jones MA, Smirnoff N, Talbot NJ (2007) Generation of reactive oxygen species by fungal NADPH oxidases is required for rice blast disease. Proc Natl Acad Sci USA 104:11772–11777

Farah ME, Amberg DC (2007) Conserved actin cysteine residues are oxidative stress sensors that can regulate cell death in yeast. Mol Biol Cell 18:1359–1365

Fiddy C, Trinci AP (1976) Mitosis, septation, branching and the duplication cycle in *Aspergillus nidulans*. J Gen Microbiol 97:169–184

Fischer R, Zekert N, Takeshita N (2008) Polarized growth in fungi – interplay between the cytoskeleton, positional markers and membrane domains. Mol Microbiol 68:813–826

Ge W, Balasubramanian MK (2008) Pxl1p, a paxillin-related protein, stabilizes the actomyosin ring during cytokinesis in fission yeast. Mol Biol Cell 19:1680–1692

Girbardt M (1957) Der Spitzenkorper von *Polystictus versicolor*. Planta 50:47–59

Guirland C, Suzuki S, Kojima M, Lu B, Zheng JQ (2004) Lipid rafts mediate chemotropic guidance of nerve growth cones. Neuron 42:51–62

Hachet O, Simanis V (2008) Mid1/anillin and the septation initiation network orchestrate contractile ring assembly for cytokinesis. Genes Dev 22:3205–3216

Hamer JE, Valent B, Chumley FG (1989) Mutations at the smo genetic locus affect the shape of diverse cell types in the rice blast fungus. Genetics 122:351–361

Han KH, Seo JA, Yu JH (2004) Regulators of G-protein signaling in *Aspergillus nidulans*: RgsA downregulates stress response and stimulates asexual sporulation through attenuation of GanB (Galpha) signaling. Mol Microbiol 53:529–540

Harris SD (1999) Morphogenesis is coordinated with nuclear division in germinating *Aspergillus nidulans* conidiospores. Microbiology 145:2747–2756

Harris SD (2010) Hyphal growth and polarity. In: Borkovich KA, Ebbole DJ (eds) Cellular and molecular biology of filamentous fungi. ASM, Washington, DC

Harris SD (2011) Cdc42/Rho GTPases in fungi: variations on a common theme. Mol Microbiol 79:1123–1127

Harris SD, Momany M (2004) Polarity in filamentous fungi: moving beyond the yeast paradigm. Fungal Genet Biol 41:391–400

Harris SD, Hofmann AF, Tedford H, Lee MP (1999) Identification and characterization of genes required for hyphal morphogenesis in the filamentous fungus *Aspergillus nidulans*. Genetics 151:1015–1025

Harris SD, Read ND, Roberson RW, Shaw B, Seiler S et al (2005) Polarisome meets spitzenkörper: microscopy, genetics, and genomics converge. Eukaryot Cell 4:225–229

Harris SD, Turner G, Meyer V, Espeso EA, Specht T et al (2009) Morphology and development in *Aspergillus nidulans*: a complex puzzle. Fungal Genet Biol 46:S82–S92

Johnson AE, Gould KL (2011) Dma1 ubiquitinates the SIN scaffold, Sid4, to impede the mitotic localization of the Plo1 kinase. EMBO J 30:341–354

Justa-Schuch D, Heilig Y, Richthammer C, Seiler S (2010) Septum formation is regulated by the RHO4-specific exchange factors BUD3 and RGF3 and by the landmark protein BUD4 in *Neurospora crassa*. Mol Microbiol 76:220–235

Keaton MA, Lew DJ (2006) Eavesdropping on the cytoskeleton: progress and controversy in the yeast morphogenesis checkpoint. Curr Opin Microbiol 9:540–546

Kim JM, Zeng CJT, Nayak T, Shao R, Huang AC et al (2009) Timely septation requires SNAD-dependent spindle pole body localization of the septation initiation network components in the filamentous fungus *Aspergillus nidulans*. Mol Biol Cell 20:2874–2884

Knaus M, Pelli-Gulli MP, van Drogen F, Springer S, Jaquenoud M et al (2007) Phosphorylation of Bem2p and Bem3p may contribute to local activation of Cdc42p at bud emergence. EMBO J 26:4501–4513

Konzack S, Rischitor PE, Enke C, Fischer R (2005) The role of the kinesin motor KipA in microtubule organization and polarized growth of *Aspergillus nidulans*. Mol Biol Cell 16:497–506

Kozubowski L, Saito K, Johnson JM, Howell AS, Zyla TR et al (2008) Symmetry-breaking polarization driven by a Cdc42p GEF-PAK complex. Curr Biol 18:1719–1726

Krapp A, Simanis V (2008) An overview of the fission yeast septation initiation network (SIN). Biochem Soc Trans 36:411–415

Kubler E, Riezman H (1993) Actin and fimbrin are required for the internalization step of endocytosis in yeast. EMBO J 12:2855–2862

Lambeth JD (2004) NOX enzymes and biology of reactive oxygen. Nat Rev Immunol 4:181–189

Layton AT, Savage NS, Howell AS, Carroll SY, Drubin DG et al (2011) Modeling vesicle traffic reveals unexpected consequences for Cdc42p-mediated polarity establishment. Curr Biol 21:184–194

Lew DJ, Reed SI (1995) Cell cycle control of morphogenesis in budding yeast. Curr Opin Genet Dev 5:17–23

Li S, Du L, Yuen G, Harris SD (2006) Distinct ceramide synthases regulate polarized growth in the filamentous fungus *Aspergillus nidulans*. Mol Biol Cell 17:1218–1227

Lindsey R, Cowden S, Hernández-Rodríguez Y, Momany M (2010) Septins AspA and AspC are important for normal development and limit the emergence of new growth foci in the multicellular fungus *Aspergillus nidulans*. Eukaryot Cell 9:155–163

Liu J, Wang H, McCollum D, Balasubramanian MK (1999) Drc1/Cps1p, a 1,3-beta-glucan synthase subunit, is essential for division septum assembly in *Schizosaccharomyces pombe*. Genetics 153:1193–1203

Liu J, Tang X, Wang H, Oliferenko S, Balasubramanian MK (2002) The localization of the integral membrane protein Cps1p to the cell division site is dependent on the actomyosin ring and the septation-inducing network in *Schizosaccharomyces pombe*. Mol Biol Cell 13:989–1000

Lo HJ, Kohler JR, DiDomenico B, Cacciapuoti A, Fink GR (1997) Nonfilamentous *C. albicans* mutants are avirulent. Cell 90:939–949

Lopez-Franco R, Bracker CE (1996) Diversity and dynamics of the Spitzenkorper in growing hyphal tips of higher fungi. Protoplasma 195:90–111

Malinska K, Malinsky J, Opekarova M, Tanner W (2003) Visualization of protein compartmentation within the plasma membrane of living yeast cells. Mol Biol Cell 14:4427–4436

Malinsky J, Opekarova M, Tanner W (2010) The lateral compartmentation of the yeast plasma membrane. Yeast 27:473–478

Martin SG (2009) Microtubule-dependent cell morphogenesis in the fission yeast. Trends Cell Biol 19:447–454

Martin SG, Berthelot-Grosjean M (2009) Polar gradients of the DYRK-family kinase Pom1 couple cell length with the cell cycle. Nature 459:852–856

Martin SG, Chang F (2005) New end take off: regulating cell polarity during the fission yeast cell cycle. Cell Cycle 4:1046–1049

Martin SW, Konopka JB (2004) Lipid raft polarization contributes to hyphal growth in *Candida albicans*. Eukaryot Cell 3:675–684

McCusker D, Denison C, Anderson S, Egelhofer TA, Yates JR III et al (2007) Cdk1 coordinates cell-surface growth with the cell cycle. Nat Cell Biol 9:506–515

Miller PJ, Johnson DI (1994) Cdc42p GTPase is involved in controlling polarized cell growth in *Schizosaccharomyces pombe*. Mol Cell Biol 14:1075–1083

Momany M, Hamer JE (1997) Relationship of actin, microtubules, and crosswall synthesis during septation in *Aspergillus nidulans*. Cell Motil Cytoskeleton 38:373–384

Momany M, Westfall PJ, Abramowsky G (1999) *Aspergillus nidulans swo* mutants show defects in polarity establishment, polarity maintenance and hyphal morphogenesis. Genetics 151:557–567

Morris NR (1975) Mitotic mutants of *Aspergillus nidulans*. Genet Res 26:237–254

Moseley JB, Nurse P (2009) Cdk1 and cell morphology: connections ans directions. Curr Opin Cell Biol 21:82–88

Moseley JB, Mayeux A, Paoletti A, Nurse P (2009) A spatial gradient coordinates cell size and mitotic entry in fission yeast. Nature 459:857–860

Nelson WJ (2003) Adaptation of core mechanisms to generate cell polarity. Nature 422:766–774

Park HO, Bi E (2007) Central roles of small GTPases in the development of cell polarity in yeast and beyond. Microbiol Mol Biol Rev 71:48–96

Pearson CL, Xu K, Sharpless KE, Harris SD (2004) MesA, a novel fungal protein required for the stabilization of polarity axes in *Aspergillus nidulans*. Mol Biol Cell 15:3658–3672

Pelham RJ Jr, Chang F (2001) Role of actin polymerization and actin cables in actin-patch movement in *Schizosaccharomyces pombe*. Nat Cell Biol 3:235–244

Perez P, Rincon SA (2010) Rho GTPases: regulation of cell polarity and growth in yeasts. Biochem J 426:243–253

Regalado CM, Crawford JW, Ritz K, Sleeman BD (1996) The origins of spatial heterogeneity in vegetative mycelia: a reaction-diffusion model. Mycol Res 100:1473–1480

Rittenour WR, Harris SD (2008) Characterization of *Fusarium graminearum* Mes1 reveals roles in cell-surface organization and virulence. Fungal Genet Biol 45:933–946

Rittenour WR, Chen M, Cahoon EB, Harris SD (2011) Control of glucosylceramide production and morphogenesis by the Bar1 ceramide synthase in *Fusarium graminearum*. PLoS One 6: e19385

Semighini C, Harris SD (2008) Regulation of apical dominance in *Aspergillus nidulans* hyphae by reactive oxygen species. Genetics 179:1919–1932

Sharpless KE, Harris SD (2002) Functional characterization and localization of the *Aspergillus nidulans* formin SEPA. Mol Biol Cell 13:469–479

Shimada Y, Gulli MP, Peter M (2000) Nuclear sequestration of the exchange factor Cdc24 by Far1 regulates cell polarity during yeast mating. Nat Cell Biol 2:117–124

Si H, Justa-Schuch D, Seiler S, Harris SD (2010) Regulation of septum formation by the Bud3-Rho4 GTPase module in *Aspergillus nidulans*. Genetics 185:165–176

Simanis V (2003) The mitotic exit and septation initiation networks. J Cell Sci 116:4261–4262

Simons K, Gerl MJ (2010) Revitalizing membrane rafts: new tools and insights. Nat Rev Mol Cell Biol 11:688–699

Slaughter BD, Smith SE, Li R (2009) Symmetry breaking in the life cycle of the budding yeast. Cold Spring Harb Perspect Biol 1:a003384

Sopko R, Huang D, Smith JC, Figeys D, Andrews BJ (2007) Activation of the Cdc42p GTPase by cyclin-dependent protein kinases in budding yeast. EMBO J 26:4487–4500

Steinberg G (2007) Hyphal growth; a tale of motors, lipids, and the Spitzenkorper. Eukaryot Cell 6:351–360

Taheri-Talesh N, Horio T, Araujo-Bazán L, Dou X, Espeso EA et al (2008) The tip growth apparatus of *Aspergillus nidulans*. Mol Biol Cell 19:1439–1449

Takeda S, Gapper C, Kaya H, Bell E, Kuchitsu K et al (2008) Local positive feedback regulation determines cell shape in root hair cells. Science 319:1241–1244

Takemoto D, Tanaka A, Scott B (2006) A p67Phox-like regulator is recruited to control hyphal branching in a fungal-grass mutualistic symbiosis. Plant Cell 18:2807–2821

Takemoto D, Tanaka A, Scott B (2007) NADPH oxidases in fungi: diverse roles of reactive oxygen species in fungal cellular differentiation. Fungal Genet Biol 44:1065–1076

Takemoto D, Kamakura S, Saikia S, Becker Y, Wrenn R et al (2011) Polarity proteins Bem1 and Cdc24 are components of the filamentous fungal NADPH oxidase complex. Proc Natl Acad Sci USA 108:2861–2866

Takeshita N, Higashitsuji Y, Konzack S, Fischer R (2008) Apical sterol-rich membranes are essential for localizing cell end markers that determine growth directionality in the filamentous fungus *Aspergillus nidulans*. Mol Biol Cell 19:339–351

Tanaka A, Takemoto D, Hyon GS, Park P, Scott B (2008) NoxA activation by the small GTPase RacA is required to maintain a mutualistic symbiotic association between *Epichloe festucae* and perennial ryegrass. Mol Microbiol 68:1165–1178

Timberlake WE (1990) Molecular genetics of *Aspergillus* development. Annu Rev Genet 24:5–36

Toret CP, Drubin DG (2006) The budding yeast endocytic pathway. J Cell Sci 119:4585–4587

Upadhyay S, Shaw BD (2008) The role of actin, fimbrin and endocytosis in growth of hyphae in *Aspergillus nidulans*. Mol Microbiol 68:690–705

Valdez-Taubus J, Pelham HR (2003) Slow diffusion of proteins in the yeast plasma membrane allows polarity to be maintained by endocytic cycling. Curr Biol 13:1636–1640

Viestica A, Tang XZ, Oliferenko S (2008) The actomyosin ring recruits early secretory compartments to the division site in fission yeast. Mol Biol Cell 19:1125–1138

Wendland J (2003) Analysis of the landmark protein Bud3 of *Ashbya gossypii* reveals a novel role in septum construction. EMBO Rep 4:200–204

Wolkow TD, Harris SD, Hamer JE (1996) Cytokinesis in *Aspergillus nidulans* is controlled by cell size, nuclear positioning and mitosis. J Cell Sci 109:2179–2188

Ye X, Lee SL, Wolkow TD, McGuire SL, Hamer JE et al (1999) Interaction between developmental and cell cycle regulators is required for morphogenesis in *Aspergillus nidulans*. EMBO J 18:6994–7001

Yoshida S, Kono K, Lowery DM, Bartolini S, Yaffe MB et al (2006) Polo-like kinase Cdc5 controls the local activation of Rho1 to promote cytokinesis. Science 313:108–111

# Chapter 2
# Tropic Orientation Responses of Pathogenic Fungi

**Alexandra Brand and Neil A.R. Gow**

**Abstract** Cellular orientation allows growth, differentiation and behaviour to respond to vectorial cues generated in the environment and in relation to cells of the same organisms or different organisms that exist in proximity to one another. In the case of fungal pathogens, the orientation of hyphae may allow the fungus to detect a host and to make strategic penetrations at points of weakness on the host surface. Within a host, tropic orientation may facilitate colonisation, ramification and dispersal within the host tissues. To achieve this, cells have to be able to coordinate their cell cycles, growth and expansion of their margins with directional growth responses. In this chapter, we review the tropic orientation responses of fungi and, with an emphasis on fungal pathogenesis, discuss and speculate on the underlying molecular mechanisms that regulate cellular tropisms. Examples are taken across the fungal kingdom, including from work on saprophytes, plant and animal pathogens, to construct a working model that speculates how a wide range of tropisms may be controlled by a more-or-less common tropic mechanism that regulates the orientation of the hyphal tip.

## 2.1 Introduction

Pathogenic fungi have developed efficient growth strategies that allow them to penetrate and infiltrate through host tissue. A common mechanism in this process is the formation of elongating structures, such as hyphal filaments, penetration pegs, shmoos and rhizoids, which translocate fungal growth within the host or environment, for example, through leaf cuticles or between epidermal cell layers. Each host offers its own unique environment, so fungi have evolved hard-wired host-specific sensing and response mechanisms (tropisms) that regulate their vectorial

A. Brand • N.A.R. Gow (✉)
Institute of Medical Sciences, University of Aberdeen, Aberdeen AB25 2ZD, UK
e-mail: n.gow@abdn.ac.uk

J. Pérez-Martín and A. Di Pietro (eds.), *Morphogenesis and Pathogenicity in Fungi*, Topics in Current Genetics 22, DOI 10.1007/978-3-642-22916-9_2, 

growth. The study of tropic responses *in vitro* has demonstrated how subtle the interplay between environmental cues and fungal response can be and potentially involves the integration of mechanical, chemical and electrical signals. The challenge now is to develop a holistic understanding of the process during infection *in vivo*, with a view to destabilising the relationship between host environment and pathogen growth response. In this chapter, we review the tropic, or pre-programmed, growth responses of fungal pathogens which are involved in disease progression in plants and humans.

Thigmotropism is the movement or orientation of an organism or cell in relation to the topography, shape and physical properties of the underlying substrate on which it is growing. In both the macrobiotic and microscopic worlds, there are many well-recognised examples of thigmotropism. Roots of plants grow around stones and obstacles or along the crevices of rocks, ivy infiltrates the wall facing of houses and grows along drain pipes and seaweeds penetrate suitable holdfasts on the sea floor to maintain their position in the tide. At the cellular level, pollen tubes enter the stigma and head straight through the style to locate and fertilise the plant ovaries, and nerve cells execute complex orientation responses according to the tissues through which they are growing and generate complex neural networks that somehow work together to generate the bewildering complexity of the brain. The hyphae of endophytes trace the perimeters of the plant cells they associate with (Fig. 2.1d), and mating gametes of fungi grow directly towards each other and, with exquisite precision, fuse precisely at their tips to facilitate karyogamy. All these orientation responses require growth to be orchestrated and directed.

The best evidence for thigmotropism facilitating fungal pathogenesis comes from work on a range of plant pathogenic species. On the outer surfaces of plants, some fungi trace the junctions between plant cortical cells, while others grow across the junctions at right angles (Fig. 2.1a, b). Good examples of the former behaviour come from fungi growing on dicotyledenous plants where the cells are arranged as a mosaic, while examples of growth perpendicular to plant epithelial cells are exhibited by some monocot pathogens (Fig. 2.1). In both cases, this behaviour seems to be adapted to facilitate the searching out of guard cells, which are often the natural infection site targets. In monocots, the guard cells are usually in staggered rows; therefore, crossing the cells at right angles will maximise the chance of a guard cell encounter. If the guard cells are scattered between a mosaic of cells, a more effective strategy is to follow the peripheries of the cortical cells. In some fungi, the guard cell lip may then trigger thigmo-differentiation of the appressorium (Fig. 2.1c) (Allen et al. 1991; Collins and Read 1997; Read et al. 1992) (see later).

Human fungal pathogens also exhibit hypha thigmotropism (Fig. 2.2) (Gow 1993, 2004; Gow et al. 1994; Perera et al. 1997), although it has not yet been established whether this is a *bone fide* virulence attribute. It has been shown that non-thigmotropic mutants (described below) are less good at causing tissue damage (Brand et al. 2008), but such mutations are normally pleiotrotopic and this tropism phenotype cannot, therefore, be uniquely attributed to alterations in the tropic behaviour of the fungus (Davies et al. 1999). However, we can speculate that the ability to grow between the cells of a tissue or within the strata of a cornified

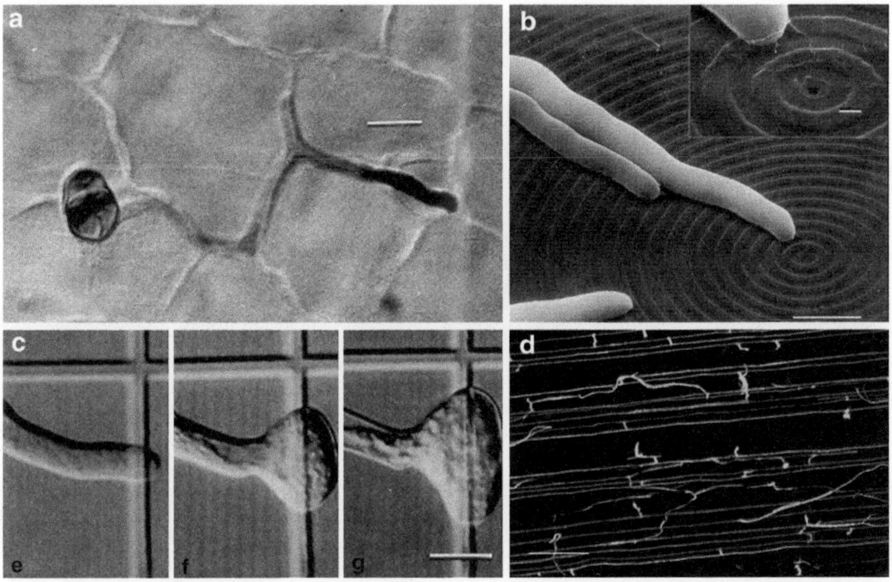


**Fig. 2.1** Thigmotropism and thigmo-differentiation of plant-associated fungi. Fungi that penetrate the host leaf via stomata have evolved host-specific hyphal growth strategies to locate and recognise stomatal guard cells. (**a**) A hypha of the fungus *Cymodothea trifolii* follows the depressions surrounding adjoining epidermal cells of the host plant *Trifolium repens* (white clover) in order to locate a stoma, on which an appressorium is formed (Roderick 1993) (*Bar* = 10 μm). (**b**) Hyphae of *Uromyces appendiculatus* (bean rust fungus) cross cortical cells of the leaf surface at right angles to maximise the likelihood of locating a stoma, which are arranged in staggered rows in its host. This tropism appears to be elicited entirely through mechanical contact sensing because it can be replicated by growth on an inert microfabricated surface, resulting in the targeting of the bullseye of a surface consisting of concentric circles (Hoch et al. 1993, C IEEE). (**c**) Thigmo-differentiation – the formation of appressoria by *U. appendiculatus* grown on a polystyrene substrate can be triggered on contact with 0.5-μm-high ridges that precisely mimic the height of guard cells of the host plant (Kwon and Hoch 1990) (*Bar* = 11.8 μm). (**d**) Hyphae of *Epichloë* endophytes, here shown tagged with fluorescent green protein, extend along the longitudinal axis of the host leaf. Hyphal extension precisely matches that of the leaf. It is thought that, through their tight association with the intercellular space, hyphae are subjected to mechanical stretching as the host cells expand, and this stress is mitigated by the onset of intercalary growth (Christensen et al. 2008) (*Bar* = 100 μm) (image courtesy of C. Voisey)

epithelium may confer an advantage on such pathogens (Perera et al. 1997; Hutton et al. 1978; Kumamoto and Vinces 2005). It is also clear that *Candida albicans* hyphae sometimes directly enter human cells, sometimes move between them and sometimes induce their own phagocytosis. Therefore, thigmotropism should be regarded as an adjunct penetration mechanism, and not the only one. It is also quite clear that non-pathogenic fungi also exhibit strong thigmotropic responses. It may be that similar advantages are conferred if hyphae of a saprophyte or symbiont can sense the dead or living cells or tissues on which they are living (Kumamoto 2008; Brand and Gow 2009; Christensen et al. 2008).

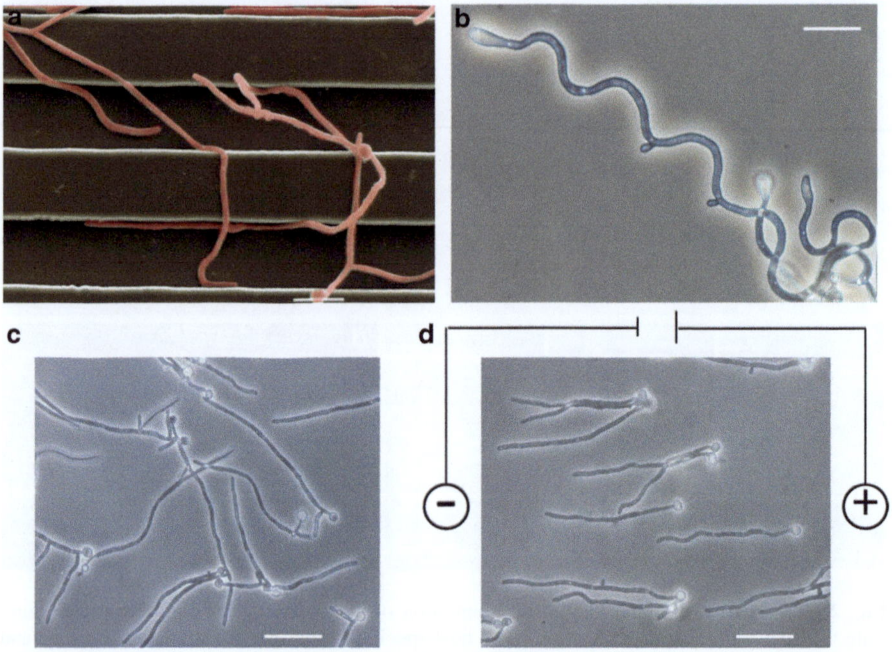


**Fig. 2.2** Tropic growth of the human fungal pathogen *Candida albicans*. (**a**) Hyphae adhered to a microfabricated quartz slide (etched with ridges of 3.25 μm) follow the contours of the substrate by re-orienting their tip growth (*Bar* = 15 μm) (image courtesy A Brand and K Mackenzie). (**b**) When *C. albicans* is grown on a semi-solid surface with poor nutrients, hyphae form regular sinusoidal waves, with a strong tendency of septa to form at apices of the curves (Brand et al. 2008) (*Bar* = 12 μm) (image courtesy A Brand and K Lee). (**c**) The random direction of growth by hyphae germinated after adhesion to a glass slide can be strikingly overridden by the application of a DC electric field prior to germination (**d**), which causes growth to be cathode oriented (Crombie et al. 1990; Brand et al. 2007) (*Bars* = 25 μm) (images courtesy A Brand)

## 2.2 Mechanism of Growth and Tropic Orientation

Considering how fundamental cellular orientation is in nature, there is really rather little information available about how this is achieved and regulated. The mechanism is assumed to be part of the machinery that is involved in cell growth and extension. But is there in addition a subset of proteins and processes that operate in parallel to those controlling growth? To create a metaphor, is there a separate molecular steering wheel that is separately regulated and distinct from the engine that drives the cell tip forward during growth? In this chapter, we consider this question in the context of thigmotropic orientations and in turn other tropic mechanisms that are exhibited by fungal hyphae. Fungi are excellent models to address such fundamental issues. Their hyphae often grow rapidly and their trajectories can be readily mapped and measured. They respond to a wide range of environmental cues by reorienting growth and can often be manipulated physiologically and genetically, and mutants that have altered growth and orientation behaviours can be studied.

Thigmotropism can be broken down into a number of component steps that can be considered separately or as part of an integral mechanism. To respond to the undulation of a surface, the surface must first be sensed directly. This requires intimate contact to be made with the surface, and so adhesion mechanisms are a component of the information chain that results in tropic growth. Once the surface is bound and sensed, the vectorial information that is defined by the surface contours must be translated into a signal that articulates with the cell biological processes that bring about polarised apical growth. This implies that orientation mechanisms involve a very large number of proteins with signalling, scaffolding and mechanical properties so that cell polarity establishment and maintenance are regulated in the context of ambient environmental cues. The cellular apparatus for tip growth in fungi is in itself an enormous field involving the microfilament and microtubular cytoskeleton and their motor proteins, the secretory pathway that provides vesicles for membrane expansion at the tip, enzymes to catalyse apical cell wall growth, and regulatory proteins such as those found in the polarisome, Arp2/3 and exocyst protein complexes (Virag and Harris 2006; Steinberg 2007; Sudbery and Court 2007; Machesky and Gould 1999; Lipschutz and Mostov 2007). In filamentous fungi, the apex has an assemblage of vesicles within a structure called the Spitzenkörper, whose dynamic properties and position are critical in defining the growth axis of the cell. In yeast, the growth axis is determined by cell-cycle regulated cortical markers that form adjacent to septin rings. They are regulated by a complex genetic circuit involving *BUD* and other genes that encode a developmental programme that determines where the site of outgrowth will occur (Casamayor and Snyder 2002; Fischer et al. 2008). The mechanism of cell orientation is, therefore, governed in part by endogenous cues, such as the cortical proteins, whilst remaining responsive to exogenous ones that have the potential to override these endogenous cues.

Therefore, the biology of cell tropic behaviour is a highly integrative field comprising aspects of cell biology, cytoskeletal function, secretion, polarity, the cell cycle and cell wall growth. A number of general reviews on the aspects of these various component fields have been suggested, but here we will focus on the studies that deal most specifically and directly with hyphal orientation and responses to topography in the context of fungal pathogenesis.

### *2.2.1 Requirement for Initial Adhesion*

For the fungal spore or cell, the naked host surface, such as the leaf of a plant or epithelial mucosa, is an unstable and even hostile environment. On plants, vibration and the shearing effect of abrasion, water flow and raindrop splash are immediate threats to successful colonisation. In the mammalian host, mechanical abrasion, the sloughing off of keratinised surface epithelial cells, blood shear flow and attack by components of the innate immune system threaten survival of the fungus. The fungal strategy is, therefore, to transfer as quickly as possible from the hostile host surface into a safer environment. This involves differentiation to a morphology specialised for penetration of the underlying host tissue (leaf and root cuticles, or endothelial cell layers), intercalation between keratinised cells or establishment of a protective biofilm layer on mucosal or plastic surfaces. The primary function of the infective fungal particle is, therefore, to remain adhered to the host long enough for this process to take place.

Surprisingly, little is known about the molecules that mediate the initial adhesion of fungi to the host, although some general concepts are well understood. Dispersive forces and non-specific interactions such as electrostatic or hydrophobic attraction provide avidity rather than affinity, and it is likely that they play a greater role in the infection of plants than mammals. Although topographically variable, plant leaves offer a relatively uniform surface of epicuticular waxes, and plant pathogens are thought to adhere via surface hydrophobin rodlets. These amphiphilic molecules not only act during spore dissemination and initial adhesion, but are also subsequently shed onto the leaf surface to mediate adhesion of the developing appressorium (Wösten et al. 1994; Talbot et al. 1993). In the complex mammalian environment, heterogeneity in surface molecules appears to be the key to adhesion success in *C. albicans*. Unlike *A. fumigatus*, which primarily infects the lung, *C. albicans* adheres to multiple host sites, including hydrophobic in-dwelling medical plastic devices, and to itself through flocculation in biofilms. We and others have shown that *C. albicans* yeast consistently binds more avidly to collagen IV, found in the kidney and the epithelial and endothelial basal lamina, than to the general extracellular matrix protein collagen I, suggesting a degree of binding specificity by the *C. albicans* yeast cell surface (Yan et al. 1998). Initial adhesion of yeast cells to extracellular matrix proteins was proposed to be mediated by as-yet unidentified promiscuous fungal receptors in a calcium-sensitive manner, where charge interactions are important (Klotz et al. 1993). Charge is strongly influenced by pH, which differs dramatically by host body site, ranging from pH 1–4 in the gut, to pH 4.2 in the vaginal mucosa and pH 7.4 in the mouth. *C. albicans* expresses a family of 8 *ALS* (*A*gglutinin-*L*ike *S*equence) genes that encode proteins with hypervariable N-termini. Together they present a range of physico-chemical properties that are involved in non-specific adhesion – hydrophobicity, electrostatic charge and hydrogen-bonding interactions. Some *ALS* genes are expressed specifically in hyphae, e.g. *ALS3*, and so are not involved in the initial adhesion of yeast cells, but when expressed heterologously in *Saccharomyces cerevisiae*, distinct adhesion profiles were identified for individual Als proteins, suggesting that this gene family contributes significantly to the predicted overall heterogeneity of surface adhesive molecules in *C. albicans* (Hoyer and Hecht 2001; Sheppard et al. 2004; Hoyer et al. 2008; Liu and Filler 2011). In addition, the amyloid-like properties of the Als proteins could allow them to increase binding avidity through the formation of adhesive plaques (Alsteens et al. 2010). Experiments using Als5 showed the plaques were initiated through the mechanical stretching of individual surface molecules. Stretching of the protein fibril unfolds the N-terminal β-sheet domains, exposing hydrophobic patches that recruit by-stander Als5 molecules to form self-propagating adhesive patches. The process is independent of cell viability, but it is conceivable that such dramatic surface changes in living cells could initiate a signal to the cell interior to indicate that adhesion had occurred.

### *2.2.2 Initial Adhesion and Mechanosensing*

Cell attachment to a surface triggers a "differentiate to survive" response in micro-organisms. The energetic cost of this transition is high, so a requirement for the integration of multiple signals ensures that this irreversible step is taken only when conditions are favourable. For example, in the bacterium *Vibrio parahaemolyticus*, adhesion is sensed by the steric hindrance to rotation of the polar flagellum by the proximity of the surface, but the switch to swarming growth also requires a nutrient signal (i.e. a lack of iron) (Gode-Potratz et al. 2011). Surface hardness is a key inducer of differentiation in many plant pathogens and even transient contact induces *Magnaporthe grisea* conidia to produce appressoria (Liu et al. 2007). How do fungi sense contact through rigid cell walls? Unlike mammalian cells, there is no evidence in fungi that the sensing of adhesion is receptor mediated via *trans*-wall fungal molecular signalling. Instead, sensing is probably mediated by perturbation of the cell wall and the consequences of this on the stretch of the underlying cell membrane. The cell wall counters a high internal turgor pressure (0.17–0.24 MPa) in the mammalian pathogen, *Pythium insidiosum* (Ravishankar et al. 2001), but it is not known whether the weight of the spore leads to wall deformation. By extrapolation from studies carried out with plant cells, tangential shear forces generated within the wall–plasma membrane–cytoskeleton linkages may be sensed by the adhered spore when it is jostled relative to the attachment surface. Shear forces are sensed in plant cells by such a mechanism and result in an immediate increase in cytosolic $Ca^{2+}$ via plasma membrane $Ca^{2+}$ channels (Pickard 1992; Ding and Pickard 1993). Hechtian strands, thin linkages composed of plasma membrane, actin and integrin-like peptides, have been observed at attachment points at the internal face of the cell wall in plasmolysed plant cells and the hyphal oomycete, *Saprolegnia ferax*, and it has been suggested that mechanosensing $Ca^{2+}$ channels are clustered at these focal points to co-ordinate the response to perturbations in cell shape (Lang et al. 2004; Volgger et al. 2010; Jaffe et al. 2002; Kaminskyj and Heath 1995). To our knowledge, Hechtian strands have not been visualised in fungi, but two proteins have been identified in *C. albicans* that are predicted to be involved in conveying information from the wall to the membrane. Dfi1 is a small transmembrane protein that extends into the cell wall, where it is cross-linked to the primary cell wall structural polymer, β-glucan. Deletion of Dfi1 renders the cell sensitive to cell wall stresses such as treatment with Congo Red (Zucchi et al. 2010). *C. albicans* also contains an orthologue of Wsc1, the cell wall integrity sensor first identified in *S. cerevisiae* (Verna et al. 1997). The distal portion of ScWsc1 protrudes from the membrane into the cell wall, and molecular engineering experiments have shown that it displays the properties of a nanospring (Dupres et al. 2009). CaWsc1 lies upstream in an orthologous cell wall integrity signalling pathway which is activated through the phosphorylation of the kinase CaMkc1 when cells are in contact with a semi-solid medium or plastic (Kumamoto 2005). Thus, adhesion seems to be sensed as cell wall perturbation, but cell responses are likely to also require integration of this signal with others from the external environment.

### 2.2.3 *Initial Adhesion and Physico-Chemical Interactions*

The requirement for additional signals for spore differentiation has been primarily established during the study of plant pathogens, but the characteristics of the inductive surfaces identified are only understood at the macro level. Many plant pathogens require a hard, hydrophobic surface and limited nutrients to activate differentiation (Warwar and Dickman 1996; Apoga et al. 2004; Shaw et al. 2006). Even though inert hard surfaces such as plastic can induce differentiation in *Uromyces appendiculatus* and *M. grisea*, the precise interplay between the fungus and an inert surface may nevertheless involve complex physico-chemical signals that operate at the nano-environmental level. Differentiation of these spores requires that the cells come into contact with moisture. At the cell surface, the "unstirred" aqueous diffusion boundary layer can extend from the cell surface by tens of micrometres. Within this zone, fungal surface molecules and effluxed ions are likely to create a localised chemical signature that could itself subsequently feedback to mediate cell behaviour, as is observed in mammalian cells (Smith et al. 2010). At the molecular level, some infective fungal particles are pre-coated with contact-activated molecular signals. Spores of the plant pathogens *Uromyces viclai-fabae* and *Blumeria graminis* carry esterases, cutinases and lipases on their surfaces. On contact with the host plant, the enzymes from *U. viclai-fabae* form an adhesive pad within seconds, and in both species, released enzyme activity generates specific breakdown products from the long-chain waxes of the host leaf cuticle. The chemical signal generated subsequently activates differentiation (Deising et al. 1992; Feng et al. 2009). Even inert hydrophobic surfaces could generate a chemical signal. It has been proposed that contact with a hydrophobic surface allows the diffusion of a surface-borne inhibitor, pyriculol, away from conidiospores of *M. grisea*, thereby relieving inhibition of differentiation (Hegde and Kolatukudy 1997). This strategy of priming the spore in a host-specific manner, therefore, serves three functions – surface adhesion, host recognition, and cell differentiation activation. A similar system may be employed by the human pathogen *C. albicans*, which features cell surface-bound proteases that operate at a variety of pH optima, giving the fungus a potential means to identify which human surface it has become adhered to by interpreting the peptide signature that is generated (Schild et al. 2011).

### 2.2.4 Requirement for Polarised Growth

The infective particles, spores and yeasts of most "professional" pathogenic fungi generate specialised polarised morphologies such as hyphae, germ tubes, appressorial penetration pegs and branched haustoria, which are designed to achieve directional mobility and to penetrate the host interior. In *U. appendiculatus*, *U. maydis*, *M. griseae* and oomycetes such as *P. infestans*, a polarised germ tube of approximately 20 μm emerges from the spore prior to the formation of the dome-like appressorium. In *C. albicans*, hyphae evaginate from the infecting yeast cell. Regulated directional growth is apparent in these morphologies. In hyphae and germ tubes, initial emergence is observed to be planar, while haustoria and the appressorial penetration peg form at an interface with the plant, indicating that surface sensing elicits a pre-programmed directional growth response in these pathogens.

A feature of differentiation in *C. albicans* is a change in the profile of surface adhesion molecules. The complement of wall adhesins changes to include, for example, *H*yphal *W*all *P*rotein 1 (Hwp1), which is a substrate for host transglutaminases (Staab et al. 1999). Where the requirement during initial adhesion is to remain attached until cell differentiation could occur, successful host penetration demands an adhesive force that enables the fungus to apply sufficient pressure to penetrate and infiltrate host tissue. Adhesion is just one part of the equation. In invasive hyphae, hydrolytic enzymes produced at the tip, and even the shape of the tip itself, aid direct penetration of the host cell membrane, thereby lowering the critical level of adhesive force required to anchor the hypha. In appressoria, penetration is achieved by generating a turgor pressure of 8 MPa (80 bar) to power the penetration peg through the leaf cuticle (Howard et al. 1991). The pressure required to penetrate the host must be adequately countered by the adhesive force so that the appressorium does not merely push itself away from the leaf. In *M. grisea*, the adhesive force has been estimated at a minimum of 500 J/m$^2$, i.e. in the value region for sticky tape rather than superglue (Goriely and Tabor 2006; Gent and Kaang 1986). Treatments that can tip the balance between host-specific fungal avidity and the mechanical resistance offered by the host might be an effective method to reduce infection, particularly by plant pathogens.

### 2.2.5 Directionality in Polarity Establishment

Directional growth requires the establishment of a single growth site to which vesicles and new wall material are delivered. In *S. cerevisiae* and *C. albicans* yeast, polarity is established and maintained by the small GTPase Cdc42 module. This module is essential in these two fungi and is highly conserved in all eukaryotes, but can be substituted by other related small GTPases in some plant pathogens (Harris and Momany 2004). The site at which Cdc42 is activated is determined in *S. cerevisiae* by cortical landmark proteins that are pre-anchored within the plasma membrane adjacent to previous bud sites (Casamayor and Snyder 2002). In *C. albicans* hyphae, Cdc42-GFP localises to the hyphal apex and immediate sub-apical region, but its activation is limited to the positioning of its guanine exchange factor (GEF), Cdc24, which in turn is localised by another small GTPase, Rsr1/Bud1 (Fig. 2.3a). Thus, it is the positioning of Rsr1 that determines the site at which polarised growth is established. Deletion of Rsr1 or its GTPase-activating protein Bud2 causes erratic wandering of the polarisome complex, which maintains

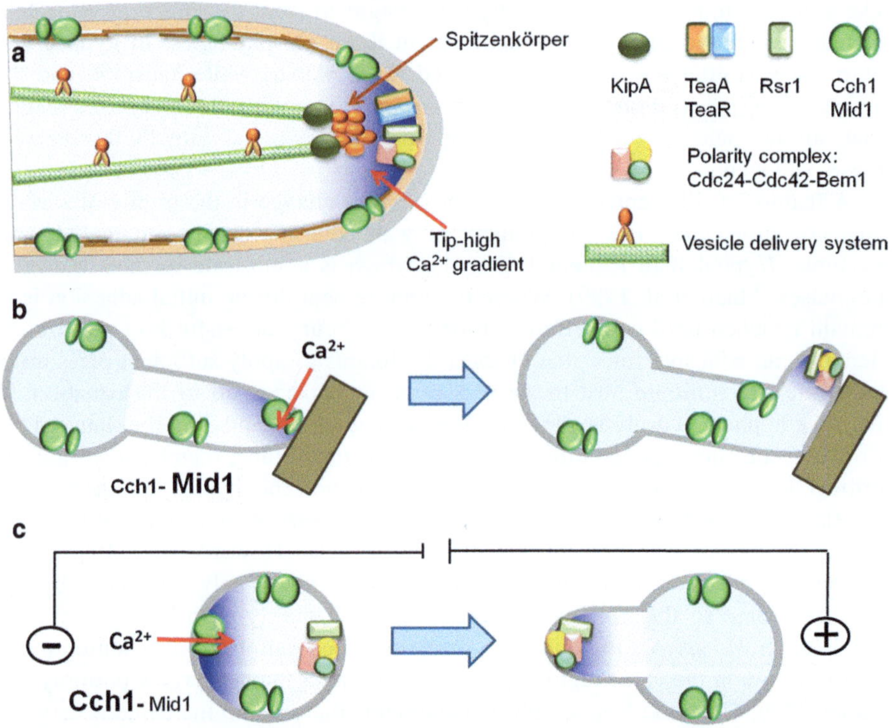


**Fig. 2.3** Model for tip growth and re-orientation of hyphae. (**a**) Fungal hyphae share conserved molecular complexes that drive polarised growth. These include a vesicle delivery system, a vesicle supply centre (Spitzenkörper), plasma membrane calcium channels and a tip-high calcium gradient. A system of cell-end markers and small GTPases marks the site of growth and directs delivery of the necessary materials (Brand and Gow 2009). (**b**) Model for thigmotropic response in *C. albicans*. Hyphal tips that contact an obstacle undergo wall and plasma membrane deformation and stress. This is sensed by the stretch-activated calcium channel regulator Mid1, causing localised calcium influx via Cch1. The asymmetry of the calcium gradient results in repositioning of the active cell polarity Cdc24–Cdc42–Bem1 complex and the establishment of a new growth axis. (**c**) Model for the galvanotropic response in *C. albicans*. In an applied electric field, the anodal face of the yeast cell becomes hyperpolarised, while the cathodal face becomes depolarised. Membrane depolarisation activates the L-type voltage-gated calcium channel Cch1, which permits entry of calcium at the cathodal side of the cell. The presence of a localised calcium gradient overrides the signal generated by other internal polarity markers, and hyphae subsequently emerge and grow towards the cathode

polarised growth, leading to hyphal evagination at random sites on the cell surface (Hausauer et al. 2005). In *C. albicans*, the site of hyphal emergence from the mother yeast cell is reported as 50% random (lateral) and 50% determined by the position of the previous bud site (Herrero et al. 1999), which allows environmental cues to play a role in hyphal directionality. The early positional markers and molecular organisers of cell polarity have been studied in depth in *S. cerevisiae* and are generally conserved in *C. albicans* but not well characterised in filamentous fungi. However, directionality of germ tube emergence is particularly important in fungal conidia and appressoria that are heavily melanized, where wall strength is an impedence for the emergence of new growth. Hyphal germination in *Podospora anserina*, a pathogen that does not form an appressorium, occurs through a specific pore in the melanized ascospore (Lambou et al. 2008). Similarly, the penetration peg emerges from appressoria within an area defined by a heavily melanized ring that surrounds a polysaccharide bilayer (Wolkow et al. 1983; Bourett and Howard 1992). The sites of these emergence zones are probably laid down during wall biosynthesis prior to melanisation. How this spatial regulation is achieved in *P. anserina* ascospores is unknown, but in *de novo* appressoria its position is likely to be determined by detection of the host surface. Yeast-like cortical landmark proteins (BUDs) are either lacking or are poorly conserved in these fungi, and alternative mechanisms that mark the site of polarity establishment remain to be identified. The Cdc42 module, essential in *C. albicans* for organising the cytoskeleton and polarised vesicle transport, is not essential for growth in *M. grisea*, *P. marneffei*, *U. maydis* or *W. dermatitidis* (Bassilana et al. 2003; Zheng et al. 2006; Boyce et al. 2001; Ye and Szaniszlo 2000). In *M. grisea*, *Colletotrichum lindemuthianum* and *Botrytis cinerea*, a role for a tetraspannin-like protein has been suggested in the establishment of polarised outgrowth. According to a model proposed by Harris and Momany, positional signals in filamentous fungi could be conveyed by membrane receptors that recognise the host plant (Harris and Momany 2004). Tetraspannins generally function to cluster membrane proteins, and mutants (*pls1*$^{-}$) were observed to have "abortive and mislocalised" penetration pegs, suggesting that Pls1 may be important for the organisation of positional signalling complexes or cell-end markers (Clergeot et al. 2001; Gourgues et al. 2004; Veneault-Fourrey et al. 2006). In *C. lagenarium*, deletion of the orthologue of the *Schizosaccharomyces pombe* cell-end marker, Tea1, resulted in abnormal appressoria that could not penetrate cellulose, but the mutant phenotype was rescued by growth on the host plant or in the presence of calcium ions. Thus, the all-important emergence of a penetration peg into the host plant provides a clear example of a contact sensing-regulated directional growth response.

## 2.3 Thigmotropic Sensing

The host specificity of fungal thigmotropic responses has best been characterised in the rust fungus *U. appendiculatus*, which undergoes thigmo-differentiation on contact with the stomatal ridge of the host leaf (Hoch et al. 1987). The response ensures that the appressorium forms at the correct host penetration site and could be elicited in vitro by ridges in an inert plastic surface (Fig. 2.1c). It was concluded that the fungus could detect the precise height of the inductive host feature (5 μm) and that the process was purely mechanical. The mechanisms whereby hyphae sense perturbations in the substratum are as yet unknown, but there is some evidence that sensing could occur through mechanosensing (MS) $Ca^{2+}$-permeable channels. Such channels have been identified in *U. appendiculatus* and the hyphae of *C. albicans*, which also displays thigmotropic turning in vitro (Zhou et al. 1991; Watts et al. 1998). Similar to *U. appendiculatus*, *C. albicans* hyphae (2 μm diameter) respond to low ridges (0.8 μm high) in the substratum, which cause them to re-orient and grow along the ridge. An orthologue of the *S. cerevisiae* $Ca^{2+}$-permeable MS channel, Mid1, has been identified in *C. albicans*. Its deletion or that of the large L-type voltage-gated channel it putatively regulates, CaCch1; another putative $Ca^{2+}$ channel, CaFig1; or the Crz1 transcription factor that regulates expression of Cch1 reduces *C. albicans* hypha re-orientation by approximately 50% (Brand et al. 2007; Karababa et al. 2006). Thigmotropism was also attenuated in the presence of a $Ca^{2+}$ chelator, gadolinium or verapamil, blockers of stretch-activated and L-type channels, respectively, but not on deletion of CaYvc1, which releases $Ca^{2+}$ from intracellular stores (Brand, unpublished). Abnormally high intracellular $Ca^{2+}$ also abolishes sinusoidal growth, another contact-dependent response of wild-type hyphae in *C. albicans* (Fig. 2.1b). Hyphae and branches of the *pmr1*Δ mutant, which cannot pump $Ca^{2+}$ into the Golgi, grew as remarkably straight rods instead of developing regular oscillating waves on low-nutrient, high-concentration agar (Bates et al. 2005; Brand et al. 2009). Thus, normal $Ca^{2+}$ flux and signalling are involved in hyphal tip directionality in *C. albicans*. The current model for contact sensing in hyphae is, therefore, one whereby wall deformation causes perturbation of the membrane, which is sensed by stretch-activated $Ca^{2+}$ channels. Localised $Ca^{2+}$ influx could then act as a signal to influence the polarity machinery, but the pathway between the two has not been characterised (Fig. 2.3c) (Brand et al. 2007).

Nevertheless, by compromising the function of components of the Cdc42 polarity complex, several of which are essential in *C. albicans*, some interesting cell polarity and tropism phenotypes have emerged. On deletion of the cell polarity Ras-like GTPase, CaRsr1/Bud1 or its GAP, Bud2, polarised growth was established and maintained, but the hyphal trajectory became increasingly erratic and the polarisome (visualised using Spa2-YFP) moved randomly within the tip (Hausauer et al. 2005). Rsr1 GTP–GDP cycling is, therefore, required to anchor the polarisome stably within the apex, and without this linkage, hyphae were completely unable to respond to any of the known external tropism cues. The mutants were also attenuated in their ability to penetrate and damage cells in a model of oral epithelial infection, suggesting that normal tip directionality could be important for tissue invasion (Brand et al. 2008). Conversely, in mutants where the loss of GTPase activity was in the Rho-like GTPase Cdc42, the thigmotropic response was reduced, although hyphal trajectories, which meandered slightly as normal, were maintained. This suggests that GTP–GDP cycling may be required to "unlock" the position of the Cdc42 module so that it can relocate within the apex in response to external cues (Brand, unpublished).

In obligately filamentous fungi, mictrotubules play a prominent role alongside actin cables in vesicle transport and polarity maintenance. Correct association of microtubules with the hyphal apex is mediated by a cell-end marker, a cortical receptor and a kinesin (TeaA, TeaR and KipA in *A. nidulans*, respectively) (reviewed by Fischer et al. 2008) (Fig. 2.3a). The involvement of the cell marker system in the control of tip orientation is evidenced by the zig-zag trajectories of the deletion mutants. How the balance of power is regulated or co-ordinated between the microtubule- and actin-based polarity systems in hypha tip directionality is not known.

Work by Bowen et al. suggests that the point of contact between the obstacle and the growth zone in the hyphal tip influences the directional response. Experiments where hyphae were grown on ridges with shallow inclines led to the conclusion that the apical growth zone in hyphal tips, estimated to describe an arc of approximately 60°, was insensitive to touch because cross-linking between the immature cell wall, plasma membrane and cytoskeleton was incomplete (Bowen et al. 2007). One could speculate that the zone of tip sensitivity might, therefore, lie sub-apically to the growth zone but forward of the zone of endocytosis where polarity effectors are internalised for recycling to the tip. This concept could explain the observations of *C. albicans* thigmotropism in vitro. When *C. albicans* hyphae are tightly adhered to the substratum, tips re-orient on contact with ridges of 0.8 μm, less than half the diameter of a hypha (2 μm). Moreover, tip re-orientation increases with reducing ridge height (Brand et al. 2007). A refinement of this experiment with a wider range of ridge heights could help to determine the region of sensitivity in the hyphal tip. In contrast to tightly adhered cells, in in vitro assays, the undulating growth of *C. albicans* hyphae seen in infection models allows the tip to approach host cells orthogonally. Instead of deflecting on contact with the host cell membrane, the hyphal growth trajectory is maintained, resulting in host cell penetration (Dalle et al. 2010) and supporting the view that the hyphal growth zone is insensitive to touch.

## 2.4 Modulation of the Thigmotropic Response

In light of in vitro investigations, numerous and complex factors that could influence a pathogen's ability to sense and respond to contact have been identified. Nutrient availability, temperature and other environmental factors affect hyphal diameter, growth trajectory, turgor pressure, tip shape, cell wall elasticity and the expression of surface adhesins (Brand et al. 2009; Bowen et al. 2007, Ravishankar et al. 2001; Money and Harold 1992; Bastidas et al. 2009). The thigmotropic sensitivity of *Aspergillus niger* increased in low nutrient conditions due to a change in hyphal tip shape, which became more closely apposed to the substratum (Bowen et al. 2007). In *C. albicans*, low nutrient availability and surface hardness caused wavy and undulating growth, away from contact and the confines of the substratum

(Brand et al. 2009, V. Veses, personal communication). In *P. insidiosum*, a temperature increase from 24 to 37°C caused hyphae to increase their diameter and exert a greater force on skin, but although the hyphal diameter remained the same, when the cells were grown in serum, the effect of temperature was reversed (Ravishankar et al. 2001). Thus, pathogenic behaviour and success are the products of genetic interactions with a number of environmental variables. The role of thigmotropism in fungal infections within the human host will, therefore, be difficult to examine, but the different orientation responses observed in *C. albicans* hyphae could reflect niche-dependent tip regulation. Re-orientation in response to small obstacles might be important in biofilm formation or during superficial mucosal infection, where hyphae intercalate between keratinised cell layers. In contrast, in blood stream infections, tissue penetration and escape from internalisation by macrophages require that hyphae do not deflect on contact.

## 2.5 Lessons from Other Tropic Responses

### *2.5.1 Galvanotropism*

Galvanotropism is the directional growth response of cells to an imposed D.C. electrical field. A wide range of cell types respond tropically or tactically to exogenous electrical fields, and most eukaryotic cells and tissue generate electrical fields (Gow 1987, 1989, 1994; Bowling et al. 1986). Fungi grow directionally in electrical fields in the order of >2–5 V/cm, but zoospores of oomycetes swim towards the positive (anode) or negative (cathode) poles of much weaker fields. In the case of the latter, this is of interest since plant roots and other tissue generate sufficiently large electrical fields to orient the trajectory of the swimming cells to specific sites on the plant surface (Miller and Gow 1989; Morris et al. 1992; van West et al. 2002). In the case of fungal hyphae, the galvanotropism response is not sufficiently sensitive to believe that they respond to the relatively weak electrical fields generated by plant or animal tissues which they may inhabit.

There are several possible consequences of cells experiencing an external electrical field. The field may influence the behaviour, activity or mobility of *trans*-membrane proteins (Gow et al. 1984; Gow 1989). At the cathode-facing end of a cell, the internally negative membrane potential will become depolarised, while at the anode end of the cell, the membrane will be hyperpolarised. Membrane depolarisation can lead to triggering of action potentials by voltage-sensing proteins and channels (Gow 1987, 2004). Imposed electrical fields can also lead to membrane electrophoresis or electro-osmosis – the counter-flow of water due to the displacement of counter-ions around the charged groups of proteins within the field.

The application of electric fields is an useful experimental system for inducing marked and uniform hyphal alignments. Such experiments can assess what environmental conditions influence the ability of a hypha to respond to an imposed tropic stimulus and to test whether a specific mutation influences the galvanotropic response (Crombie et al. 1990; McGillivray and Gow 1986, 1987). In an electric field (10 V/cm), germ tube emergence of *C. albicans* hyphae is 70–80% cathodal and can be further increased by the presence of extracellular $Ca^{2+}$ in a dose-dependent manner (Crombie et al. 1990; Brand et al. 2007). The model generated from these experiments postulates that the cathodal face of the yeast cell is depolarised by the electric field, activating the voltage-gated calcium channel CaCch1. The resulting calcium influx overrides the existing positional information within the cell and indicates a new growth site (Fig. 2.3c). Such experiments have demonstrated that certain aspects of the regulation of the galvanotropic response are shared with those required for thigmotropism. For example, both tropic responses are attenuated in media of very low calcium ion concentration (Lever et al. 1994; Brand et al. 2007) and in both cases, Rsr1, a Ras-like GTPase component of the bud-site selection mechanism, is required for tropic orientation (Brand et al. 2008). These findings suggest that the underlying mechanism that is responsible for tropic orientation contains elements that are shared and are common for a wide range of tropic responses. However, it is interesting to note that deletion of the gene encoding the voltage-sensitive calcium ion channel Cch1 strongly attenuated galvanotropism, but had less of an effect on thigmotropism. Reciprocally, the membrane stretch sensor Mid1, which is thought to modulate the properties of Cch1, is critical for thigmotropism, but not for galvanotropism (Brand et al. 2007). These experiments all point to the importance of calcium regulation in the control of tropic orientation responses.

It has also been shown that the galvanotropic response of *Neurospora crassa* is strongly dependent on the pH of the medium – and indeed the response has an isolelectric point where there is no galvanotropism (Lever et al. 1994). This suggests that the pH may affect the mobility of key polarity-determining proteins embedded in the cell membrane, which in turn can be moved laterally in an electrical field, resulting in turn in directionality of hypha growth. Putting these phenomena together, it can be speculated that electrical fields induce tropic responses by influencing both the distribution of and ionic fluxes through calcium ion pumps and calcium channels in the cell membrane. Many aspects of cellular physiology resulting in tip growth are calcium dependent. For example, actin polymerisation and depolymerisation are both $Ca^{2+}$ dependent. Galvanotropic mechanisms often implicate actin organisation as the ultimate target of the changes induced by electrical field exposure (Gow 1994). However, it should also be pointed out that electrical fields can also result in the orientation of cell division plane of bacteria (Rajnicek et al. 1994), and although bacteria are now known to have some elements that are orthologues of eukaryotic cytoskeletal proteins, they do not have actin-dependent cell growth. A second actin-dependent cell biological process is that of vesicle fusion with the cell membrane. It is, therefore, feasible that these orientation mechanisms induce gradients and asymmetries in calcium ions that result in preferential vesicle fusion events and the spatial organisation of actin assembly and disassembly – and hence, directed cell extension and tropic growth.

### 2.5.2 Aerotropism and Chemotropism to Nutrients and Pheromones

Mycelia of filamentous fungi form well-spaced, usually non-overlapping hyphae, and individual hyphae often exhibit obvious negative autotropisms (Aoki et al. 1998). The environmental cues that explain these tropic responses have not been established – but they have been hypothesised to be due to either positive aerotropism towards regions of high oxygen tension or a form of negative chemotropism away from secreted fungal staling products. However, the evidence for *bone fide* chemotropism to soluble nutrients such as glucose and amino acids in most classes of fungi is not strong (Gooday 1975; Jansson et al. 1988). It is quite easy to use point sources of solutes at the ends of filled capillary tubes to reveal chemotaxis of bacteria, oomycete zoospores and chemotropic tropic alignments of zygomycete rhizoids, such as for the fungus *Allomyces macrogynus* or *Blastocladiella emersonii* (Youatt et al. 1988). In contrast, chemotropism of ascomycete and basidiomycete hyphae has been very difficult to demonstrate (Gooday 1975). The notable exception to this is the very strong and noticeable tropic response seen to sex pheromones of gametes of all the major classes of fungi (Gooday and Adams 1993). Sometimes such encounters occur in complex environments with competing environmental tropic cues. It is possible that the formation of biofilms stabilises such matrices and facilitates chemotropic orientations (Daniels et al. 2006). It is not known as yet if such sexual orientations are also subject to the tropic regulation mechanisms underpinning thigmotropic and galvanotropic orientation.

## 2.6 Models and Speculations

The studies summarised above suggest a number of common elements and themes emerging from studies of fungal tropic growth responses. Calcium ion flux across the apical cell membrane of the hypha mediated by Mid1 and Cch1 influences a number of tropic responses including thigmotropism and galvanotropism. Proteins that regulate calcium transport, such as Fig. 1 in *C. albicans*, are also implicated in the growth of mating shmoos (Yang et al. 2011), which undergo chemotropism in response to sex pheromone. Elements within the RAC/Rho GTPase signalling pathways such as Rsr1/Bud1 and Cdc42 that influence bud-site selection in yeast are also clearly involved in tropic growth, as are components of the actin-based and microtubule-based cytoskeleton (e.g. Tea1). Connecting these observations is the possibility that standing gradients of calcium ion gradients in the apical dome of the hypha could modulate the activities of components of the cortical polarity establishment and maintenance apparatus, which in turn regulates the site of vesicle secretion and cytoskeleton function. Both these latter processes are strongly influenced by $Ca^{2+}$, and some of the upstream signalling proteins contain calcium-binding motifs, such as EF hands, and it remains to be tested whether these domains are required for the integrity of the tropism signalling mechanism to be maintained. This model has certain attractions in so far as it would account for the ability of hyphal cells to respond to both exogenous (nutrients, pheromones, obstacles and electric fields) and endogenous signals (cortical markers for budding, germ tube evagination sites and septal junctions). What seems clear from the analysis of some of the mutants listed above is that orientation mechanisms can be dissociated from core growth mechanisms. Therefore, it seems that hyphae do have a cellular steering wheel and a range of sensors that enable them to explore, respond to and exploit their environments and to refine strategies for pathogenesis efficiently.

**Acknowledgements** Our research in this area has been funded by the MRC, BBSRC, Wellcome trust and Royal Society. AB is the recipient of an MRC New Investigator Award and a Royal Society University Research Fellowship.

## References

Allen EA, Hoch HC, Stavely JR, Steadman JR (1991) Uniformity among races of *Uromyces appendiculatus* in response to topographical signalling for appressorium formation. Phytopathology 81:883–887

Alsteens D, Garcia MC, Lipke PN, Dufresne YF (2010) Force-induced formation and propagation of adhesion nanodomains in living fungal cells. Proc Natl Acad Sci USA 107:20744–20749

Aoki A, Ito-Kuwa S, Nakamura K, Vidotta V, Takeo K (1998) Oxygen as a possible tropic factor in hyphal growth of *Candida albicans*. Mycoscience 39:231–238

Apoga D, Barnard J, Craighead HG, Hoch HC (2004) Quantification of substratum contact required for initiation of *Colletotrichum graminicola* appressoria. Fungal Genet Biol 41:1–12

Bassilana M, Blyth J, Arkowitz RA (2003) Cdc24, the GDP-GTP exchange factor for Cdc42, is required for invasive hyphal growth of *Candida albicans*. Eukaryot Cell 2:9–18

Bastidas RJ, Heitman J, Cardenas ME (2009) The protein kinase Tor1 regulates adhesin gene expression in *Candida albicans*. PLoS Pathog 5:e1000294

Bates S, MacCallum DM, Bertram G, Munro CA, Hughes HB, Buurman ET, Brown AJP, Odds FC, Gow NAR (2005) *Candida albicans* Pmr1p, a secretory pathway P-type $Ca^{2+}/Mn^{2+}$-ATPase, is required for glycosylation and virulence. J Biol Chem 280:23408–23415

Bourett TM, Howard RJ (1992) Actin in penetration pegs of the fungal rice blast pathogen, *Magnaporthe grisea*. Protoplasma 168:20–26

Bowen AD, Davidson FA, Keatch R, Gadd GM (2007) Induction of contour sensing in *Aspergillus niger* by stress and its relevance to fungal growth mechanics and hyphal tip structure. Fungal Genet Biol 44:484–491

Bowling DFJ, Edwards MC, Gow NA, Bowling DFJ, Edwards MC, Gow NAR (1986) Electrical currents at the leaf surface of *Commelina communis* and their relationships to stomatal activity. J Exp Bot 179:876–882

Boyce KJ, Hynes MJ, Adrianopoulos A (2001) The CDC42 homolog of the dimorphic fungus *Penicillium marneffei* is required for correct cell polarization during growth but not development. J Bacteriol 183:3447–3457

Brand A, Gow NAR (2009) Mechanisms of hypha orientation of fungi. Curr Opin Microbiol 12:1–8

Brand A, Shanks S, Duncan VMS, Yang M, Mackenzie K, Gow NAR (2007) Hyphal orientation of *Candida albicans* is regulated by a calcium-dependent mechanism. Curr Biol 17:347–352

Brand A, Vacharaksa A, Bendel C, Norton J, Haynes P, Henry-Stanley M, Wells C, Ross K, Gow NAR, Gale CA (2008) An internal polarity landmark is important for externally induced hyphal behaviours in *Candida albicans*. Eukaryot Cell 7:712–720

Brand A, Lee K, Veses B, Gow NAR (2009) Calcium homeostasis is required for contact-dependent helical and sinusoidal tip growth in *Candida albicans* hyphae. Mol Microbiol 71:1155–1164

Casamayor A, Snyder M (2002) Bud-site selection and cell polarity in budding yeast. Curr Opin Microbiol 5:179–186

Christensen MJ, Bennett RJ, Ansari HA, Koga H, Johnson RD, Bryan GT, Simposon WR, Koolarard JP, Nickless EM, Voisey CR (2008) Epichloe endophytes grow by intercalary hyphal extension in elongating grass leaves. Fungal Genet Biol 45:84–93

Clergeot PH, Gourgues M, Cots J, Laurans F, Latorse MP, Pepin R, Tharreau D, Notteghem JL, Lebrun MH (2001) PLS1, a gene encoding a tetraspanin-like protein, is required for penetration of rice leaf by the fungal pathogen *Magnaporthe grisea*. Proc Natl Acad Sci USA 98:6963–6968

Collins TJ, Read ND (1997) Appressorium induction by topographical signals from six cereal rusts. Physiol Mol Plant Pathol 51:169–179

Crombie T, Gow NAR, Gooday GW (1990) Influence of applied electrical fields on yeast and hyphal growth of *Candida albicans*. J Gen Microbiol 136:311–317

Dalle F, Wächtler B, L'Ollivier C, Holland G, Bannert N, Wilson D, Labruère C, Bonnin A, Hube B (2010) Cellular interactions of *Candida albicans* with human oral epithelial cells and enterocytes. Cell Microbiol 12:248–271

Daniels KJ, Srikantha T, Lockhart SR, Pujol C, Soll DR (2006) Opaque cells signal white cells to form biofilms in *Candida albicans*. EMBO J 25:2240–2252

Davies JM, Stacey AJ, Gilligan CA (1999) *Candida albicans* hyphal invasion: thigmotropism or chemotropism? FEMS Microbiol Lett 171:245–249

Deising H, Nicholson RL, Haug M, Howard RJ, Mendgen K (1992) Adhesion pad formation and the involvement of cutinase and esterases in the attachment of uredospores to the host cuticle. Plant Cell 4:1101–1111

Ding JP, Pickard BG (1993) Mechanosensory calcium-selective cation channels in epidermal cells. Plant J 3:83–110

Dupres V, Alsteens D, Wilk S, Hansen B, Heinisch JJ, Dufrene YF (2009) The yeast Wsc1 cell surface sensor behaves like a nanospring in vivo. Nat Chem Biol 5:857–862

Feng J, Wang F, Liu G, Greenshields D, Shen W, Kaminskyj S, Hughes GR, Peng Y, Selvaraj G, Zou J, Wei Y (2009) Analysis of a *Blumeria graminis*-secreted lipase reveals the importance of host epicuticular wax components for fungal adhesion and development. Mol Plant Microbe Interact 22:1601–1610

Fischer R, Zekert N, Takeshita N (2008) Polarized growth in fungi – interplay between the cytoskeleton, positional markers and membrane domains. Mol Microbiol 68:813–826

Gent AN, Kaang S (1986) Pull-off forces for adhesive tapes. J Appl Polym Sci 32:4689–4700

Gode-Potratz CJ, Kustusch RJ, Breheny PJ, Weiss DS, McCarter LL (2011) Surface sensing in *Vibrio parahaemolyticus* triggers a programme of gene expression that promotes colonization and virulence. Mol Microbiol 79:240–263

Gooday GW (1975) Chemotaxis and chemotropism in fungi and algae. In: Carlile MJ (ed) Primitive sensory and communication systems. Academic, London, pp 155–204

Gooday GW, Adams DJ (1993) Sex hormones and fungi. Adv Microb Physiol 34:69–145

Goriely A, Tabor M (2006) Estimates of biomechanical forces in *Magnaporthe grisea*. Mycol Res 110:755–759

Gourgues M, Brunet-Simon A, Lebrun MH, Levis C (2004) The tetraspanin BcPls1 is required for appressorium-mediated penetration of *Botrytis cinerea* into host plant leaves. Mol Microbiol 51:619–629

Gow NAR (1987) Polarity and branching induced by electrical fields. In: Poole RK, Trinci APJ (eds) Spatial organisation in eukaryotic microbes. Special publications of the society for general microbiology, vol 23. IRL Press, Oxford, pp 25–41
Gow NAR (1989) The circulating ionic currents of microorganisms. Adv Microb Physiol 30:89–123
Gow NAR (1993) Non-chemical signals used for host location and invasion by fungal pathogens. Trends Microbiol 1:45–50
Gow NAR (1994) Growth and guidance of the hyphal apex. Microbiology 140:3193–3205, Fleming Lecture
Gow NAR (2004) New angles in mycology: studies in directional growth and directional motility. Mycol Res 108:5–13
Gow NAR, Kropf DL, Harold FM (1984) Growing hyphae of *Achlya bisexualis* generate a longitudinal pH gradient in the surrounding medium. J Gen Microbiol 130:2967–2974
Gow NAR, Perera THS, Sherwood-Higham J, Gooday GW, Gregory DW, Marshall D (1994) Investigation of touch-sensitive responses by hyphae of the human pathogenic fungus *Candida albicans*. Scanning Microsc 8:705–710
Harris SD, Momany M (2004) Polarity in filamentous fungi: moving beyond the yeast paradigm. Fungal Genet Biol 41:391–400
Hausauer DL, Gerami-Nejad M, Kistler-Anderson C, Gale CA (2005) Hyphal guidance and invasive growth in *Candida albicans* require the Ras-Like GTPase Rsr1p and its GTPase-activating protein Bud2p. Eukaryot Cell 4:1273–1286
Hedge Y, Kollatukudy PE (1997) Cuticular waxes relieve self-inhibition of germination and appressorium formation by conidia of *Magnaporthe grisea*. Physiol Mol Plant Pathol 51:75–84
Herrero AB, Lopez MC, Fernandez-Lago L, Dominguez A (1999) *Candida albicans* and *Yarrowia lipolytica* as alternative models for analysing budding patterns and germ tube formation in dimorphic fungi. Microbiology 145:2727–2737
Hoch HC, Bojko RJ, Comeau GL, Allen EA (1993) Integrating microfabrication and biology. Circuits and Devices 9:16–22
Hoch H, Staples R, Whitehead B, Comeau J, Wolf E (1987) Signalling for growth orientation and cell differentiation by surface topography in *Uromyces*. Science 235:1659–1662
Howard RJ, Ferrari MA, Roach DH, Money NP (1991) Penetration of hard substrates by a fungus employing enormous turgor pressures. Proc Acad Sci USA 88:11281–11284
Hoyer LL, Hecht JE (2001) The ALS5 gene of *Candida albicans* and analysis of the Als5p N-terminal domain. Yeast 18:49–60
Hoyer LL, Green CB, Oh SH, Zhao X (2008) Discovering the secrets of the *Candida albicans* agglutinin-like sequence (ALS) gene family – a sticky pursuit. Med Mycol 46:1–15
Hutton RD, Kerbs S, Yee K (1978) Scanning electron microscopy of experimental *Trichophyton mentagrophytes* infections in guinea pig skin. Infect Immun 21:247–253
Jaffe MJ, Leopold AC, Staples RC (2002) Thigmo responses in plants and fungi. Am J Bot 89:375–382
Jansson H-B, Johansson T, Nordbring-Herts B, Tunlid A, Odham G (1988) Chemotropic growth of germ tubes of *Cochliobolus sativus* to barley roots or root exudates. Trans Br Mycol Soc 90:647–650
Kaminskyj SG, Heath IB (1995) Integrin and spectrin homologues, and cytoplasm-wall adhesion in tip growth. J Cell Sci 108:849–856
Karababa M, Valentino E, Pardini G, Coste AT, Bille J, Sanglard D (2006) CRZ1, a target of the calcineurin pathway in *Candida albicans*. Mol Microbiol 59:1429–1451
Klotz SA, Rutten MJ, Smith RL, Babcock SR, Cunningham MD (1993) Adherence of *Candida albicans* to immobilized extracellular matrix proteins is mediated by calcium-dependent surface glycoproteins. Microb Pathog 14:133–147
Kumamoto CA (2005) A contact-activated kinase signals *Candida albicans* invasive growth and biofilm development. Proc Natl Acad Sci USA 102:5576–5581

Kumamoto CA (2008) Molecular mechanisms of mechanosensing and their roles in fungal contact sensing. Nat Rev Microbiol 6:667–673

Kumamoto CA, Vinces MD (2005) Alternative *Candida albicans* lifestyles: growth on surfaces. Annu Rev Microbiol 59:113–133

Kwon YH, Hoch HC (1990) Temporal and spatial dynamics of appressorium formation in *Uromyces appendiculatus*. Exp Mycol 15:116–131

Lambou K, Malagnac F, Barbisan C, Tharreau D, Lebrun MH, Silar P (2008) The crucial role of the Pls1 tetraspanin during ascospore germination in *Podospora anserina* provides an example of the convergent evolution of morphogenetic processes in fungal plant pathogens and saprobes. Eukaryot Cell 7:1809–1818

Lang I, Barton DA, Overall RL (2004) Membrane – wall attachments in plasmolysed plant cells. Protoplasma 224:231–243

Lever M, Robertson B, Buchan ADB, Gooday GW, Gow NAR (1994) pH and $Ca^{2+}$ dependent galvanotropism of filamentous fungi: implications and mechanisms. Mycol Res 98:301–306

Lipschutz J, Mostov K (2007) Exocytosis: the many masters of the exocyst. Curr Biol 1:212–214

Liu Y, Filler SG (2011) *Candida albicans* Als3, a multifunctional adhesin and invasin. Eukaryot Cell 10:168–173

Liu H, Suresh A, Willard FS, Siderovski DP, Lu S, Naqvi NI (2007) Rgs1 regulates multiple Gα subunits in *Magnaporthe* pathogenesis, asexual growth and thigmotropism. EMBO J 26:690–700

Machesky LM, Gould KL (1999) The Arp2/3 complex: a multifunctional actin organizer. Curr Opin Cell Biol 11:117–121

McGillivray AM, Gow NAR (1986) Applied electrical fields polarize the growth of mycelial fungi. J Gen Microbiol 132:2515–2525

McGillivray AM, Gow NAR (1987) The transhyphal electrical current of *Neurospora crassa* is carried principally by protons. J Gen Microbiol 133:2875–2881

Miller AL, Gow NAR (1989) Correlation between profile of ion-current circulation and root development. Physiol Plant 75:102–108

Money NP, Harold FM (1992) Extension growth of the water mold *Achlya*: interplay of turgor and wall strength. Proc Natl Acad Sci USA 89:4245–4249

Morris BM, Reid B, Gow NAR (1992) Electrotaxis of zoospores of *Phytophthora palmivora* at physiologically relevant field strengths. Plant Cell Environ 15:645–653

Perera THS, Gregory DW, Marshall D, Gow NAR (1997) Contact sensing in hyphae of dermatophytic and saprophytic fungi. J Med Vet Mycol 35:289–294

Pickard BG (1992) Wall to membrane linkers, stretch activated channels, and the detection of tension, voltage, temperature, auxin, and pH. ASGSB Bull 6:31

Rajnicek AM, McCaig CD, Gow NAR (1994) Electric fields induce curved growth of *Enterobacter cloacae*, *Escherichia coli* and *Bacillus subtilus* cells: implications for mechanisms of galvanotropism and bacterial growth. J Bacteriol 176:702–713

Ravishankar JP, Davis CM, Davis DJ, MacDonald E, Makselan SD, Millward L, Money NP (2001) Mechanics of solid tissue invasion by the mammalian pathogen *Pythium insidiosum*. Fungal Genet Biol 34:167–175

Read ND, Kellock LK, Knight H, Trewavas AJ (1992) Contact sensing during infection by fungal pathogens. In: Callow JA, Green JR (eds) Perspectives in plant cell recognition, vol 48. Cambridge University Press, Cambridge, pp 137–172

Roderick HW (1993) The infection of white clover (*Trifolium repens*) by conidia of *Cymadothea trifolii*. Mycol Res 97:227–232

Schild L, Heyken A, de Groot PWJ, Hiller E, Mock M, de Koster C, Horn U, Rupp S, Hube B (2011) Proteolytic cleavage of covalently linked cell wall proteins by *Candida albicans* Sap9 and Sap10. Eukaryot Cell 10:98–109

Shaw BD, Carroll GC, Hoch HC (2006) Generality of the prerequisite of conidium attachment to a hydrophobic substratum as a signal for germination among *Phyllosticta* species. Mycologia 98:186–194

Sheppard DC, Yeaman MR, Welch WH, Phan QT, Fu Y, Ibrahim AS, Filler SG, Zhang M, Waring AJ, Edwards JE Jr (2004) Functional and structural diversity in the Als protein family of *Candida albicans*. J Biol Chem 279:30480–30489

Smith PJS, Collis LP, Messerli MA (2010) Windows to cell function and dysfunction: signatures written in the boundary layers. Bioessays 32:514–523

Staab JF, Bradway SD, Fidel PL, Sundstrom P (1999) Adhesive and mammalian transglutaminase substrate properties of *Candida albicans* Hwp1. Science 283:1535–1538

Steinberg G (2007) Hyphal growth: a tale of motors, lipids, and the Spitzenkörper. Eukaryot Cell 6:351–360

Sudbery P, Court H (2007) Polarised growth in fungi. In: Howard RJ, Gow NAR (eds) The Mycota VIII, 2nd edn. Springer, Berlin, pp 137–166

Talbot NJ, Ebbole DJ, Hamer JE (1993) Identification and characterization of *MPG1*, a gene involved in pathogenicity from the rice blast fungus *Magnaporthe grisea*. Plant Cell 5:1575–1590

van West P, Morris BM, Reid B, Appiah AA, Osborne MC, Campbell TA, Shepherd SJ, Gow NAR (2002) Plant pathogens use electric fields to target plant roots. Mol Plant Microbe Interact 15:790–798

Veneault-Fourrey C, Lambou K, Lebrun MH (2006) Fungal Pls1 tetraspanins as key factors of penetration into host plants: a role in re-establishing polarized growth in the appressorium? FEMS Microbiol Lett 256:179–184

Verna J, Lodder A, Lee K, Vagts A, Ballester R (1997) A family of genes required for maintenance of cell wall integrity and for the stress response in *Saccharomyces cerevisiae*. Proc Natl Acad Sci USA 94:13804–13809

Virag A, Harris SD (2006) The Spitzenkörper: a molecular perspective. Mycol Res 110:4–13

Volgger M, Lang I, Ove-ìka M, Lichtscheidl I (2010) Plasmolysis and cell wall deposition in wheat root hairs under osmotic stress. Protoplasma 243:51–62

Warwar V, Dickman M (1996) Effects of calcium and calmodulin on spore germination and appressorium development in *Colletotrichum trifolii*. Appl Microbiol Environ 62:74–79

Watts H, Very A-A, Perera THS, Davies J, Gow NAR (1998) Thigmotropism and stretch-activated channels in the pathogenic fungus *Candida albicans*. Microbiology 144:689–695

Wolkow PM, Sisler HD, Vigil EL (1983) Effect of inhibitors of melanin biosynthesis on structure and function of appressoria of *Colletotrichum lindemuthianum*. Physiol Plant Pathol 23:55–71

Wösten HA, Schuren FH, Wessels JG (1994) Interfacial assembly of a hydrophobin into an amphipathic protein membrane mediates fungal attachment to hydrophobic surfaces. EMBO J 13:5848–5854

Yan S, Rodrigues RG, Cahn-Hidalgo D, Walsh TJ, Roberts DD (1998) Hemoglobin induces binding of several extracellular matrix proteins to *Candida albicans*. J Biol Chem 273:5638–5644

Yang M, Brand A, Srikantha T, Daniels K, Soll DR, Gow NAR (2011) Fig1 facilitates calcium influx and localises to membranes destined to undergo fusion during mating in *Candida albicans*. Eukaryot Cell 10:435–444

Ye X, Szaniszlo PJ (2000) Expression of a constitutively active Cdc42 homologue promotes development of sclerotic bodies but represses hyphal growth in the zoopathogenic fungus *Wangiella (Exophiala) dermatitidis*. J Bacteriol 182:4941–4950

Youatt J, Gow NAR, Gooday GW (1988) Bioelectric and biosynthetic aspects of cell polarity in *Allomyces macrogynus*. Protoplasma 146:118–126

Zheng W, Zhao Z, Chen J, Liu W, Ke H, Zhou J, Lu G, Darvill AG, Albersheim P, Wu S, Wang Z (2006) A Cdc42 ortholog is required for penetration and virulence of *Magnaporthe grisea*. Fungal Genet Biol 46:450–460

Zhou XL, Stumpf MA, Hoch HC, Kung C (1991) A mechanosensitive channel in whole cells and in membrane patches of the fungus *Uromyces*. Science 253:1415–1417

Zucchi PC, Davis TR, Kumamoto CA (2010) A *Candida albicans* cell wall-linked protein promotes invasive filamentation into semi-solid medium. Mol Microbiol 76:733–748

# Chapter 3
# Hyphal Fusion

**André Fleißner**

**Abstract** Since the early days of mycology, hyphal fusion or anastomosis has been recognized as a common feature of colony establishment and development in filamentous fungi. However, the role and function of this process remained mostly unclear. In recent years, much progress in understanding the molecular basis of anastomosis has been made, and numerous genes and proteins essential for fusion were identified. Insights emerging from these studies include the notion that hyphal fusion employs conserved signaling pathways, but adopts them in unusually dynamic fashions. In addition, increasing evidence suggests that anastomosis formation and pathogenic hyphal development share common machineries to some extent. Future challenges in studying hyphal fusion include deciphering the molecular networks controlling this complex cellular process and understanding the biological function of anastomosis.

## 3.1 Introduction

In the summer of 1888, the white lilies in the garden of Marshall Ward, a professor of Botany at the Royal Indian College in Cooper's Hill, met their fate. A disease apparently caused by some filamentous fungus carried off more than 90% of these beautiful plants. Ward's subsequent studies identified *Botrytis* as the culprit, which prompted him to further analyze the growth and development of this parasite. A certain feature of the mycelial colony caught his attention, namely, the constant occurrence of cross-connections or fusions between hyphae. Formation of these anastomoses was initiated by small branches, which mutually attracted each other and subsequently fused. Intrigued by these observations, Ward hypothesized

A. Fleißner (✉)
Institut für Genetik, Technische Universität Braunschweig, Braunschweig, Germany
e-mail: A.Fleissner@tu-braunschweig.de

J. Pérez-Martín and A. Di Pietro (eds.), *Morphogenesis and Pathogenicity in Fungi*, Topics in Current Genetics 22, DOI 10.1007/978-3-642-22916-9_3,

that fusion hyphae communicate and tropically interact through the means of secreted signals (Ward 1888).

Today we know that hyphal fusion – also known as anastomosis – is a common feature of colony establishment and development in many filamentous fungi (Glass et al. 2000; Read et al. 2009). Fusion occurs at different developmental stages and involves various types of hyphae (Fleissner et al. 2008). Vegetative spores of many different fungal species develop specialized fusion structures, the so-called conidial anastomosis tubes (CATs) (Roca et al. 2005a; Read et al. 2009). Through CATs, individual germlings connect and form functional units of higher order, which further develop into mycelial colonies. Within such a colony, various types of hyphae differentiate, exhibiting diverse growth behaviors. Leading hyphae at the periphery of the colony typically avoid each other by growing straight out from the colony and away from their neighbors (Buller 1931; Trinci 1984). In contrast, within the inner, older parts of the mycelium, hyphal branches attract each other and fuse, as observed and described by Ward. Thereby additional hyphal cross-connections are formed and mycelial interconnectedness increases (Ward 1888; Hickey et al. 2002).

During sexual reproduction of heterothallic species, mating partners fuse either via vegetative hyphae, establishing a transient heterokaryotic stage, or via specialized reproductive hyphae called trichogynes (Bistis 1981, 1996; Bruggeman et al. 2003). After fertilization, the formation of a structure containing the sexual spores, the ascus, also occurs via a cell fusion event in ascomycete fungi (Beckett and Wilson 1968; Raju 1980).

While the early mycologists described the morphological features of these fusion events, the role and function of anastomosis formation for fungal vegetative growth and development remain mysterious. Ward hypothesized that fusion might serve in maintaining homeostasis within the colony such that anastomoses are formed "to nourish the whole mycelium more equably, or to equilibrate certain differences which have unavoidably made themselves apparent in the metabolic process" (Ward 1888). After more than 120 years of research, this hypothesis is generally accepted (Glass et al. 2004; Fleissner et al. 2008; Read et al. 2009), but strong data supporting this idea are still lacking.

In contrast, much progress in understanding the molecular basis of germling and hyphal fusion has been made in recent years. The majority of the respective studies was conducted in the saprophytic fungi *Neurospora crassa* and *Sordaria macrospora* (Fleissner et al. 2008; Read et al. 2009); however, lately a growing number of investigations regarding the role and mechanism of anastomosis employ pathogenic species such as *Fusarium oxysporum* or *Alternaria brassicicola* (Craven et al. 2008; Prados Rosales and Di Pietro 2008; Rispail and Di Pietro 2009; Lopez-Berges et al. 2010; Ruiz-Roldan et al. 2010). Forward and reverse genetics approaches combined with biochemical analysis and live cell imaging identified several genes and proteins essential for anastomosis formation. Intriguingly, hyphal fusion and pathogenic development are under the control of similar signaling pathways (Craven et al. 2008; Prados Rosales and Di Pietro 2008; Lopez-Berges et al. 2010), and germling fusion involves an unusual mode of coordinated cell–cell signaling (Fleissner et al. 2009b).

The purpose of this chapter is to summarize our current knowledge of the molecular basis of vegetative germling and hyphal fusion and to introduce hypotheses on the function of anastomoses for fungal development, including their potential relationship to the parasitic lifestyle.

## 3.2 Types of Vegetative Fusion

### *3.2.1 Hyphal Fusion*

The colonies of many filamentous fungi are divided into morphologically distinct regions. In the colony periphery, hyphae typically exhibit negative autotropism such that the leading hyphae and their branches actively avoid each other (Trinci 1984). In contrast, in the inner older parts of the colony, branches are formed, which fill the spaces between the hyphae they originated from. These hyphal extensions frequently show positive autotropism such that they attract each other and fuse, thereby forming cross-connections (Buller 1931; Hickey et al. 2002).

While two existing hyphal tips commonly attract each other, fusion branches are also able to induce the formation of new pegs at the sides of neighboring hyphae. These new tips subsequently fuse with the initiating hyphae. In both cases, two hyphal tips establish physical contact and fuse. This tip-to-tip fusion mode appears to be common in filamentous fungi, including *Fusarium* spp., *Rhizoctonia solani*, *Cryphonectria parasitica*, *Pyricularia oryzae*, and *N. crassa* (Naito 1978; Chen and Wu 1977; Newhouse and MacDonald 1991; McCabe et al. 1999; Hickey et al. 2002).

Changes in growth direction during tropic responses are usually associated with repositioning of the Spitzenkörper (Riquelme et al. 1998; Hickey et al. 2002). The Spitzenkörper is thought to function as a vesicle supply center through which vesicles containing enzymes and other growth components are transported to the growing hyphal tip (Gierz and Bartnicki-Garcia 2001; Riquelme et al. 2007). After physical contact of two fusion tips in *N. crassa*, the Spitzenkörper of both hyphae become oriented directly opposite to each other and mark the point, where the fusion pore is formed. These vesicle-rich structures remain at the fusion point until fusion is completed, suggesting some potential functions during the fusion process, such as the delivery of cell wall degrading enzymes (Hickey et al. 2002).

### *3.2.2 Germling Fusion*

Besides hyphal fusion, the early mycologists observed another type of fungal cell merger: fusion between germinated or ungerminated vegetative spores. In 1930,

Köhler described the formation of small fusion bridges between conidia, germ tubes, or both (Köhler 1930) in *Botrytis alii*. These fusion bridges appeared notably different from germ tubes such that they were significantly narrower. To distinguish them from general hyphae, Köhler chose the term "Fusionshyphen" (fusion hyphae). More recently, Roca and coworkers coined the term conidial anastomosis tubes ("CATs") for similar structures observed in *Colletotrichum* and *N. crassa* (Roca et al. 2003, 2004, 2005b). A literature survey found that fusion between spores and germlings via CATs is common in filamentous fungi and that it is reported for more than 70 species covering more than 20 genera (Roca et al. 2005a).

In the saprophyte *N. crassa*, CATs are formed directly between conidia and germ tubes. Frequently, also the tips of the germ tubes fuse (Fig. 3.1a), followed by the formation of a newly growing hyphal tip (Roca et al. 2005b; Fleissner et al. 2005). In the latter case, determining if the germ tube is still a germ tube or if its apex has differentiated into a CAT is technically difficult. Detailed characterization of a

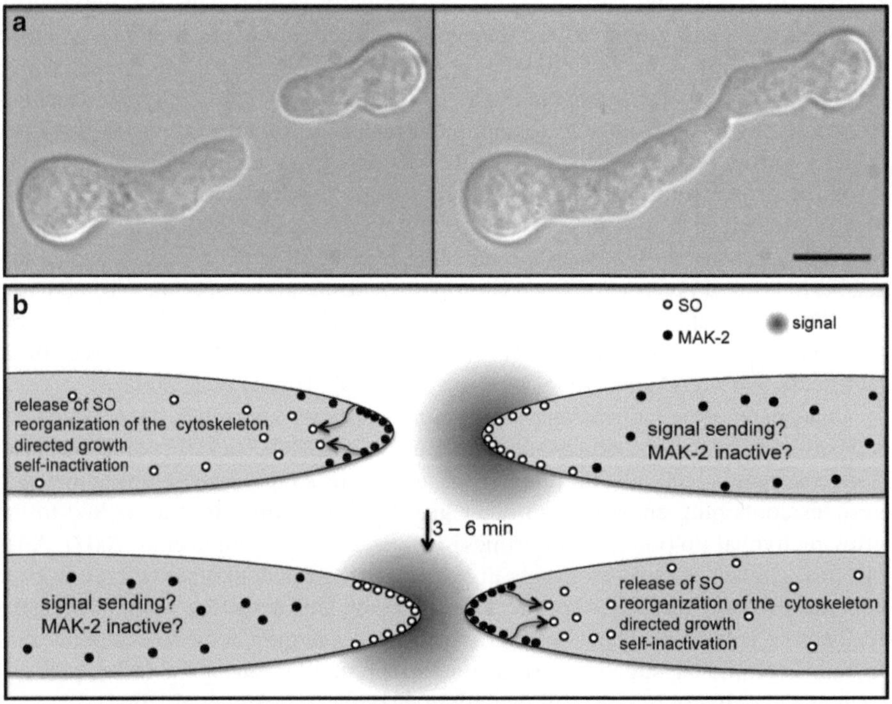


**Fig. 3.1** Germling fusion in *N. crassa* (**a**) Conidial germ tubes of *N. crassa* mutually attract each other (*left*) and subsequently fuse (*right*). Size bar: 5 μm. (**b**) Working model of the cell–cell signaling mechanism involved in *N. crassa* germling fusion. The MAP kinase MAK-2 and the filamentous ascomycete-specific SO protein are recruited to fusion tips in an alternating manner. Receiving the signal results in activation of MAK-2, reorganization of the cytoskeleton, and changes in growth direction. Subsequent inactivation of the signaling pathway might involve a negative feedback mechanism. Functional MAK-2 is required for the release of SO from the plasma membrane. Recruitment of SO might be related to signal release

recently identified mutant, which still exhibits CAT formation and fusion but appears to be defective in germ tube fusion, will further our understanding of this issue (Schürg and Fleißner, unpublished).

## 3.3 Molecular Basis of Fusion

From the cytological point of view, germling and hyphal fusion are carefully orchestrated multistep processes, whose individual stages depend on specific spatiotemporal adaptation of the cellular program. Initiation of fusion seems to require a certain competence, since not all hyphal branches within a colony and all germinating spores fuse (Hickey et al. 2002; Buller 1931). Once competency is established, fusion hyphae need to either identify a partner tip or induce its formation. The subsequent tropic interaction requires the involved external signals to be permanently translated into repositioning of the tip growth machinery toward the partner cell. As soon as the two tips touch, they have to recognize the cell–cell contact and shift their cellular program from "directed growth" toward "adhesion and fusion."

In recent years, the analysis of mutants with defects in one or more of these aspects of germling and hyphal fusion provided insight into genes and proteins required for these complex processes. All mutants affected in germling fusion also possess defects in hyphal fusion (as far as both aspects were tested), suggesting that both processes share a common molecular machinery (Xiang et al. 2002; Pandey et al. 2004; Fleissner et al. 2005; Prados Rosales and Di Pietro 2008; Simonin et al. 2010; Aldabbous et al. 2010).

### *3.3.1 Competency and Induction*

The early mycologists studying vegetative cell fusion in filamentous fungi observed a connection between culture conditions and anastomosis frequency. By cultivating *Leptosphaeria coniothyrium* and *Monilia fructigena* on agar containing different concentrations of malt extract, Laibach found a significant negative correlation between the amount of available nutrients and the number of observed fusions (Laibach 1928). Similar results were obtained for *B. allii* and *Fusarium* species (Köhler 1930). In the bean pathogen, *Colletotrichum lindemuthianum* fusion of CATs is fully repressed in axenic culture by nutrients and is only observed in water (Ishikawa et al. 2010). Recently, an influence of the nitrogen source on the fusion frequency of *F. oxysporum* germlings was reported. In the presence of ammonium nitrate, fusion was significantly reduced compared to that in cultures growing on sodium nitrate. Interestingly, this repressive effect of ammonium could be overwritten in the presence of rapamycin, a specific inhibitor of the Ser/Thr kinase TOR (Lopez-Berges et al. 2010). TOR is highly conserved in eukaryotes and controls growth and development in response to the availability of nutrients (Martin and Hall 2005; Soulard et al. 2009). Furthermore, fusion repression by the presence of ammonium ions depends on MeaB and AreA (Lopez-Berges et al. 2010), two central regulators of nitrogen metabolism (Wong et al. 2007; Caddick et al. 1994). Thus, the influence of nutrient availability on the fusion frequency seems to be mediated by the general pathways controlling nutrient response.

## *3.3.2 Recognition and Directed Growth*

### 3.3.2.1 An Unsual Signaling Mechanism

In *N. crassa*, CAT induction requires the NRC-1/MEK-2/MAK-2 MAP kinase module, which is homologous to the Fus3 pheromone response pathway of *Saccharomyces cerevsiae* (Pandey et al. 2004; Roca et al. 2005a, b; Maerz et al. 2008). CAT formation is also significantly reduced in strains lacking SO, a WW domain containing protein, found only in the filamentous ascomycete lineage. The molecular function of SO is unknown, but Δ*so* CATs are unable to chemotropically interact with each other or with wild-type germlings. Similarly, fusion between mature hyphae is also absent in this mutant (Fleissner et al. 2005). Inactivation of *so* homologs led to comparable fusion deficiencies in *S. macrospora*, *F. oxysporum*, and *A. brassicicola* (Engh et al. 2007; Prados Rosales and Di Pietro 2008; Craven et al. 2008).

Analysis of the subcellular dynamics of SO and MAK-2 during germling fusion in *N. crassa* recently revealed that the two fusion partners coordinately alternate between two physiological stages, while growing toward each other (Fleissner et al. 2009b). These physiological switches include the alternating recruitment of SO and MAK-2 to the plasma membrane of the hyphal tip region, where they accumulate in complexes of about 300 nm in diameter. Intriguingly, this oscillating localization of the two proteins occurs exactly in antiphase such that SO localizes to the tip of one cell, while MAK-2 resides at the plasma membrane of the fusion partner. This unusual cellular behavior requires the partners to be less than 15 μm apart. Once initiated, the oscillation period is between 6 and 12 min, and four to six switches in protein recruitment are observed until the cells make physical contact. Briefly after the tips touch, SO and MAK-2 accumulate at the contact region, where MAK-2 resides until fusion is completed.

Interpretations of these surprising findings include the hypothesis that the two fusion germlings coordinately alternate between signal sending and receiving (Fig. 3.1b). Cell–cell signaling and fusion often require physiological or genetic differences between the partners, such that one cell is the signal sender and the other the receiver. In *S. cerevisiae*, only cells of opposite mating type can mate. Each of the two cells secretes a mating type-specific pheromone and expresses a surface receptor specific for the recognition of the signal from the partner. Thus, two distinct signal gradients form, which allow the mating partners to use the same downstream machinery for signal transduction and directed growth, while avoiding self-stimulation. In contrast, fusion germlings in *N. crassa* are genetically identical. Employment of two different signaling compounds appears unlikely in this case. However, simultaneous production of the same signaling compound by both cells would prevent the formation of signaling gradients and would possibly result in self-activation. Coordinated alternation between signal sending and receiving, however, would solve this conundrum. Signal secretion in this model would have to be rather pulsative than permanent. Based on known MAP kinase functions, the tip accumulating MAK-2 would belong to the signal-receiving cell. It is common for these signaling proteins to translocate within the cell in response to external stimuli. In *S. cerevisiae*, Fus3 accumulates at shmoo tips where it promotes reorganization of the cytoskeleton and subsequent tropic growth (van Drogen et al. 2001; Matheos et al. 2004). Similar functions seem likely for MAK-2 during germling fusion. If the cells switch between signal sending and receiving and plasma membrane recruitment of the kinase indicates signal reception, SO recruitment would be related to signal release.

Many aspects concerning the molecular basis and biological function of this process await further investigation. One key question is the nature of the involved signaling molecule and its cognate receptor. Homologs of the yeast pheromones and pheromone receptors are present in filamentous ascomycete species and are essential for sexual mating partner interactions, such as trichogyne attraction by male conidia (Bobrowicz et al. 2002; Kim et al. 2002; Poggeler and Kuck 2001). In *N. crassa*, pheromones and their G-protein-coupled receptors are dispensable for germling or hyphal fusion (Kim and Borkovich 2004, 2006), indicating that sexual and vegetative fusion employ different signals and receptors. In addition, *N. crassa* mutant strains deficient in G-protein subunits Gβ, Gγ, or all three Gα (Kays and Borkovich 2004; Krystofova and Borkovich 2005) are still vegetative fusion competent, demonstrating that anastomosis formation does not rely on G-protein signaling (Fleißner and Glass, unpublished data).

In addition to identifying the upstream components of the hyphal fusion machinery, topics for further investigation should include the molecular function of SO, the structure of potential positive and negative feedback loops promoting the switching behavior, and the cross-talk of the MAK-2 cascade with other signaling pathways.

#### 3.3.2.2 Other Signaling Pathways

Besides the MAK-2 MAP kinase cascade, additional signal transduction pathways are essential for vegetative hyphal fusion. *N. crassa* possesses three distinct MAP kinase modules: the described MAK-2 module, the OS-2 cascade involved in osmostress signaling, and the MAK-1 cell wall remodeling pathway (Borkovich et al. 2004). Similar to MAK-2, OS-2 and MAK-1 as well as their upstream kinases are essential for vegetative hyphal fusion in *N. crassa* (Maerz et al. 2008). Similarly, a *Fusarium graminearum* mutant affected in the *mak-1* homolog *mgv1* is unable to form heterokaryons, suggesting a potential defect in hyphal fusion

(Hou et al. 2002). In *N. crassa*, defects of the Δ*mak-2* mutant are associated with increased activity of MAK-1 and are partially suppressed by defects in the NDR kinase COT-1 (Maerz et al. 2008). Further studies are needed to unravel the contributions of each of these pathways to the fusion process and to decipher their potential roles in fusion competency, cell communication, and the actual cell and membrane fusion process.

Recent studies in *S. macrospora* and *N. crassa* identified homologs of the Far multiprotein complex of *S. cerevisiae* as essential for vegetative hyphal fusion. In yeast, the proteins Far3, Far7, Far8, Far9, Far10, and Far11 form a complex, which is required for G1 cell cycle arrest after mating pheromone stimulation (Kemp and Sprague 2003). In the genome of *N. crassa*, only homologs of Far11, Far8, and Far9/10 are present (*ham-2*, *ham-3*, and *ham-4*, respectively) (Simonin et al. 2010). Deletion of *ham-2*, *ham-3*, or *ham-4* rendered the mutants anastomosis deficient (Simonin et al. 2010), while sexual trichogyne–conidium fusion was unaffected. Similarly, *S. macrospora* mutants affected in the *ham-2* and *ham-3* homologs *pro22* and *pro11* exhibit deficiencies in vegetative hyphal fusion (Rech et al. 2007; Bloemendal et al. 2010; Bernhards and Poggeler 2011). Similar to yeast mating, germling fusion in *N. crassa* also seems to be accompanied by mitotic arrest (Roca et al. 2010). Potential defects in coordinating nuclear behavior and cell fusion might cause the phenotype observed in these mutants. The human homologs of *ham-3* and *ham-4* (striatin and SLMAP) are parts of the so-called STRIPAK complex, a large multiprotein assembly. Interestingly, homologs of another component of this complex – Mob3 – are also essential for hyphal fusion in *N. crassa* and *S. macrospora* (Maerz et al. 2009; Bernhards and Poggeler 2011). SLMAP is essential for myoblast fusion during muscle development (Guzzo et al. 2004), suggesting potential conserved functions from fungi to animals. Based on studies in different systems, striatin complexes are thought to function as locally assembled signalosomes involved in the coordination of different signal transduction pathways (Bernhards and Poggeler 2011; Benoist et al. 2006).

#### 3.3.2.3 Other Fusion Mutants

During the last years, several other genes/proteins have been identified as essential factors for vegetative germling/hyphal fusion. In *N. crassa*, these include components involved in the formation of glycosylphosphatidylinositol-(GPI) anchors which attach proteins to the outer leaflet of the plasma membrane (Bowman et al. 2006, 2009), the transcription factors RCM-1 and RCO-1, and HAM-5, a WD40 domain containing protein potentially involved in modulating MAP kinase responses (Aldabbous et al. 2010). In *Aspergillus fumigatus* mutants lacking the GATA-type transcriptional activator NsdD are impaired in undergoing heterokaryon formation, indicating defects in hyphal fusion (Szewczyk and Krappmann 2010). Future studies have to identify the function and position of these components in the protein interaction network mediating germling/hyphal fusion.

### 3.3.3 Membrane Fusion

Most of the hyphal fusion mutants isolated so far are defective in early stages of the fusion process such as fusion competency or cell–cell communication. However, cell–cell recognition and establishing of tropic responses are just half the battle. After achieving physical contact, complex and highly regulated cellular processes are essential for the completion of cell fusion. Fusion tips have to recognize contact, tightly adhere, break down their cell walls, and finally form a fusion pore by merging of their plasma membranes. On the molecular level, these reactions must be tightly controlled to avoid potentially lethal malfunctions, such as cellular leakage. So far, no components mediating cell–cell adhesion and cell wall destruction during fusion have been identified in filamentous fungi.

The final stage of cell–cell fusion is initiated by plasma membrane merger. So far only a very limited number of proteins mediating plasma membrane fusion have been identified in different model systems (Oren-Suissa and Podbilewicz 2010). *S. cerevisiae* mutants carrying a deletion of *prm1*, a gene activated in response to pheromone, exhibit normal mating interactions preceding fusion; however, the actual cell merger is blocked in about 50% of the fusion pairs (Heiman and Walter 2000). Transmission electron microscopy analysis revealed that in *prm1* fusion pairs failing to fuse, cell wall breakdown still occurs. However, the plasma membranes between the cells remain intact. In *N. crassa*, fusion takes place at different developmental stages of the life cycle. Absence of its *Prm1* homolog affects vegetative germling fusion and sexual trichogyne–conidium fusion in a comparable manner. In addition, defects during the development of asci in Δ*Prm1* crosses suggest a similar deficiency in fusion of the crozier cell, a prerequisite for ascus formation (Fleissner et al. 2009a). Together these observations suggest that PRM1 promotes plasma membrane merger as part of a general membrane fusion machinery of fungi.

## 3.4 Why Do Hyphae Fuse?

Although anastomosis formation is a common feature of growth and development of filamentous fungi, its biological function remains unclear. Fusion between spores and/or germlings during colony establishment is believed to increase fitness and competitiveness of the forming mycelium (Fleissner et al. 2008). Pooling the resources of many individuals might result in higher germination and/or growth rates, and thus faster colonization of a given substrate (Hay 1995; Roca et al. 2003). Also, just increasing the colony size by connecting many smaller individuals might increase the chances of successful establishment. In basidiomycete fungi, larger colonies can have advantages over smaller units. In confrontation assays testing the interaction of the basidiomycete wood decomposers *Steccherinum fimbriatum* and *Hypholoma fasciculare*, one species eventually displaces the other. Which species succeeded over the other strongly depended on the inoculum size (Dowson et al. 1988). Similarly, confrontations of different sized colonies of the same species revealed that larger units generally dominated smaller ones (Holmer and Stenlid 1993). Whether a similar correlation between colony size and fitness exists in ascomycete species, whose colonies are generally much shorter-lived than those of basidiomycete fungi, still awaits exploration.

Anastomosis formation within mature colonies is thought to promote homeostasis within the mycelial network. Different portions of the colony can show different metabolic activities such that the region of growth can be different from the part where energy and building material are generated. Comparison of the transcriptional profiles of inner and outer regions of actively growing colonies of *N. crassa* revealed that genes involved in protein biosynthesis and energy production were more active in the middle section of the mycelium, while factors involved in polar growth, membrane biosynthesis, and signaling were enriched at the periphery of the colony (Kasuga and Glass 2008). Thus, uniform radial colony growth is likely to depend on redistribution of nutrients and molecules from sources to sinks. *N. crassa* wild-type colonies typically feature very even hyphal growth fronts consisting of individual hyphae with similar growth speed. In contrast, peripheral hyphae of the anastomosis-deficient *so* mutant exhibit significant differences in linear extension, resulting in uneven and frayed-appearing colony edges (Fleissner et al. 2005). An attractive hypothesis is that the lack of hyphal cross-connections in the mutant results in an uneven distribution of nutrients and growth molecules between individual hyphae, thereby causing uneven growth of hyphae.

A better-defined function of hyphal fusion consists in its promotion of hyphal integrity after injury. The colonies of filamentous ascomycete fungi typically grow as multinucleate syncytia. While this structural organization provides developmental advantages, it puts the entire colony at risk after wounding of individual compartments. To prevent an excessive loss of cytoplasm, septal pores are quickly plugged by a specialized hexagonally shaped organelle, the Woronin body (Woronin 1864; Collinge and Markham 1987; Markham and Collinge 1987; Tenney et al. 2000). After the injured hyphal segment is plugged, new tip growth is initiated from the neighboring intact compartments. These newly formed tips grow through the dead section and can fuse, thereby patching the hyphal interruption (Rothert 1892; Buller 1933).

Anastomosis formation between genetically different individuals is thought to promote genetic variability through parasexual recombination in species lacking sexual reproduction (Pontecorvo 1956; Read and Roca 2006). While this process has been highly valuable for genetic analysis of asexual species, such as *Aspergillus niger* (Swart et al. 2001; Bos et al. 1988), its role in nature remains unclear. In natural environments, heterokaryon formation is restricted by vegetative incompatibility, which limits or even prevents gene flow and recombination (Begueret et al. 1994; Glass et al. 2000; Loubradou and Turcq 2000).

Not only the function but also the consequences of hyphal fusion for colony growth remain poorly defined. Fusion between compatible hyphae often results in dramatic changes in cytoplasmic flow such that the cytoplasm of one hyphae quickly flows into the fusion partner, or that the direction of cytoplasmic flow is inverted within the hyphae. As a consequence, nuclei, organelles, metabolites, signaling factors, and other hyphal components quickly translocate throughout the mycelial colony (Hickey et al. 2002). The influence of this subcellular behavior on the physiology of the colony is so far unclear.

When genetically identical hyphae fuse in the pathogenic fungus *F. oxysporum*, an interesting nuclear dynamic process occurs. Formation of a fusion bridge between two uninuclear compartments is followed by a mitotic nuclear division in the fused segment. One of the daughter nuclei subsequently migrates through the fusion bridge into the adjacent hyphal compartment, which is followed by degradation of the resident nucleus (Ruiz-Roldan et al. 2010). This sophisticated mechanism might ensure that the number of nuclei remains stable in the typically uninucleate hyphal compartments of *F. oxysporum*. In contrast, similar nuclear behavior related to compatible fusion is not observed in *N. crassa* (Roca et al. 2010), whose hyphae consist of multinucleate segments.

## 3.5 Fusion and Pathogenicity

A tantalizing emerging theme is the potential relationship between hyphal fusion and pathogenic hyphal development. Future investigation of this relationship is worthwhile for several reasons. First, germling and hyphal fusion occur in many pathogenic species during host colonization, suggesting some function related to pathogenic development. Second, some molecular factors central for anastomosis formation are also essential for or contribute to infectious development, suggesting some potential evolutionary link between these two processes.

Fusion appears to be a common part of pathogenic colony growth. Examples include fusion of *Colletotrichum* spec. germlings on the surface of cowpea leaves (Latunde-Dada et al. 1999) or hyphal fusion between invading hyphae of *Magnaporthe grisea* in young lesions on rice (Chen and Wu 1977). Prados Rosales and coworkers found that *F. oxysporum* conidia germinating on tomato roots readily fuse via CATs (Prados Rosales and Di Pietro 2008). Mutants of this fungus affected in the *so* homolog or the fus3 (MAP kinase) homolog (Δ*fso1* and Δ*fmk1*, respectively) are unable to form such hyphal networks on the host surface. As a consequence, the developing mycelia are more easily washed away from the plant roots. However, Δ*fso1* is still able to infect, indicating that the ability to form anastomoses is not a prerequisite for, but promotes, pathogenic development (Prados Rosales and Di Pietro 2008). In contrast, in *A. brassicicola*, mutants affected in the homologous genes (Δ*Aso1* and Δ*amk1*) are both fusion defective and nonpathogenic on the host plant cabbage (Craven et al. 2008).

These examples also illustrate that the signaling pathways of hyphal fusion and pathogenic development are shared. In *N. crassa*, all mutants affected in the kinases of the homologous signaling modules of the yeast Fus3/Kss1, Slt2, and Hog1 pathways are fusion defective (Pandey et al. 2004; Roca et al. 2005b; Maerz et al. 2008).

Homologs of Fus3/Kss1 and Slt2 are essential for the pathogenic development of many fungal parasites on plants, including biotrophic, hemibiotrophic, and necrotrophic species (for review, see Zhao et al. 2007). Homologs of the Hog1 module are essential for infection in some pathogens, but dispensable in others. While it is not surprising that central regulators such as MAP kinases are involved in a variety of different developmental processes, comparing fusion and pathogenicity from the mechanistic point of view might be worthwhile. During both developmental processes, fungal hyphae have to perceive external signals in order to direct their growth: during germling/hyphal fusion, the fusion partners have to identify each other and establish physical contact. During pathogenic development, fungi have to recognize the host cell and direct their growth through potential entry sites into the host tissue or actively penetrate the surface, which also requires regulated orientation of the hypha. Analysis of the subcellular dynamics of the MAP kinase MAK-2 during germling fusion suggests that this MAP kinase might link signal perception with reorientation of the cytoskeleton. A similar function could be proposed for its homologs in pathogenic fungi.

The formation of appressoria, specialized infection structures that tightly adhere to the host cells, has several features reminiscent to germling and hyphal fusion. Both distinct developmental programs require cell–cell recognition, followed by cessation of hyphal elongation. In both cases, secretion of growth material such as cell wall components or enzymes involved in tip extension has to be replaced by factors deconstructing either the plant-derived polymers or the fungal outer casing. A prerequisite for trouble-free fusion or appressorial function is tight adherence of the fusion partners or the parasitic structure to the host surface. Further comparison of the molecular factors controlling both different developmental programs might prove valuable for our understanding of the evolution of hyphal growth and behavior.

While a potential relationship between fusion and plant infection appears highly speculative, a direct relationship between anastomosis and pathogenicity is observed in a mycoparasitic host–pathogen interaction of two zygomycete fungi. Infection of *Absidia glauca* by its parasite *Parasitella parasitica* includes tropic interactions and subsequent hyphal fusion of host and pathogen (Kellner et al. 1993). The observation of mating type dependency of this interaction led to the hypothesis that the parasite employs mechanisms of sexual propagation to infect its host (Satina and Blakeslee 1926; Jeffries 1985). This example illustrates how basic hyphal behavior can be adapted to serve novel functions.

## 3.6 Conclusion

Although hyphal fusion is a very common and basic feature of growth and development of filamentous fungi, its biological role and underlying molecular mechanisms remain only poorly understood. In recent years, there has been a revival of interest in studying anastomosis formation. There is emerging evidence that fusion employs conserved signaling pathways and molecular networks, some of them adapted in unique ways. Studying hyphal fusion will further our understanding of fungal biology. In addition, it has the potential to make significant contributions to the broader subjects of eukaryotic cell biology and development, including plasma membrane fusion, cell–cell communication, directed growth, subcellular dynamics of signaling factors, and cell–cell adhesion. Intriguingly, hyphal fusion and pathogenic development appear to share some molecular machineries. Both research fields can mutually benefit each other; for example, looking for fusion defects as a general part of characterizing pathogenicity mutants or testing the role of additional known pathogenicity factors for promoting hyphal fusion in saprophytes.

**Acknowledgments** I thank Dr. Carolyn Rasmussen and Timo Schürg for critical reading of the manuscript.

## References

Aldabbous MS, Roca MG, Stout A, Huang IC, Read ND, Free SJ (2010) The ham-5, rcm-1 and rco-1 genes regulate hyphal fusion in *Neurospora crassa*. Microbiology 156:2621–2629

Beckett A, Wilson IM (1968) Ascus cytology of *Podospora anserina*. J Gen Microbiol 53:81–87

Begueret J, Turcq B, Clave C (1994) Vegetative incompatibility in filamentous fungi: het genes begin to talk. Trends Genet 10:441–446

Benoist M, Gaillard S, Castets F (2006) The striatin family: a new signaling platform in dendritic spines. J Physiol Paris 99:146–153

Bernhards Y, Poggeler S (2011) The phocein homologue SmMOB3 is essential for vegetative cell fusion and sexual development in the filamentous ascomycete *Sordaria macrospora*. Curr Genet 57(2):133–149

Bistis GN (1981) Chemotropic interactions between trichogynes and conidia of opposite mating-type in *Neurospora crassa*. Mycologia 73:959–975

Bistis GN (1996) Trichogynes and fertilization in uni- and bimating type colonies of *Neurospora tetrasperma*. Fungal Genet Biol 20:93–98

Bloemendal S, Lord KM, Rech C, Hoff B, Engh I, Read ND, Kuck U (2010) A mutant defective in sexual development produces aseptate ascogonia. Eukaryot Cell 9:1856–1866

Bobrowicz P, Pawlak R, Correa A, Bell-Pedersen D, Ebbole DJ (2002) The *Neurospora crassa* pheromone precursor genes are regulated by the mating type locus and the circadian clock. Mol Microbiol 45:795–804

Borkovich KA, Alex LA, Yarden O, Freitag M, Turner GE, Read ND, Seiler S, Bell-Pedersen D, Paietta J, Plesofsky N, Plamann M, Goodrich-Tanrikulu M, Schulte U, Mannhaupt G, Nargang FE, Radford A, Selitrennikoff C, Galagan JE, Dunlap JC, Loros JJ, Catcheside D, Inoue H, Aramayo R, Polymenis M, Selker EU, Sachs MS, Marzluf GA, Paulsen I, Davis R, Ebbole DJ, Zelter A, Kalkman ER, O'Rourke R, Bowring F, Yeadon J, Ishii C, Suzuki K, Sakai W, Pratt R (2004) Lessons from the genome sequence of *Neurospora crassa*: tracing the path from genomic blueprint to multicellular organism. Microbiol Mol Biol Rev 68:1–108

Bos CJ, Debets AJ, Huybers A, Kobus G, Slakhorst SM (1988) Genetic analysis and the construction of master strains for assignment of genes to six linkage groups in *Aspergillus niger*. Curr Genet 14:437–443

Bowman SM, Piwowar A, Al Dabbous M, Vierula J, Free SJ (2006) Mutational analysis of the glycosylphosphatidylinositol (GPI) anchor pathway demonstrates that GPI-anchored proteins are required for cell wall biogenesis and normal hyphal growth in *Neurospora crassa*. Eukaryot Cell 5:587–600

Bowman SM, Piwowar A, Arnone ED, Matsumoto R, Koudelka GB, Free SJ (2009) Characterization of GPIT-1 and GPIT-2, two auxiliary components of the *Neurospora crassa* GPI transamidase complex. Mycologia 101:764–772

Bruggeman J, Debets AJ, Swart K, Hoekstra RF (2003) Male and female roles in crosses of *Aspergillus nidulans* as revealed by vegetatively incompatible parents. Fungal Genet Biol 39:136–141

Buller A (1931) Researches on fungi. Longman, London

Buller A (1933) Researches on fungi. Longman, London

Caddick MX, Peters D, Platt A (1994) Nitrogen regulation in fungi. Antonie Van Leeuwenhoek 65:169–177

Chen J, Wu H (1977) Hyphal anastomosis in *Pyricularia oryzae*. Protoplasma 92:281–287

Collinge AJ, Markham P (1987) Woronin bodies rapidly plug septal pores of severed *Penicillium chrysogenum* hyphae. Exp Mycol 9:80–85

Craven KD, Velez H, Cho Y, Lawrence CB, Mitchell TK (2008) Anastomosis is required for virulence of the fungal necrotroph *Alternaria brassicicola*. Eukaryot Cell 7:675–683

Dowson CG, Rayner ADM, Boddy L (1988) The form and outcome of mycelial interactions involving cord-forming decomposer basidiomycetes in homogeneous and heterogeneous environments. New Phytol 109:423–432

Engh I, Wurtz C, Witzel-Schlomp K, Zhang HY, Hoff B, Nowrousian M, Rottensteiner H, Kuck U (2007) The WW domain protein PRO40 is required for fungal fertility and associates with Woronin bodies. Eukaryot Cell 6:831–843

Fleissner A, Sarkar S, Jacobson DJ, Roca MG, Read ND, Glass NL (2005) The so locus is required for vegetative cell fusion and postfertilization events in *Neurospora crassa*. Eukaryot Cell 4:920–930

Fleissner A, Simonin AR, Glass NL (2008) Cell fusion in the filamentous fungus, *Neurospora crassa*. Methods Mol Biol 475:21–38

Fleissner A, Diamond S, Glass NL (2009a) The *Saccharomyces cerevisiae* PRM1 homolog in *Neurospora crassa* is involved in vegetative and sexual cell fusion events but also has postfertilization functions. Genetics 181:497–510

Fleissner A, Leeder AC, Roca MG, Read ND (2009b) Oscillatory recruitment of signaling proteins to cell tips promotes coordinated behavior during cell fusion. Proc Natl Acad Sci USA 106:19387–19392

Gierz G, Bartnicki-Garcia S (2001) A three-dimensional model of fungal morphogenesis based on the vesicle supply center concept. J Theor Biol 208:151–164

Glass NL, Jacobson DJ, Shiu PK (2000) The genetics of hyphal fusion and vegetative incompatibility in filamentous ascomycete fungi. Annu Rev Genet 34:165–186

Glass NL, Rasmussen C, Roca MG, Read ND (2004) Hyphal homing, fusion and mycelial interconnectedness. Trends Microbiol 12:135–141

Guzzo RM, Wigle J, Salih M, Moore ED, Tuana BS (2004) Regulated expression and temporal induction of the tail-anchored sarcolemmal-membrane-associated protein is critical for myoblast fusion. Biochem J 381:599–608

Hay FS (1995) Unusual germination of spores of *Arthrobotrys conoides* and *A. cladodes*. Mycol Res 99:981–982

Heiman MG, Walter P (2000) Prm1p, a pheromone-regulated multispanning membrane protein, facilitates plasma membrane fusion during yeast mating. J Cell Biol 151:719–730

Hickey PC, Jacobson D, Read ND, Louise Glass NL (2002) Live-cell imaging of vegetative hyphal fusion in *Neurospora crassa*. Fungal Genet Biol 37:109–119

Holmer L, Stenlid J (1993) The importance of inoculum size for the competitive ability of wood decomposing fungi. FEMS Microbiol Ecol 12:169–176

Hou Z, Katan T, Kistler HC, Xu JR (2002) A mitogen-activated protein kinase gene (MGV1) in *Fusarium graminearum* is required for female fertility, heterokaryon formation, and plant infection. Mol Plant Microbe Interact 15:1119–1127

Ishikawa FH, Souza EA, Read ND, Roca MG (2010) Live-cell imaging of conidial fusion in the bean pathogen, *Colletotrichum lindemuthianum*. Fungal Biol 114:2–9

Jeffries P (1985) Mycoparasitism within the zygomycetes. Bot J Linnean Soc 91:135–150

Kasuga T, Glass NL (2008) Dissecting colony development of *Neurospora crassa* using mRNA profiling and comparative genomics approaches. Eukaryot Cell 7:1549–1564

Kays AM, Borkovich KA (2004) Severe impairment of growth and differentiation in a *Neurospora crassa* mutant lacking all heterotrimeric G alpha proteins. Genetics 166:1229–1240

Kellner M, Burmester A, Wostemeyer A, Wostemeyer J (1993) Transfer of genetic information from the mycoparasite *Parasitella parasitica* to its host *Absidia glauca*. Curr Genet 23:334–337

Kemp HA, Sprague GF Jr (2003) Far3 and five interacting proteins prevent premature recovery from pheromone arrest in the budding yeast *Saccharomyces cerevisiae*. Mol Cell Biol 23:1750–1763

Kim H, Borkovich KA (2004) A pheromone receptor gene, pre-1, is essential for mating type-specific directional growth and fusion of trichogynes and female fertility in *Neurospora crassa*. Mol Microbiol 52:1781–1798

Kim H, Borkovich KA (2006) Pheromones are essential for male fertility and sufficient to direct chemotropic polarized growth of trichogynes during mating in *Neurospora crassa*. Eukaryot Cell 5:544–554

Kim H, Metzenberg RL, Nelson MA (2002) Multiple functions of mfa-1, a putative pheromone precursor gene of *Neurospora crassa*. Eukaryot Cell 1:987–999

Köhler E (1930) Zur Kenntnis der vegetativen Anastomosen der Pilze (II. Mitteilung). Planta 10:495–522

Krystofova S, Borkovich KA (2005) The heterotrimeric G-protein subunits GNG-1 and GNB-1 form a Gbetagamma dimer required for normal female fertility, asexual development, and galpha protein levels in *Neurospora crassa*. Eukaryot Cell 4:365–378

Laibach F (1928) Über Zellfusionen bei Pilzen. Planta 5:340–359

Latunde-Dada A, O'Conell R, Nash C, Lucas J (1999) Stomatal penetration of cowpea (*Vigna unguiculata*) leaves by a *Colletotrichum* species causing latent anthracnose. Plant Pathology 48:777–785

Lopez-Berges MS, Rispail N, Prados-Rosales RC, Di Pietro A (2010) A nitrogen response pathway regulates virulence functions in *Fusarium oxysporum* via the protein kinase TOR and the bZIP protein MeaB. Plant Cell 22:2459–2475

Loubradou G, Turcq B (2000) Vegetative incompatibility in filamentous fungi: a roundabout way of understanding the phenomenon. Res Microbiol 151:239–245

Maerz S, Ziv C, Vogt N, Helmstaedt K, Cohen N, Gorovits R, Yarden O, Seiler S (2008) The nuclear Dbf2-related kinase COT1 and the mitogen-activated protein kinases MAK1 and MAK2 genetically interact to regulate filamentous growth, hyphal fusion and sexual development in *Neurospora crassa*. Genetics 179:1313–1325

Maerz S, Dettmann A, Ziv C, Liu Y, Valerius O, Yarden O, Seiler S (2009) Two NDR kinase-MOB complexes function as distinct modules during septum formation and tip extension in *Neurospora crassa*. Mol Microbiol 74:707–723

Markham P, Collinge AJ (1987) Woronin bodies of filamentous fungi. FEMS Microbiol Rev 46:1–11

Martin DE, Hall MN (2005) The expanding TOR signaling network. Curr Opin Cell Biol 17:158–166

Matheos D, Metodiev M, Muller E, Stone D, Rose MD (2004) Pheromone-induced polarization is dependent on the Fus3p MAPK acting through the formin Bni1p. J Cell Biol 165

McCabe P, Gallagher M, Deacon J (1999) Microscopic observation of perfect hyphal fusion in *Rhizoctonia solani*. Mycol Res 103

Naito H (1978) Hyphal fusion in *Fusarium* leaf spot fungus of rice plants. Trans Mycol Soc Japan 19:11–21

Newhouse J, MacDonald W (1991) The ultrastructure of hyphal anastomoses between vegetatively compatible and incompatible virulent and hypovirulent strains of *Cryphonectria parasitica*. Can J Bot 69:602–614

Oren-Suissa M, Podbilewicz B (2010) Evolution of programmed cell fusion: common mechanisms and distinct functions. Dev Dyn 239:1515–1528

Pandey A, Roca MG, Read ND, Glass NL (2004) Role of a mitogen-activated protein kinase pathway during conidial germination and hyphal fusion in *Neurospora crassa*. Eukaryot Cell 3:348–358

Poggeler S, Kuck U (2001) Identification of transcriptionally expressed pheromone receptor genes in filamentous ascomycetes. Gene 280:9–17

Pontecorvo G (1956) The parasexual cycle in fungi. Annu Rev Micobiol 10:393–400

Prados Rosales RC, Di Pietro A (2008) Vegetative hyphal fusion is not essential for plant infection by *Fusarium oxysporum*. Eukaryot Cell 7:162–171

Raju NB (1980) Meiosis and ascospore genesis in *Neurospora*. Eur J Cell Biol 23:208–223

Read ND, Roca MG (2006) Vegetative hyphal fusion in filamentous fungi. In: Baluska F, Volkmann D, Barlow PW (eds) Cell–Cell channels. Landes Bioscience, Georgetown, TX, pp 87–98

Read ND, Lichius A, Shoji JY, Goryachev AB (2009) Self-signalling and self-fusion in filamentous fungi. Curr Opin Microbiol 12:608–615

Rech C, Engh I, Kuck U (2007) Detection of hyphal fusion in filamentous fungi using differently fluorescence-labeled histones. Curr Genet 52:259–266

Riquelme M, Reynaga-Pena CG, Gierz G, Bartnicki-Garcia S (1998) What determines growth direction in fungal hyphae? Fungal Genet Biol 24:101–109

Riquelme M, Bartnicki-Garcia S, Gonzalez-Prieto JM, Sanchez-Leon E, Verdin-Ramos JA, Beltran-Aguilar A, Freitag M (2007) Spitzenkorper localization and intracellular traffic of green fluorescent protein-labeled CHS-3 and CHS-6 chitin synthases in living hyphae of *Neurospora crassa*. Eukaryot Cell 6:1853–1864

Rispail N, Di Pietro A (2009) *Fusarium oxysporum* Ste12 controls invasive growth and virulence downstream of the Fmk1 MAPK cascade. Mol Plant Microbe Interact 22:830–839

Roca MG, Davide LC, Mendes-Costa MC, Wheals A (2003) Conidial anastomosis tubes in *Colletotrichum*. Fungal Genet Biol 40:138–145

Roca MG, Davide LC, Davide LM, Mendes-Costa MC, Schwan RF, Wheals AE (2004) Conidial anastomosis fusion between *Colletotrichum* species. Mycol Res 108:1320–1326

Roca M, Read ND, Wheals AE (2005a) Conidial anastomosis tubes in filamentous fungi. FEMS Microbiol Lett 249:191–198

Roca MG, Arlt J, Jeffree CE, Read ND (2005b) Cell biology of conidial anastomosis tubes in *Neurospora crassa*. Eukaryot Cell 4:911–919

Roca MG, Kuo HC, Lichius A, Freitag M, Read ND (2010) Nuclear dynamics, mitosis, and the cytoskeleton during the early stages of colony initiation in *Neurospora crassa*. Eukaryot Cell 9:1171–1183

Rothert W (1892) Über Sclerotium hydrophilum Sacc. einen sporenlosen Pilz. Botanische Zeitung 50:358–370

Ruiz-Roldan MC, Kohli M, Roncero MI, Philippsen P, Di Pietro A, Espeso EA (2010) Nuclear dynamics during germination, conidiation, and hyphal fusion of *Fusarium oxysporum*. Eukaryot Cell 9:1216–1224

Satina S, Blakeslee AF (1926) The Mucor parasite Parasitella in relation to sex. Proc Natl Acad Sci USA 12:202–207

Simonin AR, Rasmussen CG, Yang M, Glass NL (2010) Genes encoding a striatin-like protein (ham-3) and a forkhead associated protein (ham-4) are required for hyphal fusion in *Neurospora crassa*. Fungal Genet Biol 47:855–868

Soulard A, Cohen A, Hall MN (2009) TOR signaling in invertebrates. Curr Opin Cell Biol 21:825–836

Swart K, Debets AJ, Bos CJ, Slakhorst M, Holub EF, Hoekstra RF (2001) Genetic analysis in the asexual fungus *Aspergillus niger*. Acta Biol Hung 52:335–343

Szewczyk E, Krappmann S (2010) Conserved regulators of mating are essential for *Aspergillus fumigatus* cleistothecium formation. Eukaryot Cell 9:774–783

Tenney K, Hunt I, Sweigard J, Pounder JI, McClain C, Bowman EJ, Bowman BJ (2000) Hex-1, a gene unique to filamentous fungi, encodes the major protein of the Woronin body and functions as a plug for septal pores. Fungal Genet Biol 31:205–217

Trinci APJ (1984) Regualtion of hyphal branching and hyphal orientation. In: Jennings DH, Rayner ADM (eds) The ecology and physiology of the fungal mycelium. Cambridge University Press, Cambridge, UK, pp 23–52

van Drogen F, Stucke VM, Jorritsma G, Peter M (2001) MAP kinase dynamics in response to pheromones in budding yeast. Nat Cell Biol 3:1051–1059

Ward H (1888) A lily disease. Ann Bot 2:319–382

Wong KH, Hynes MJ, Todd RB, Davis MA (2007) Transcriptional control of nmrA by the bZIP transcription factor MeaB reveals a new level of nitrogen regulation in *Aspergillus nidulans*. Mol Microbiol 66:534–551

Woronin M (1864) Zur Entwicklungsgeschichte der Ascobolus pulcherrimus Cr. und einiger Pezizen. Abh Senkenb Naturforsch 5:333–344

Xiang Q, Rasmussen C, Glass NL (2002) The ham-2 locus, encoding a putative transmembrane protein, is required for hyphal fusion in *Neurospora crassa*. Genetics 160:169–180

Zhao X, Mehrabi R, Xu JR (2007) Mitogen-activated protein kinase pathways and fungal pathogenesis. Eukaryot Cell 6:1701–1714

# Chapter 4
# Signaling of Infectious Growth in *Fusarium oxysporum*

**Elena Pérez-Nadales and Antonio Di Pietro**

**Abstract** Infection-related development in fungal pathogens is regulated by a complex network of signaling pathways. In the vascular wilt fungus *Fusarium oxysporum*, the highly conserved mitogen-activated protein kinase (MAPK) Fmk1 controls invasive growth and virulence via the homeodomain transcription factor Ste12. Recently, the transmembrane mucin Msb2 was identified as a new component functioning upstream of this MAPK cascade in the process of host recognition and invasive growth. Signaling mucins contain a highly glycosylated extracellular domain, a single transmembrane region, and a short cytoplasmic tail. They have been extensively studied in mammalian cells as cell receptors functioning upstream of cancer-related MAPK pathways. In fungal pathogens of plants and humans, transmembrane mucins have recently emerged as novel virulence factors with a role in MAPK signaling and infection-related morphogenesis. In this chapter, we highlight the importance of these findings and discuss open questions on the mechanism of signal sensing and transmission by membrane mucins, as well as their relevance in the regulation of pathogenic development of fungi.

## 4.1 The *Trans*-Kingdom Pathogen *Fusarium oxysporum*

Fungi of the genus *Fusarium* are ubiquitous in soil, plant debris, and other organic substrates (Booth 1971). The widespread distribution of the genus relies on its ability to grow on a wide range of substrates and on its efficient mechanisms for

E. Pérez-Nadales • A. Di Pietro (✉)
Departamento de Genética, Universidad de Córdoba, Córdoba, Spain
e-mail: ge2dipia@uco.es

J. Pérez-Martín and A. Di Pietro (eds.), *Morphogenesis and Pathogenicity in Fungi*, Topics in Current Genetics 22, DOI 10.1007/978-3-642-22916-9_4, 

dispersal. *Fusarium oxysporum* is the causal agent of vascular wilt disease in a wide variety of economically important crops. *Fusarium* wilt is a major limiting factor in the production of many agricultural and horticultural crops, including tomato (*Lycopersicon* spp.), banana (*Musa* spp.), cabbage (*Brassica* spp.), onion (*Allium* spp.), cotton (*Gossypium* spp.), flax (*Linum* spp.), muskmelon (Cucumis spp.), pea (*Pisum* spp.), watermelon (*Citrullus* spp.), carnation (*Dianthus* spp.), chrysanthemum (*Chrysanthemum* spp.), gladiolus (*Gladiolus* spp.), and tulip (*Tulipa* spp.) (Armstrong and Armstrong 1981). *F. oxysporum* can survive for extended time periods as a saprophyte on infected plant debris in the soil, either in the form of mycelium, conidia or, most commonly, thick-walled chlamydospores (Agrios 2005). The presence of the host roots induces conidial germination through unknown signals, followed by fungal adhesion and differentiation of infection hyphae that penetrate the root preferentially through natural openings at the junctions between epidermal cells (Bishop and Cooper 1983; Lagopodi et al. 2002; Perez-Nadales and Di Pietro 2011; Rodriguez-Galvez and Mendgen 1995). The following steps of infection are the invasion of the root cortex and colonization of the xylem vessels that eventually leads to the expression of the characteristic vascular wilt symptoms (Agrios 2005).

Besides its well-studied activity as a plant pathogen, *F. oxysporum* is also known as an emerging pathogen of humans that can cause a broad spectrum of clinical infections, ranging from corneal keratitis (Hua et al. 2010) and onychomycosis to disseminated multiorgan infections in immunocompromised patients that frequently have fatal outcomes (Nucci and Anaissie 2002; Ortoneda et al. 2004). *F. oxysporum*, together with *F. solani* and *F. verticillioides*, are responsible for practically all cases of invasive fusariosis in humans (Guarro and Gene 1995; Nucci and Anaissie 2007). Due to its capacity to infect both plants and mammals, *F. oxysporum* has been successfully used as a model for the analysis of trans-kingdom pathogenicity on plant and mammalian hosts (Ortoneda et al. 2004; Martinez-Rocha et al. 2008).

The complete genome sequence of a tomato pathogenic strain of *F. oxysporum* f. sp. *lycopersici* was recently published (Ma et al. 2010), and provides an invaluable tool for the identification of new pathogenicity genes, leading to a deeper understanding of the molecular basis of infection. In this chapter, we review the current knowledge on the molecular pathways implicated in the regulation of infectious growth in *F. oxysporum* f. sp. *lycopersici*, with particular emphasis on a mitogen-activated protein kinase (MAPK) pathway, which is conserved and essential for virulence in a wide array of fungal plant pathogens (Lee et al. 2003; Lengeler et al. 2000; Rispail et al. 2009; Zhao et al. 2007). We then focus on the role of signaling mucins, a class of proteins that have recently emerged as novel virulence factors with a role in MAPK signaling and infection-related morphogenesis. We highlight the importance of these findings for understanding the molecular mechanisms that regulate pathogenic development fungi.

## 4.2 Signaling Pathways Regulating Infection-Related Morphogenesis and Virulence

Fungal plant pathogens have evolved strategies to recognize suitable hosts, penetrate and invade the plant tissue, overcome the host defenses, and optimize growth within the plant. To perform these tasks, they must process chemical and physical signals from the host through distinct cellular signal transduction pathways which coordinate the morphogenetic changes associated with pathogenic development. This includes directed hyphal growth, adhesion to the plant surface, differentiation of specialized infection structures, and secretion of effectors. Key stimuli sensed by phytopathogenic fungi include environmental parameters such as nutritional status, surface hardness, topography or hydrophobicity, plant compounds, and others (Mendoza-Mendoza et al. 2009; Ohtake et al. 1999; Uchiyama and Okuyama 1990; Xiao et al. 1997). Signal transduction pathways participate in the perception of these stimuli via cognate sensor molecules and propagate the signal intracellularly, leading to the synthesis of specific gene products and modulating fundamental cellular processes such as polarity, cell cycle, adherence, growth, and secondary metabolism. Signal transduction in fungal plant pathogens involves, among others, G proteins, components of cAMP signaling and MAPK cascades (Lee et al. 2003; Lengeler et al. 2000; Qi and Elion 2005; Wilson and Talbot 2009; Zhao et al. 2007).

### *4.2.1 The Pathogenicity MAPK Cascade*

MAPK cascades comprise a conserved module of three kinases: the MAPK, the MAPK kinase (MAPKK or MEK), and the MAPKK kinase (MAPKKK or MEKK) that sequentially activate each other by phosphorylation (Chang and Karin 2001). The upstream signals are sensed by specific receptors that trigger the MAPK module directly or through intermediate signaling components. MAPKs phosphorylate a diverse set of substrates, including transcription factors, translational regulators, protein kinases, phosphatases, and other classes of proteins, thereby regulating metabolism, cellular morphology, cell cycle progression, and gene expression in response to a variety of extracellular stresses and molecular signals.

Among the MAPKs implicated in fungal virulence, the yeast and fungal extracellular signal-regulated kinase (YERK1) subfamily (Kultz and Burg 1998) plays a key role in infection-related morphogenesis and pathogenicity. In *Saccharomyces cerevisiae* there are two members of this subfamily, Fus3 which regulates the response to mating pheromone and Kss1 which controls a morphogenetic switch from budding to filamentous growth in response to nutrient limitation. In contrast to yeast, most filamentous fungi have only one MAPK orthologous to Fus3 and Kss1. Xu and Hamer (1996) first reported that the ortholog of *S. cerevisiae* Fus3 in the rice blast fungus *Magnaporthe oryzae*, designated Pmk1 (for pathogenicity MAP kinase 1), was required for infection of rice plants. Subsequently, orthologs of

Pmk1 were shown to be essential for pathogenicity in a wide range of biologically diverse plant pathogens (Di Pietro et al. 2001; Jenczmionka et al. 2003; Lev and Horwitz 2003; Mey et al. 2002; Takano et al. 2000; Zheng et al. 2000; Lev et al. 1999; Ruiz-Roldan et al. 2001). In appressorium-forming species such as *M. oryzae*, *Colletotrichum lagenarium*, *Cochliobolus heterostrophus,* and *Pyrenophora teres*, mutants lacking the Fus3/Kss1 ortholog fail to differentiate appressoria (Lev et al. 1999; Takano et al. 2000; Xu and Hamer 1996; Ruiz-Roldan et al. 2001). However, these mutants also fail to grow on the plant tissue when inoculated into wound sites, suggesting that this MAPK regulates additional basic virulence functions beside appressorium development. In support of this view, species that penetrate their plant hosts directly without the need for appressoria such as *F. oxysporum*, *F. graminearum,* or the necrotrophic pathogen *Botrytis cinerea*, also require this MAPK for infection (Di Pietro et al. 2001; Jenczmionka et al. 2003; Zheng et al. 2000). In *F. oxysporum*, the Pmk1 ortholog Fmk1 shares over 90% identity with the orthologous MAPKs from the leaf pathogens *M. oryzae*, *C. lagenarium*, and *C. heterostrophus* (Di Pietro et al. 2001; Xu and Hamer 1996; Lev et al. 1999; Takano et al. 2000). Fmk1 is not essential for germination, but required for infection of tomato plants. Fmk1 controls infection-related morphogenesis, as it is required for correct differentiation of infection hyphae in the presence of tomato roots, as well as for attachment to and penetration of the roots (Di Pietro et al. 2001). Additionally, *F. oxysporum* Δ*fmk1* mutants have a strongly reduced ability to grow invasively on tomato fruit tissue, suggesting alterations in the expression profile of cell-wall degrading enzymes (Di Pietro et al. 2001). Indeed, transcript levels of *pl1* encoding an endopectate lyase are reduced in Δ*fmk1* mutants (Di Pietro et al. 2001). Fmk1 is also required for vegetative hyphal fusion, a ubiquitous process in filamentous fungi whose biological function is poorly understood (Prados Rosales and Di Pietro 2008). While hyphal fusion is not essential for plant infection, the establishment of hyphal networks may contribute to optimize virulence-related functions such as adhesion to host surfaces or exploitation of the limited nutrient resources encountered during infection (Prados Rosales and Di Pietro 2008).

### 4.2.2 G-Protein Signaling and cAMP-PKA cascade

Heterotrimeric G protein-mediated signal perception and propagation plays a central role in controlling cell growth, development, virulence, and secondary metabolite production in fungi (Bolker 1998; Lengeler et al. 2000; Kulkarni et al. 2005). Heterotrimeric G proteins consisting of Gα, Gβ, and Gγ subunits, typically transmit the signal from G protein-coupled receptors in a process that involves GDP-to-GTP exchange of the guanine nucleotide bound to the Gα subunit, followed by release of the Gα subunit from the Gβγ dimer. These subunits interact with a variety of downstream effectors such as cAMP-dependent protein kinase A (PKA), phospholipases, MAPKs, and calmodulin-dependent pathways and ion channels, featuring as candidate proteins involved in the coordinated regulation of the distinct signal transduction pathways.

G-protein signaling via the cAMP-PKA pathway plays a key role in controlling cell polarity in fungi. Constitutive activation of the cAMP cascade in *U. maydis* through deletion of the PKA regulatory subunit *ubc1* or expression of a dominant active allele of the Gα subunit Gpa3 abrogates formation of polarized filaments, whereas deletion of GPA3 leads to strongly elongated filament-like cells (Gold et al. 1994; Kruger et al. 2000). Deletion of the Gβ subunit Bpp1 in *U. maydis* also produces a hyperfilamentous phenotype that can be reversed by addition of exogenous cAMP (Muller et al. 2004).

Downstream of heterotrimeric G proteins, components of the cAMP signaling pathway regulate key aspects of fungal pathogenicity. In a classical cAMP signaling pathway, the signal is transmitted from a transmembrane cell surface receptor via heterotrimeric G-proteins to adenylyl cyclase, which synthesizes the secondary messenger cAMP (Lee et al. 2003). cAMP signaling is required for regulation of mating, hyphal morphogenesis, infection structure formation, sclerotium formation, sporulation, and spore germination in different plant pathogenic fungi (Gao and Nuss 1996; Kasahara and Nuss 1997; Muller et al. 2004; Nishimura et al. 2003; Liu and Dean 1997; Delgado-Jarana et al. 2005; Jain et al. 2002; Jain et al. 2003; Yamauchi et al. 2004). In *M. oryzae*, exogenous cAMP stimulates appressorium formation on noninducing hydrophilic surfaces, and disruption of a gene encoding a cAMP-dependent PKA causes a delay in appressorium formation (Mitchell and Dean 1995; Tucker and Talbot 2001).

In *F. oxysporum*, the cAMP/PKA cascade is directly involved in hyphal growth and development and interacts with the Fmk1 pathway to regulate expression of virulence genes (Prados-Rosales et al. 2006; Delgado-Jarana et al. 2005; Jain et al. 2002, 2003). Activation of cAMP-PKA signaling in *F. oxysporum* is coordinately regulated by the heterotrimeric G protein α and β subunits Fga1 and Fgb1, respectively, and results in inhibition of cell elongation and apical growth, promoting differentiation and conidiation (Delgado-Jarana et al. 2005). In addition, Fgb1 signals in a separate cAMP-independent pathway involved in polarization of the actin cytoskeleton, where it promotes reorientation of the polarity axis and subapical branching (Delgado-Jarana et al. 2005). Interestingly, upon pheromone stimulation the Gβ subunit Ste4p of *S. cerevisiae* recruits the polarity determinant Far1p, the guanine-nucleotide exchange factor Cdc24p, and the GTPase Cdc42 to the plasma membrane, where they participate in the assembly of a Cdc42p-dependent signaling complex that reorients polarity of the actin cytoskeleton and directs shmoo growth (Butty et al. 1998; Nern and Arkowitz 1999; Pruyne and Bretscher 2000). During this recruitment process, Ste4p interacts directly with the small G-protein Rho1p which is known to activate the cell integrity pathway via protein kinase C and the cell integrity MAPK Mpk1 (Bar et al. 2003). In *F. oxysporum*, Rho1 is essential for normal hyphal growth and plant infection, but not for virulence on immunodepressed mice (Martinez-Rocha et al. 2008). To date, firm evidence for a direct link between a G protein subunit and components of Mpk1 in a plant pathogenic species is still lacking.

While essential for plant infection, the Fmk1 and cAMP/PKA pathways contribute only marginally to virulence of *F. oxysporum* on mammalian hosts, similarly to the orthologous pathways in the human pathogenic fungus *Candida albicans* (Davidson et al. 2003; Csank et al. 1998; Prados-Rosales et al. 2006). Thus, infection of *F. oxysporum* on plants, but not on mammals is blocked by inactivating either the Fmk1 MAPK or the Gβ subunit Fgb1 functioning in the cAMP cascade. Interestingly, however, Δ*fmk1*Δ*fgb1* double mutants are avirulent on mice, and both Fmk1 and Fgb1 coordinately contribute to adhesion of fungal hyphae to mammalian tissue (Prados-Rosales et al. 2006). The nature of the surface components that mediate host adhesion of *F. oxysporum* is currently unknown. A proteomic approach suggested the possible involvement of glycosyl-phosphatidylinositol (GPI)-linked glycoproteins present at the cell surface (Prados-Rosales et al. 2009). Interestingly, the *S. cerevisiae* MAPKs Kss1 and Fus3 control expression of GPI-linked glycoproteins such as Flo11p and Fig2p which are required for adhesion during invasive growth and mating, respectively (Guo et al. 2000). In *Candida*, a family of structurally related GPI-linked glycoproteins termed adhesins promote adherence to mammalian tissue (Cormack et al. 1999). A second type of cell surface molecules involved in fungal adhesion are O-glycosylated mannoproteins, which also influence hyphal hydrophobicity (Singleton et al. 2005). Interestingly, hyphal surface hydrophobicity is markedly reduced in *F. oxysporum* Δ*fgb1* and Δ*fmk1* mutants and even further impaired in a Δ*fmk1*Δ*fgb1* strain, suggesting a coordinated regulation of the genes relevant for maintenance of this phenotypical trait. It remains unclear at which level cAMP and MAPK signaling pathways interact to coordinate the regulation of cell polarity and hyphal development in *F. oxysporum* and other plant pathogenic species.

## 4.3 The Infectious Growth MAPK Signaling Network in *F. oxysporum*

Recently, two new components of the Fmk1 MAPK cascade have been characterized in *F. oxysporum*: the homeodomain transcription factor Ste12 (Rispail and Di Pietro 2009) and the cell surface mucin Msb2 (Fig. 4.1a) (Perez-Nadales and Di Pietro 2011). *F. oxysporum* Ste12 controls invasive growth, the major Fmk1-dependent pathogenicity function. Both Δ*fmk1* and Δ*ste12* mutants fail to penetrate cellophane membranes, colonize living plant tissue, and kill tomato plants (Fig. 4.2). By contrast, Msb2 contributes only partially to this function, as Δ*msb2* mutants exhibit a significant but not complete reduction in cellophane penetration and invasive growth on fruits. Interestingly, a second type of transmembrane receptor, the tetraspan protein Sho1, also contributes to invasive growth in *F. oxysporum* (Pérez-Nadales and Di Pietro, unpublished data). Msb2 and Ste12 also have different functions in regulating another Fmk1-controlled process, secretion of pectinolytic enzymes. While Ste12 is dispensable for clear halo

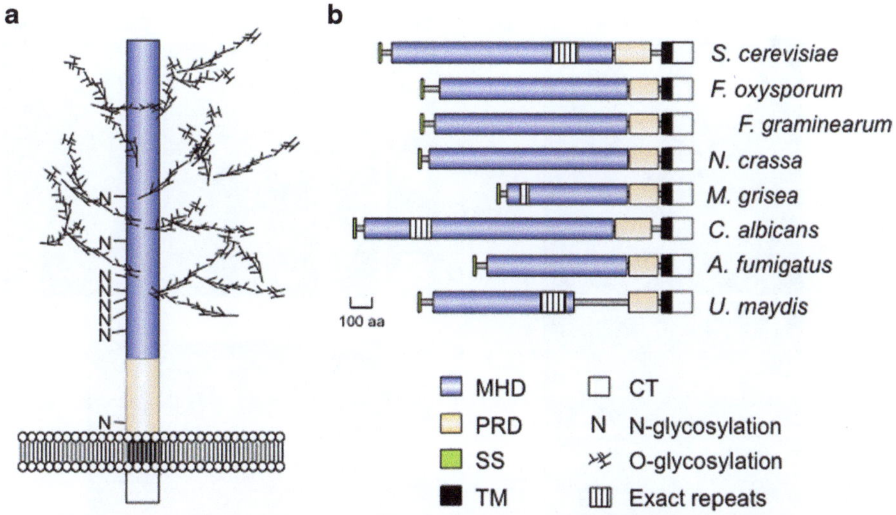


**Fig. 4.1** Structure of Msb2 mucin orthologs from fungal species. (**a**) Schematic representation of *F. oxysporum* Msb2 is shown on scale. The large number of carbohydrate side chains, combined with the large size of the MHD results in the hallmark rigid structure of tethered mucins, which is likely to extend to a remarkable distance from the cell surface. (**b**) Comparative representation of predicted mucin orthologs from the indicated fungal species. Key domains include the N-terminal signal sequence (SS, *green*); the serine/threonine rich mucin homology domain (MHD, *purple*); the positive regulatory domain (PRD, *orange*), a single transmembrane domain (TM, *black*); and the cytoplasmic tail (CT, *white*)

production on polygalacturonic acid-containing plates, Msb2 contributes partially to pectinolytic activity, since Δ*msb2* mutants show an intermediate phenotype when compared with the wt and the Δ*fmk1* mutant. Neither Msb2 nor Ste12 are required for additional Fmk1-controlled functions such as hyphal fusion and adhesion to tomato roots, suggesting a role of additional, yet unidentified receptors and transcription factors in these processes.

In addition to its function in the Fmk1 MAPK cascade, Msb2 appears to act in additional signaling pathways (Perez-Nadales and Di Pietro 2011). The Δ*msb2* mutants are more sensitive to the cell wall targeting compounds Congo red and calcofluor white than the wild type and Δ*fmk1* strains, suggesting a function of Msb2 in maintenance of cell wall integrity (Fig. 4.2b). Interestingly, a Δ*msb2* Δ*fmk1* double mutant is significantly more affected by calcofluor white than any of the single mutants, pointing to an additional role of Msb2 as a coordinator of cell wall integrity via an alternative cell stress response pathway other than Fmk1 (Fig. 4.3). One candidate could be the Mpk1 MAPK pathway which is involved in cell wall integrity (Levin 2005). Alternatively, Msb2 may contribute to cell wall integrity via components of the high osmolarity glycerol (HOG) MAPK cascade. In yeast, both the Mpk1 and the Hog1 MAPKs are necessary for the polarization of actin filaments toward weakened cell wall domains following cell wall damage or osmotic insult, respectively (Zarzov et al. 1996; Mazzoni et al. 1993; Brewster and

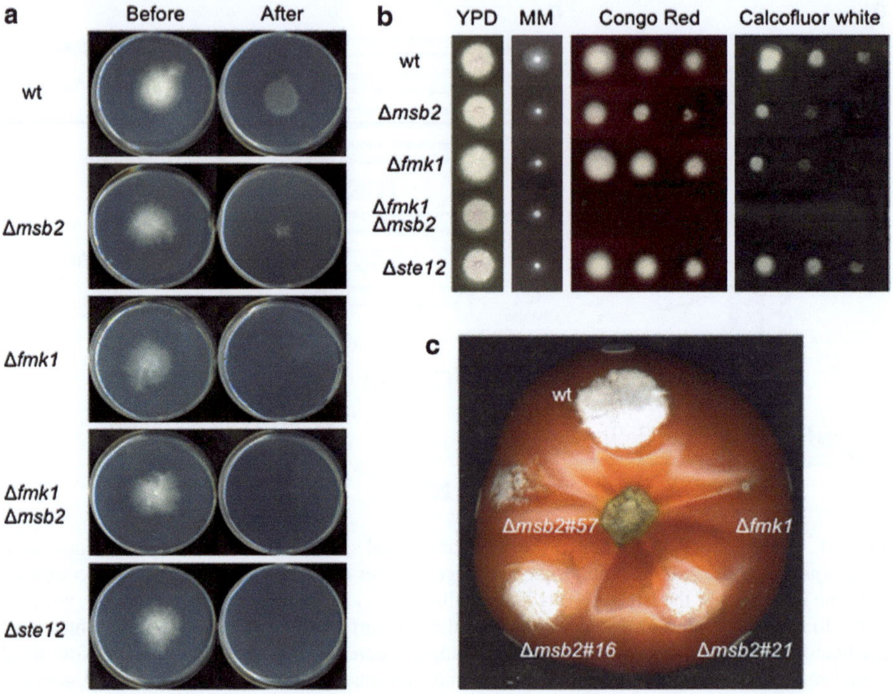


**Fig. 4.2** The Msb2/Fmk1/Ste12 pathway promotes invasive growth. (**a**) Penetration of cellophane membranes. The indicated strains were grown on a minimal medium (MM) plate covered by a cellophane membrane (before). The cellophane with the fungal colony was removed and plates were incubated for an additional day (after). (**b**) Msb2 contributes to hyphal growth under conditions of nutrient limitation and cell integrity stress. Strains were grown on yeast peptone glucose (YPD), minimal medium (MM), or YPD supplemented with the cell wall targeting compounds Congo Red or Calcofluor White. (**c**) Msb2 contributes to invasive growth on living tissue. Tomato fruits were inoculated with microconidia of the wild type strain, the Δ*fmk1* mutant and three independent Δ*msb2* mutants, and incubated at 28°C for 4 days

Gustin 1994; Yuzyuk et al. 2002). Elements of the HOG MAPK pathway have also been implicated in the response to cell wall stress via the Mpk1 pathway in *S. cerevisiae* (Bermejo et al. 2008). While Msb2 was shown to promote Hog1 phosphorylation on solid medium (Perez-Nadales and Di Pietro 2011), a direct implication of the HOG pathway in cell wall integrity of *F. oxysporum* remains to be confirmed. Interestingly, *S. cerevisiae* Msb2 was shown to interact with Cdc42, suggesting a possible link between Msb2-dependent MAPK signaling and reorganization of the actin cytoskeleton. Moreover, additional studies have demonstrated an interaction between Msb2 and other cell surface proteins, such as Bni4, a protein that targets chitin deposition to sites of polarized growth by linking chitin synthase to septins, and the kinase Cla4 (DeMarini et al. 1997; Drees et al. 2001), further supporting the idea that Msb2 may be part of the Cdc42 regulatory pathway. Further

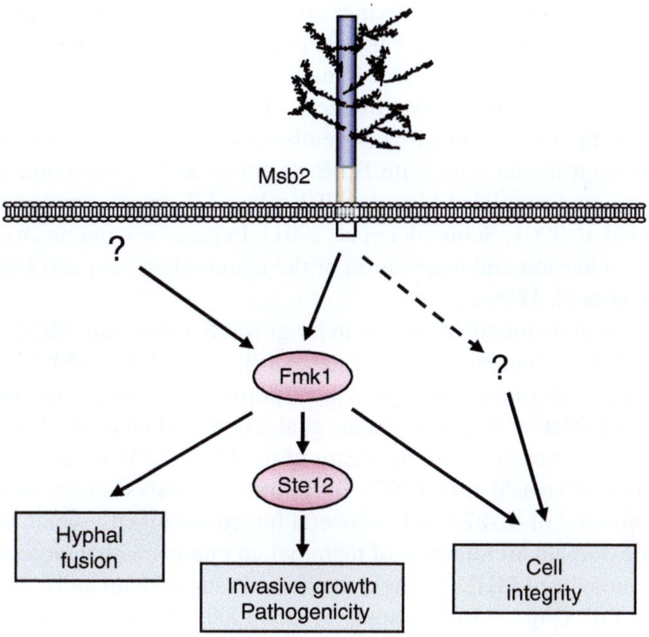


**Fig. 4.3** Model for the role of Msb2 and Fmk1 in signaling for pathogenic development of *F. oxysporum*. The cell surface mucin Msb2 functions upstream of the MAPK Fmk1 to promote invasive growth and virulence via the homeodomain transcription factor Ste12. Additional Fmk1-controlled functions such as vegetative hyphal fusion are regulated independently of Msb2. In addition, both Msb2 and Fmk1 contribute to cell integrity through distinct pathways

research is required to dissect the separate contributions of Msb2 and Fmk1 to cell wall remodeling in response to cell wall stress and, ultimately, during infectious development.

## 4.4 Fungal Signaling Mucins Are Activators of Pathogenic Development

Mucins are type I integral membrane proteins that typically have large extracellular domains containing a number of highly O-glycosylated repeat regions rich in serine and threonine residues [mucin homology domain (MHD)] and a short cytoplasmic tail (Fig. 4.1a) (Carraway et al. 2003; Agrawal et al. 1998). The size of cytoplasmic tails in cell-surface mucins varies from 22 to 80 residues (Carraway et al. 2003). Mucins have been extensively studied in mammalian cells, where they act as barriers to pathogen infection (Carson et al. 1998). It has been postulated that some cell surface mucins may serve as sensors of the extracellular environment by directly sensing changes in the external conditions such as pH, ionic composition, or physical interactions and promoting intracellular signaling in response to ligand binding or conformational changes (Carraway et al. 2003). Signaling mucins are key factors in metastasis in a variety of human cancers (Carraway et al. 2003). MUC1 and MUC4 are the prototypic human signaling mucin members which were originally identified as molecular markers of carcinoma cells (Wreschner et al. 1994). MUC1, the best characterized member, is involved in activation of MAPK pathways through interactions with ErbB receptors and downstream activation of extracellular-signal-regulated kinases (ERKs 1 and 2) in mouse mammary glands (Meerzaman et al. 2001; Schroeder et al. 2001). In parasites, mucins have also been implicated in adhesion and penetration of the mammalian host cell (Almeida et al. 1994; Di Noia et al. 1996).

The first signaling mucin identified in fungi was *S. cerevisiae* Msb2, followed by its paralog Hkr1 (Cullen et al. 2004; Tatebayashi et al. 2007). Msb2 is required for the Kss1-dependent filamentous growth response that takes place upon nutrient deprivation (Pitoniak et al. 2009; Vadaie et al. 2008; Cullen et al. 2004). Moreover, Msb2 regulates osmosensitivity upstream of the Hog1 MAPK, in concert with the Hkr1 mucin (Tatebayashi et al. 2007). Bioinformatic analysis detected the presence of a single ortholog of Msb2/Hkr1 in several fungal pathogens, which conserves the characteristic domain architecture of mammalian mucins with a large extracellular, highly O-glycosylated MHD, a single transmembrane domain and a relatively short cytoplasmic tail (Fig. 4.1b) (Rispail et al. 2009). In mammalian membrane-associated mucins, a 120-amino acid domain called SEA module is located in the extracellular region juxtaposed to the transmembrane domain and mediates autoproteolytic cleavage during posttranslational processing in the endoplasmic reticulum. This generates two subunits that remain noncovalently associated during cellular transport through the endoplasmic reticulum and Golgi complex to the cell surface (Lillehoj et al. 2003; Parry et al. 2001; Wreschner et al. 2002). Interestingly, fungal Msb2 proteins contain a conserved region of approximately 100 aminoacids located upstream of the transmembrane domain (Perez-Nadales and Di Pietro 2011; Rispail et al. 2009). This region, known as the positive regulatory domain (PRD), is essential for the signaling function of *S. cerevisiae* Msb2 and Hkr1 mucins (Cullen et al. 2004; Tatebayashi et al. 2007). It has been suggested that a mechanism similar to the SEA-mediated cleavage of mammalian mucins may operate during the processing of fungal mucins (Perez-Nadales and Di Pietro 2011). Interestingly, the extracellular domains of membrane-associated fungal mucins are released from the cell surface, a process named shedding (Vadaie et al. 2008; Perez-Nadales and Di Pietro 2011). The mechanisms that control shedding have not been clearly elucidated, although in *S. cerevisiae* it was shown that the PRD is required for this process (Vadaie et al. 2008). There is also evidence that specific proteases from the group of yapsins could mediate shedding of the extracellular mucin domains from the cell surface (Thathiah and Carson 2004; Vadaie et al. 2008). The *F. oxysporum* Msb2 protein is also shed from the surface of fungal colonies (Pérez-Nadales and Di Pietro 2011), but the relevance of shedding for mucin activation and fungal infection is currently unknown.

Increasing evidence suggests a role of Msb2 in regulating the initial steps of fungal infection upon contact with the host. In the non-appressorium-forming

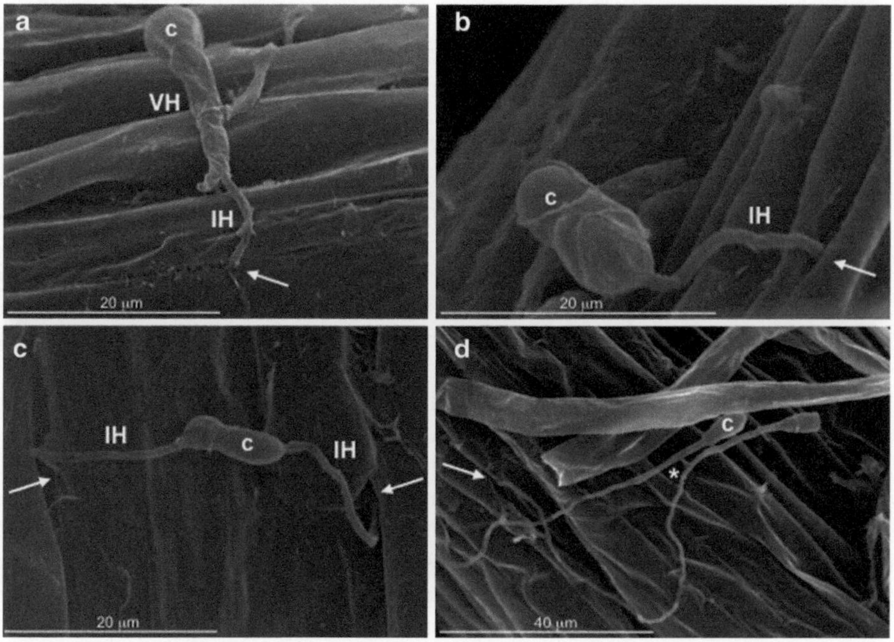


**Fig. 4.4** Scanning electron microscope analysis of penetration of tomato roots by *F. oxysporum*. (**a–d**) *F. oxysporum* microconidia (**c**) germinating on the surface of tomato roots, where they produce vegetative (VH) and/or infectious (IH) hyphae that can undergo hyphal fusion (indicated by an *asterisk* in **d**). Penetration events involve directed growth of IH toward natural openings between epidermal root cells, followed by direct penetration without the development of specialized infection structures (penetration sites indicated by *arrows*). Both the mucin sensor Msb2 and the MAPK Fmk1 contribute to sensing and penetration (Perez-Nadales and Di Pietro 2011). The detailed molecular mechanisms of signal perception remain to be elucidated

pathogen *F. oxysporum*, hyphal penetration of tomato roots takes place through natural openings between epidermal root cells (Fig. 4.4) (Perez-Nadales and Di Pietro 2011; Lagopodi et al. 2002). At present, it is unclear through which physical and/or chemical mechanisms *F. oxysporum* infection hyphae successfully locate these preexisting openings of the tomato root surface. Interestingly, Δ*msb2* and Δ*fmk1* mutants are markedly reduced in efficiency of penetration of tomato roots (40–45% for the mutants compared with >80% for the wild type), suggesting that Msb2 contributes to sensing of these penetration sites (Perez-Nadales and Di Pietro 2011). In the aerial plant pathogens *M. oryzae* and *U. maydis*, the development of specialized appresoria takes place upon contact with the leaf surface and the signals involved include ethylene, epicuticular waxes, cutin monomers, and hydrophobicity (Tucker and Talbot 2001; Kumamoto 2008; Mendoza-Mendoza et al. 2009). In these two fungi, Msb2 functions in concert with the plasma membrane receptor Sho1 to regulate appressorium development on the plant surface and on inductive hydrophobic surfaces via the MAPKs Kpp2/Kpp6 and Pmk1, respectively (Lanver et al. 2010; Liu et al. 2010). In the human pathogen

*C. albicans*, Msb2 contributes to invasion of solid surfaces, possibly by regulating the activity of the Kss1 MAPK ortholog Cek1 (Roman et al. 2009). Msb2 orthologs in *F. oxysporum* and *C. albicans* were also shown to have an additional function in regulation of the cell wall stress response, suggesting a role in cell integrity. Moreover, similar to *S. cerevisiae*, Msb2 participates in the regulation of the osmotic stress response in *C. albicans*, although this regulation appears to take place via an unknown Hog1-independent pathway (Roman et al. 2009).

One intriguing question is how mucins exert their function as sensors of stress, nutritional status, or plant surface signals. It has been speculated that the physico-chemical nature of the sugar polymers in the mucin domain may render them susceptible to alterations in extracellular parameters such as pH, ionic concentration, hydration, or interaction with ligands. In fact, organic polymer gels are highly sensitive to the solvent properties (Tanaka et al. 1980). Thus, a shift in osmolarity could cause a significant volume change in the MHD domain, thereby exposing the PRD and/or transmembrane domains to initiate signaling (Tatebayashi et al. 2007). With regards to sensing of the cellular nutritional status, it has long been known that glycosylation defects can activate a filamentous growth like response in yeast (Cullen et al. 2000; Lee and Elion 1999). Interestingly, deletion of the mucin domain of Msb2 resulted in constitutive activation of the filamentation pathway (Cullen et al. 2004). A hypothetical mechanism has been proposed in *S. cerevisiae*, whereby starvation-induced activation of the aspartyl protease Yps1 would result in specific cleavage of the Msb2 mucin domain and activation of the filamentation pathway (Vadaie et al. 2008). In support for a role of glycosylation of the extracellular region on the signaling properties of Msb2, disruption of the gene encoding the protein *O*-mannosyltransferase Pmt4 combined with N-glycosylation defects induced by tunicamycin resulted in reduced glycosylation of the Msb2 mucin domain and in induction of the filamentous growth response (Yang et al. 2009). These results, in turn, led to the speculation that the Msb2 mucin domain may function as a sensor of nutrient deprivation. Under poor nutritional conditions, underglycosylation of this region would unmask the regulatory domain, thereby initiating pathway activation (Yang et al. 2009). A recent study on the in vivo measurement of the mechanical behavior of the highly glycosylated mucin-like transmembrane sensor Wsc1, which functions upstream of the yeast cell wall integrity MAPK pathway, suggested that it behaves like a nanospring in response to cell surface stress (Dupres et al. 2009). Interestingly, underglycosylation of Wsc1 by *pmt4* deletion caused dramatic alterations in protein spring properties, supporting a pivotal role of glycosylation at the extracellular serine/threonine-rich region. In mammals, the highly branched O-linked oligosaccharides located in the large extracellular mucin domain are known to confer rigidity to the protein, resulting in a so-called bottle-brush structure (Hattrup and Gendler 2008). As a result of this configuration, mucins extend several hundred nanometers from the cell surface and are involved in specific monitoring and sensing of the molecular microenvironment. Finally, mucins may also participate in contact sensing by means of their association with the plasma membrane. Transmembrane helices of the mammalian angiotensin II type I and bradykinin G-protein-coupled receptors were shown to undergo conformational changes in response to mechanical stretching of the membrane and to other membrane perturbations such as hypotonic stress or incubation with molecules that enhance membrane fluidity (Chachisvilis et al. 2006; Yasuda et al. 2008). Further studies are required to dissect the contribution of Msb2 transmembrane mucins to contact sensing and perception of putative plant signals in phytopathogenic fungi.

## 4.5 Future Outlook

In fungal pathogens, plant-derived signals trigger germination and development of infection-related morphogenesis. The molecular details of the signal transduction programs involved in this process remain largely uncharacterized. Elucidating the nature of the activating signals and the mechanisms of transduction is one of the key challenges in the field of molecular plant pathology.

Cell surface mucins have recently been characterized as upstream activators of MAPK pathways in several fungal species. Orthologs of the Msb2 mucin protein from *S. cerevisiae*, which regulates Kss1 and Hog1 activities, have been implicated in the regulation of Kss1 ortholog pathogenicity MAPK pathways in three phytopathogenic fungal species (Lanver et al. 2010; Liu et al. 2010; Perez-Nadales and Di Pietro 2011) and one human fungal pathogen (Roman et al. 2009), suggesting a broadly conserved role for mucins in pathogenic development. In mammals, cell surface mucins are typically involved in functions such as cytoprotection and regulation of cellular signal transduction pathways. They have also been implicated in the invasive properties of cancer cells, because mucin genes are overexpressed in a variety of human cancers (Carraway et al. 2003; Corfield et al. 2000) and in several respiratory diseases such as cystic fibrosis, chronic obstructive pulmonary disease, and asthma (Hattrup and Gendler 2008). It has been suggested that tumors may use mucins to alter the local microenvironment during metastatic development (Hollingsworth and Swanson 2004). Overall, these data highlight the relevant biological function of this type of receptor molecules in eukaryotic cells and the need for understanding their mode of action.

At present, it is unclear which mechanism regulates mucin activation during fungal pathogenicity and whether this process requires shedding or proteolytic processing of the intact, membrane-bound form, or whether both mechanisms operate simultaneously. Mucins from plant pathogenic fungi have been directly implicated in the perception of physical or chemical signals present on the host surface (Lanver et al. 2010; Liu et al. 2010; Perez-Nadales and Di Pietro 2011). Thus, characterization of cell surface mucins in these species has proved a highly promising tool for elucidating an unsolved fundamental question in this field of research: the physico-chemical nature of the plant signals sensed by fungal pathogens. Additional key points that need to be addressed are: what are these sensors detecting? What is their mode of action? What are their end-targets? Future studies should attempt to address these open questions and provide new insights into the field of mucin biology. This will undoubtedly advance our understanding of fungal infection, but may also provide useful insights into other biological processes involving signaling mucins such as cancer progression and pulmonary disease.

**Acknowledgements** Our research was financially supported by the SIGNALPATH Marie Curie Research Training Network (MRTN-CT-2005-019277) and by grants BIO2008-04479-E, EUI2009-03942 and BIO2010-15505 from the Spanish Ministerio de Ciencia e Innovación (MICINN).

## References

Agrawal B, Gendler SJ, Longenecker BM (1998) The biological role of mucins in cellular interactions and immune regulation: prospects for cancer immunotherapy. Mol Med Today 4:397–403

Agrios G (2005) Plant pathology, 5th edn. Academic, San Diego, CA

Almeida IC, Ferguson MA, Schenkman S, Travassos LR (1994) Lytic anti-alpha-galactosyl antibodies from patients with chronic Chagas' disease recognize novel O-linked oligosaccharides on mucin-like glycosyl-phosphatidylinositol-anchored glycoproteins of Trypanosoma cruzi. Biochem J 304:793–802

Armstrong GM, Armstrong JK (1981) Formae speciales and races of *Fusarium oxysporum* causing wilt diseases. Fusarium: diseases, biology, and taxonomy. Pennsylvania State University Press, Philadelphia, PA

Bar EE, Ellicott AT, Stone DE (2003) Gbetagamma recruits Rho1 to the site of polarized growth during mating in budding yeast. J Biol Chem 278:21798–21804

Bermejo C, Rodriguez E, Garcia R, Rodriguez-Pena JM, Rodriguez de la Concepcion ML, Rivas C, Arias P, Nombela C, Posas F, Arroyo J (2008) The sequential activation of the yeast HOG and SLT2 pathways is required for cell survival to cell wall stress. Mol Biol Cell 19:1113–1124

Bishop CD, Cooper RM (1983) An ultrastructural study of root invasion of three vascular wilt diseases. Physiol Mol Plant Pathol 22:15–27

Bolker M (1998) Sex and crime: heterotrimeric G proteins in fungal mating and pathogenesis. Fungal Genet Biol 25(3):143–156

Booth C (1971) The genus *Fusarium*. The Eastern Press Ltd., London

Brewster JL, Gustin MC (1994) Positioning of cell growth and division after osmotic stress requires a MAP kinase pathway. Yeast 10:425–439

Butty AC, Pryciak PM, Huang LS, Herskowitz I, Peter M (1998) The role of Far1p in linking the heterotrimeric G protein to polarity establishment proteins during yeast mating. Science 282:1511–1516

Carraway KL, Ramsauer VP, Haq B, Carothers Carraway CA (2003) Cell signaling through membrane mucins. Bioessays 25:66–71

Carson DD, DeSouza MM, Kardon R, Zhou X, Lagow E, Julian J (1998) Mucin expression and function in the female reproductive tract. Hum Reprod Update 4:459–464

Chachisvilis M, Zhang YL, Frangos JA (2006) G protein-coupled receptors sense fluid shear stress in endothelial cells. Proc Natl Acad Sci USA 103:15463–15468

Chang L, Karin M (2001) Mammalian MAP kinase signalling cascades. Nature 410:37–40

Corfield AP, Myerscough N, Longman R, Sylvester P, Arul S, Pignatelli M (2000) Mucins and mucosal protection in the gastrointestinal tract: new prospects for mucins in the pathology of gastrointestinal disease. Gut 47:589–594

Cormack BP, Ghori N, Falkow S (1999) An adhesin of the yeast pathogen Candida glabrata mediating adherence to human epithelial cells. Science 285:578–582

Csank C, Schroppel K, Leberer E, Harcus D, Mohamed O, Meloche S, Thomas DY, Whiteway M (1998) Roles of the *Candida albicans* mitogen-activated protein kinase homolog, Cek1p, in hyphal development and systemic candidiasis. Infect Immun 66:2713–2721

Cullen PJ, Schultz J, Horecka J, Stevenson BJ, Jigami Y, Sprague GF Jr (2000) Defects in protein glycosylation cause SHO1-dependent activation of a STE12 signaling pathway in yeast. Genetics 155:1005–1018

Cullen PJ, Sabbagh W Jr, Graham E, Irick MM, van Olden EK, Neal C, Delrow J, Bardwell L, Sprague GF Jr (2004) A signaling mucin at the head of the Cdc42- and MAPK-dependent filamentous growth pathway in yeast. Genes Dev 18:1695–1708

Davidson RC, Nichols CB, Cox GM, Perfect JR, Heitman J (2003) A MAP kinase cascade composed of cell type specific and non-specific elements controls mating and differentiation of the fungal pathogen *Cryptococcus neoformans*. Mol Microbiol 49:469–485

Delgado-Jarana J, Martinez-Rocha AL, Roldan-Rodriguez R, Roncero MI, Di Pietro A (2005) *Fusarium oxysporum* G-protein beta subunit Fgb1 regulates hyphal growth, development, and virulence through multiple signalling pathways. Fungal Genet Biol 42:61–72

DeMarini DJ, Adams AE, Fares H, De Virgilio C, Valle G, Chuang JS, Pringle JR (1997) A septin-based hierarchy of proteins required for localized deposition of chitin in the *Saccharomyces cerevisiae* cell wall. J Cell Biol 139(1):75–93

Di Noia JM, Pollevick GD, Xavier MT, Previato JO, Mendoça-Previato L, Sánchez DO, Frasch AC (1996) High diversity in mucin genes and mucin molecules in Trypanosoma cruzi. J Biol Chem 271:32078–32083

Di Pietro A, Garcia-Maceira FI, Meglecz E, Roncero MI (2001) A MAP kinase of the vascular wilt fungus *Fusarium oxysporum* is essential for root penetration and pathogenesis. Mol Microbiol 39:1140–1152

Drees BL, Sundin B, Brazeau E, Caviston JP, Chen GC, Guo W, Kozminski KG, Lau MW, Moskow JJ, Tong A, Schenkman LR, McKenzie A III, Brennwald P, Longtine M, Bi E, Chan C, Novick P, Boone C, Pringle JR, Davis TN, Fields S, Drubin DG (2001) A protein interaction map for cell polarity development. J Cell Biol 154:549–571

Dupres V, Alsteens D, Wilk S, Hansen B, Heinisch JJ, Dufrêne YF (2009) The yeast Wsc1 cell surface sensor behaves like a nanospring in vivo. Nat Chem Biol 5:857–862

Gao S, Nuss DL (1996) Distinct roles for two G protein alpha subunits in fungal virulence, morphology, and reproduction revealed by targeted gene disruption. Proc Natl Acad Sci USA 93:14122–14127

Gold S, Duncan G, Barrett K, Kronstad J (1994) cAMP regulates morphogenesis in the fungal pathogen *Ustilago maydis*. Genes Dev 8:2805–2816

Guo B, Styles CA, Feng Q, Fink GR (2000) A *Saccharomyces* gene family involved in invasive growth, cell-cell adhesion, and mating. Proc Natl Acad Sci USA 97:12158–12163

Hattrup CL, Gendler SJ (2008) Structure and function of the cell surface (tethered) mucins. Annu Rev Physiol 70:431–457

Hollingsworth MA, Swanson BJ (2004) Mucins in cancer: protection and control of the cell surface. Nat Rev Cancer 4:45–60

Hua X, Yuan X, Di Pietro A, Wilhelmus KR (2010) The molecular pathogenicity of *Fusarium keratitis*: a fungal transcriptional regulator promotes hyphal penetration of the cornea. Cornea 29:1440–1444

Jain S, Akiyama K, Mae K, Ohguchi T, Takata R (2002) Targeted disruption of a G protein alpha subunit gene results in reduced pathogenicity in *Fusarium oxysporum*. Curr Genet 41:407–413

Jain S, Akiyama K, Kan T, Ohguchi T, Takata R (2003) The G protein beta subunit FGB1 regulates development and pathogenicity in *Fusarium oxysporum*. Curr Genet 43:79–86

Jenczmionka NJ, Maier FJ, Losch AP, Schafer W (2003) Mating, conidiation and pathogenicity of *Fusarium graminearum*, the main causal agent of the head-blight disease of wheat, are regulated by the MAP kinase gpmk1. Curr Genet 43:87–95

Kasahara S, Nuss DL (1997) Targeted disruption of a fungal G-protein beta subunit gene results in increased vegetative growth but reduced virulence. Mol Plant Microbe Interact 10:984–993

Kruger J, Loubradou G, Wanner G, Regenfelder E, Feldbrugge M, Kahmann R (2000) Activation of the cAMP pathway in *Ustilago maydis* reduces fungal proliferation and teliospore formation in plant tumors. Mol Plant Microbe Interact 13:1034–1040

Kulkarni RD, Thon MR, Pan H, Dean RA (2005) Novel G-protein-coupled receptor-like proteins in the plant pathogenic fungus *Magnaporthe grisea*. Genome Biol 6:R24

Kultz D, Burg M (1998) Evolution of osmotic stress signaling via MAP kinase cascades. J Exp Biol 201:3015–3021

Kumamoto CA (2008) Molecular mechanisms of mechanosensing and their roles in fungal contact sensing. Nat Rev Microbiol 6:667–673

Lagopodi AL, Ram AF, Lamers GE, Punt PJ, Van den Hondel CA, Lugtenberg BJ, Bloemberg GV (2002) Novel aspects of tomato root colonization and infection by *Fusarium oxysporum* f. sp. *radicis-lycopersici* revealed by confocal laser scanning microscopic analysis using the green fluorescent protein as a marker. Mol Plant Microbe Interact 15:172–179

Lanver D, Mendoza-Mendoza A, Brachmann A, Kahmann R (2010) Sho1 and Msb2-related proteins regulate appressorium development in the smut fungus *Ustilago maydis*. Plant Cell 22:2085–2101

Lee BN, Elion EA (1999) The MAPKKK Ste11 regulates vegetative growth through a kinase cascade of shared signaling components. Proc Natl Acad Sci USA 96:12679–12684

Lee N, D'Souza CA, Kronstad JW (2003) Of smuts, blasts, mildews, and blights: cAMP signaling in phytopathogenic fungi. Annu Rev Phytopathol 41:399–427

Lengeler KB, Davidson RC, D'Souza C, Harashima T, Shen WC, Wang P, Pan X, Waugh M, Heitman J (2000) Signal transduction cascades regulating fungal development and virulence. Microbiol Mol Biol Rev 64:746–785

Lev S, Horwitz BA (2003) A mitogen-activated protein kinase pathway modulates the expression of two cellulase genes in *Cochliobolus heterostrophus* during plant infection. Plant Cell 15:835–844

Lev S, Sharon A, Hadar R, Ma H, Horwitz BA (1999) A mitogen-activated protein kinase of the corn leaf pathogen *Cochliobolus heterostrophus* is involved in conidiation, appressorium formation, and pathogenicity: diverse roles for mitogen-activated protein kinase homologs in foliar pathogens. Proc Natl Acad Sci USA 96:13542–13547

Levin DE (2005) Cell wall integrity signaling in *Saccharomyces cerevisiae*. Microbiol Mol Biol Rev 69:262–291

Lillehoj EP, Han F, Kim KC (2003) Mutagenesis of a Gly-Ser cleavage site in MUC1 inhibits ectodomain shedding. Biochem Biophys Res Commun 307:743–749

Liu S, Dean RA (1997) G protein alpha subunit genes control growth, development, and pathogenicity of *Magnaporthe grisea*. Mol Plant Microbe Interact 10(9):1075–1086

Liu W, Zhou X, Li G, Li L, Kong L, Wang C, Zhang H, Xu JR (2010) Multiple plant surface signals are sensed by different mechanisms in the rice blast fungus for appressorium formation. PLoS Pathog 7:e1001261

Ma LJ, Van der Does HC, Borkovich KA, Coleman JJ, Daboussi MJ, Di Pietro A, Dufresne M, Freitag M, Grabherr M, Henrissat B, Houterman PM, Kang S, Shim WB, Woloshuk C, Xie X, Xu JR, Antoniw J, Baker SE, Bluhm BH, Breakspear A, Brown DW, Butchko RAE, Chapman S, Coulson R, Coutinho PM, Danchin EGJ, Diener A, Gale LR, Gardiner DM, Goff S, Hammond-Kosack KE, Hilburn K, Hua-Van A, Jonkers W, Kazan K, Kodira CD, Koehrsen M, Kumar L, Lee YH, Li L, Manners JM, Miranda-Saavedra D, Mukherjee M, Park G, Park J, Park SY, Proctor RH, Regev A, Ruiz-Roldan MC, Sain D, Sakthikumar S, Sykes S, Schwartz DC, Turgeon BG, Wapinski I, Yoder O, Young S, Zeng Q, Zhou S, Galagan J, Cuomo CA, Kistler HC, Rep M (2010) Comparative genomics reveals mobile pathogenicity chromosomes in *Fusarium oxysporum*. Nature 464:367–373

Martinez-Rocha AL, Roncero MI, Lopez-Ramirez A, Marine M, Guarro J, Martinez-Cadena G, Di Pietro A (2008) Rho1 has distinct functions in morphogenesis, cell wall biosynthesis and virulence of *Fusarium oxysporum*. Cell Microbiol 10:1339–1351

Mazzoni C, Zarov P, Rambourg A, Mann C (1993) The SLT2 (MPK1) MAP kinase homolog is involved in polarized cell growth in *Saccharomyces cerevisiae*. J Cell Biol 123:1821–1833

Meerzaman D, Shapiro PS, Kim KC (2001) Involvement of the MAP kinase ERK2 in MUC1 mucin signaling. Am J Physiol Lung Cell Mol Physiol 281:L86–91

Mendoza-Mendoza A, Berndt P, Djamei A, Weise C, Linne U, Marahiel M, Vranes M, Kamper J, Kahmann R (2009) Physical-chemical plant-derived signals induce differentiation in *Ustilago maydis*. Mol Microbiol 71:895–911

Mey G, Held K, Scheffer J, Tenberge KB, Tudzynski P (2002) CPMK2, an SLT2-homologous mitogen-activated protein (MAP) kinase, is essential for pathogenesis of *Claviceps purpurea* on rye: evidence for a second conserved pathogenesis-related MAP kinase cascade in phytopathogenic fungi. Mol Microbiol 46:305–318

Mitchell TK, Dean RA (1995) The cAMP-dependent protein kinase catalytic subunit is required for appressorium formation and pathogenesis by the rice blast pathogen *Magnaporthe grisea*. Plant Cell 7:1869–1878

Muller P, Leibbrandt A, Teunissen H, Cubasch S, Aichinger C, Kahmann R (2004) The Gbeta-subunit-encoding gene bpp 1 controls cyclic-AMP signaling in *Ustilago maydis*. Eukaryot Cell 3:806–814

Nern A, Arkowitz RA (1999) A Cdc24p-Far1p-Gbetagamma protein complex required for yeast orientation during mating. J Cell Biol 144:1187–1202

Nishimura M, Park G, Xu JR (2003) The G-beta subunit MGB1 is involved in regulating multiple steps of infection-related morphogenesis in *Magnaporthe grisea*. Mol Microbiol 50:231–243

Nucci M, Anaissie E (2002) Cutaneous infection by Fusarium species in healthy and immunocompromised hosts: implications for diagnosis and management. Clin Infect Dis 35:909–920

Ohtake M, Yamamoto H, Uchiyama T (1999) Influences of metabolic inhibitors and hydrolytic enzymes on the adhesion of appressoria of *Pyricularia oryzae* to wax-coated cover-glasses. Biosci Biotechnol Biochem 63:978

Ortoneda M, Guarro J, Madrid MP, Caracuel Z, Roncero MI, Mayayo E, Di Pietro A (2004) *Fusarium oxysporum* as a multihost model for the genetic dissection of fungal virulence in plants and mammals. Infect Immun 7:1760–1766

Parry S, Silverman HS, McDermott K, Willis A, Hollingsworth MA, Harris A (2001) Identification of MUC1 proteolytic cleavage sites in vivo. Biochem Biophys Res Comm 283:715–720

Perez-Nadales E, Di Pietro A (2011) The membrane mucin Msb2 controls invasive growth and plant infection in *Fusarium oxysporum*. Plant Cell 23:1171–1185

Pitoniak A, Birkaya B, Dionne HM, Vadaie N, Cullen PJ (2009) The signaling mucins Msb2 and Hkr1 differentially regulate the filamentation mitogen-activated protein kinase pathway and contribute to a multimodal response. Mol Biol Cell 20:3101–3114

Prados Rosales RC, Di Pietro A (2008) Vegetative hyphal fusion is not essential for plant infection by *Fusarium oxysporum*. Eukaryot Cell 7:162–171

Prados-Rosales RC, Serena C, Delgado-Jarana J, Guarro J, Di Pietro A (2006) Distinct signalling pathways coordinately contribute to virulence of *Fusarium oxysporum* on mammalian hosts. Microbes Infect 8:2825–2831

Prados-Rosales R, Luque-Garcia JL, Martinez-Lopez R, Gil C, Di Pietro A (2009) The *Fusarium oxysporum* cell wall proteome under adhesion-inducing conditions. Proteomics 9:4755–4769

Pruyne D, Bretscher A (2000) Polarization of cell growth in yeast. I. Establishment and maintenance of polarity states. J Cell Sci 113:365–375

Qi M, Elion EA (2005) MAP kinase pathways. J Cell Sci 118:3569–3572

Rispail N, Di Pietro A (2009) *Fusarium oxysporum* Ste12 controls invasive growth and virulence downstream of the Fmk1 MAPK cascade. Mol Plant Microbe Interact 22:830–839

Rispail N, Soanes D, Ant C, Czajkowski R, Grünlere A, Huguet R, Perez-Nadales E, Poli A, Sartorel E, Valiante V, Yang M, Beffa R, Brakhage A, Gow N, Kahmann R, Lebrun M-H,

Lenasi H, Perez-Martin J, Talbot NJ, Wendland J, Di Pietro A (2009) Comparative genomics of MAP kinase and calcium–calcineurin signalling components in plant and human pathogenic fungi. Fungal Genet Biol 46:287–298

Rodriguez-Galvez E, Mendgen K (1995) Cell wall synthesis in cotton roots after infection with *Fusarium oxysporum*. The deposition of callose, arabinogalactans, xyloglucans, and pectic components into walls, wall appositions, cell plates and plasmodesmata. Planta 197:535–545

Roman E, Cottier F, Ernst JF, Pla J (2009) Msb2 signaling mucin controls activation of Cek1 mitogen-activated protein kinase in *Candida albicans*. Eukaryot Cell 8:1235–1249

Ruiz-Roldan MC, Maier FJ, Schafer W (2001) PTK1, a mitogen-activated-protein kinase gene, is required for conidiation, appressorium formation, and pathogenicity of *Pyrenophora teres* on barley. Mol Plant Microbe Interact 14:116–125

Schroeder JA, Thompson MC, Gardner MM, Gendler SJ (2001) Transgenic MUC1 interacts with epidermal growth factor receptor and correlates with mitogen-activated protein kinase activation in the mouse mammary gland. J Biol Chem 276:13057–13064

Singleton DR, Masuoka J, Hazen KC (2005) Surface hydrophobicity changes of two *Candida albicans* serotype B mnn4delta mutants. Eukaryot Cell 4:639–648

Takano Y, Kikuchi T, Kubo Y, Hamer JE, Mise K, Furusawa I (2000) The *Colletotrichum lagenarium* MAP kinase gene CMK1 regulates diverse aspects of fungal pathogenesis. Mol Plant Microbe Interact 13:374–383

Tanaka T, Fillmore D, Sun S, Nishio I, Swislow G, Shah A (1980) Phase transitions in ionic gels. Phys Rev Lett 45:1636–1639

Tatebayashi K, Tanaka K, Yang HY, Yamamoto K, Matsushita Y, Tomida T, Imai M, Saito H (2007) Transmembrane mucins Hkr1 and Msb2 are putative osmosensors in the SHO1 branch of yeast HOG pathway. EMBO J 26:3521–3533

Thathiah A, Carson DD (2004) MT1-MMP mediates MUC1 shedding independent of TACE/ADAM17. Biochem J 382:363–373

Tucker SL, Talbot NJ (2001) Surface attachment and pre-penetration stage development by plant pathogenic fungi. Annu Rev Phytopathol 39:385–417

Uchiyama T, Okuyama K (1990) Participation of *Oryza sativa* leaf wax in appressorium formation by *Pyricularia oryzae*. Phytochemistry 29(1):91–92

Vadaie N, Dionne H, Akajagbor DS, Nickerson SR, Krysan DJ, Cullen PJ (2008) Cleavage of the signaling mucin Msb2 by the aspartyl protease Yps1 is required for MAPK activation in yeast. J Cell Biol 181:1073–1081

Wilson RA, Talbot NJ (2009) Under pressure: investigating the biology of plant infection by *Magnaporthe oryzae*. Nat Rev Microbiol 7:185–195

Wreschner DH, Zrihan-Licht S, Baruch A, Sagiv D, Hartman ML, Smorodinsky N, Keydar I (1994) Does a novel form of the breast cancer marker protein, MUC1, act as a receptor molecule that modulates signal transduction? Adv Exp Med Biol 353:17–26

Wreschner DH, McGuckin MA, Williams SJ, Baruch A, Yoeli M, Ziv R, Okun L, Zaretsky J, Smorodinsky N, Keydar I, Neophytou P, Stacey M, Lin HH, Gordon S (2002) Generation of ligand-receptor alliances by "SEA" module-mediated cleavage of membrane-associated mucin proteins. Protein Sci 11:698–706

Xiao JZ, Watanabe T, Sekido S, Choi WB, Kamakura T, Yamaguchi I (1997) An anti-hydrotactic response and solid surface recognition of germ tubes of the rice blast fungus, *Magnaporthe grisea*. Biosci Biotechnol Biochem 61:1225

Xu JR, Hamer JE (1996) MAP kinase and cAMP signaling regulate infection structure formation and pathogenic growth in the rice blast fungus *Magnaporthe grisea*. Genes Dev 10:2696–2706

Yamauchi J, Takayanagi N, Komeda K, Takano Y, Okuno T (2004) cAMP-PKA signaling regulates multiple steps of fungal infection cooperatively with Cmk1 MAP Kinase in *Colletotrichum lagenarium*. MPMI 17:1355–1365

Yang HY, Tatebayashi K, Yamamoto K, Saito H (2009) Glycosylation defects activate filamentous growth Kss1 MAPK and inhibit osmoregulatory Hog1 MAPK. EMBO J 28:1380–1391

Yasuda N, Miura S, Akazawa H, Tanaka T, Qin Y, Kiya Y, Imaizumi S, Fujino M, Ito K, Zou Y, Fukuhara S, Kunimoto S, Fukuzaki K, Sato T, Ge J, Mochizuki N, Nakaya H, Saku K, Komuro I (2008) Conformational switch of angiotensin II type 1 receptor underlying mechanical stress-induced activation. EMBO Rep 9:179–186

Yuzyuk T, Foehr M, Amberg DC (2002) The MEK kinase Ssk2p promotes actin cytoskeleton recovery after osmotic stress. Mol Biol Cell 13:2869–2880

Zarzov P, Mazzoni C, Mann C (1996) The SLT2(MPK1) MAP kinase is activated during periods of polarized cell growth in yeast. EMBO J 15:83–91

Zhao X, Mehrabi R, Xu JR (2007) Mitogen-activated protein kinase pathways and fungal pathogenesis. Eukaryot Cell 6:1701–1714

Zheng L, Campbell M, Murphy J, Lam S, Xu JR (2000) The *BMP1* gene is essential for pathogenicity in the gray mold fungus *Botrytis cinerea*. Mol Plant Microbe Interact 13:724–732

# Chapter 5
# Integrating Cdk Signaling in *Candida albicans* Environmental Sensing Networks

**Carlos R. Vázquez de Aldana and Jaime Correa-Bordes**

**Abstract** Cyclin-dependent protein kinases (Cdks) control cell cycle progression and morphological switches in eukaryotic cells. Based on recent findings concerning the evolution of Cdk phosphorylation sites in the Ascomycete linage, we shall analyze the density of Cdk motifs in the *Candida* proteome using the $S_{LR}$ algorithm, focusing on protein sequences of regulatory modules that play important roles in the environmental sensing of *Candida albicans*. Since Cdks are also involved in morphogenesis and environmental signaling, this search could help us to speculate about how Cdk signaling might be integrated in these regulatory networks that control *C. albicans* morphopathogenic determinants.

## 5.1 Introduction

In eukaryotic cells, global regulatory networks control cell physiology in response to external and internal cues. Many of these signaling networks are highly modular, making them more evolvable and providing increased fitness in competitive and changing environments (Bhattacharyya et al. 2006). Fungi are able to colonize and occupy highly divergent niches, ranging from high osmotic environments to plants and mammalian hosts. Their ability to adapt to these hostile environments depends on their capability to sense a variety of external cues, transduce the signals to specific cytoplasmic targets, and activate the appropriate responses.

Unlike the majority of fungi, the Ascomycete *Candida albicans* is normally found as a commensal in the gastrointestinal tract of humans and warm-blooded animals. Although generally asymptomatic, *C. albicans* can cause mucosal infections

C.R. Vázquez de Aldana
Instituto de Biología Funcional y Genómica, CSIC-Universidad de Salamanca, Salamanca, Spain

J. Correa-Bordes (✉)
Departamento Ciencias Biomédicas, Universidad de Extremadura, Badajoz, Spain
e-mail: jcorrea@unex.es

J. Pérez-Martín and A. Di Pietro (eds.), *Morphogenesis and Pathogenicity in Fungi*, Topics in Current Genetics 22, DOI 10.1007/978-3-642-22916-9_5, 

in healthy people. In patients with a deficient immune system, such as HIV-compromised individuals or patients treated with immunosuppressive drugs after organ transplantation, this yeast can produce systemic infections in which the fungus can spread to all major organs of the body, leading to death in around 50% of bloodstream infections (Eggimann et al. 2003).

*C. albicans* has several attributes that allow it to adapt rapidly to changing environmental signals that contribute to host colonization. One of those best studied is the ability to switch between different morphologies, such as yeast, pseudohyphae, and hyphae (Sudbery et al. 2004). The yeast-to-hypha transition, triggered by a wide range of environmental cues, is regulated by multiple signaling pathways that control the transcription of a set of hypha-specific genes (HSGs), many of which encode known virulence factors (Calderone and Fonzi 2001; Biswas et al. 2007; Whiteway and Bachewich 2007). The promoter regions of HSGs integrate the signals of multiple activators and repressors, although the interplay between them remains largely unknown. The two major transcriptional activators are Cph1 and Efg1 (Liu et al. 1994; Stoldt et al. 1997), which act downstream of the MAPK and the cAMP-PKA pathways, respectively. The double mutant *cph1 egf1* blocks hyphal transitions under most conditions tested and shows reduced virulence (Lo et al. 1997). Negative regulation of hyphal growth is achieved through the combinatorial association of the Tup1 repressor with the DNA-binding proteins (DBPs) Nrg1, Rfg1, and Mig1 (Braun and Johnson 1997; Braun et al. 2000; Murad et al. 2001a, b; Kadosh and Johnson 2005). Cells depleted of any of these repressors are able to activate hyphal growth under yeast-growth conditions. Since Mig1-Tup1 regulate a set of genes other than those regulated by Nrg1-Tup1, it has been suggested that DBPs target Tup1 to specific subsets of genes. However, Mig1 and Nrg1 can also repress the expression of other genes in a Tup1-independent manner (Murad et al. 2001a), suggesting that the DBPs associated with Tup1 (rather than Tup1 itself) are likely to be regulated during the yeast-to-hypha transition (Braun and Johnson 1997; Braun et al. 2000; Murad et al. 2001a, b; Kadosh and Johnson 2005).

Another example of the enormous plasticity of *C. albicans* cells is white–opaque (W/O) switching. This fungus undergoes an epigenetic switch between two cell types, known as white and opaque (Slutsky et al. 1987). These cell types differ in their cell morphology, metabolic state and mating behavior, and in their ability to form biofilms, their preferred niches in the host and interactions with the immune system. In recent years, new insights into the mechanisms that control this transition have allowed researchers to uncover complex relationships that relate switching, mating. and pathogenesis (Lohse and Johnson 2009; Soll 2009).

Although *Saccharomyces cerevisiae* and *C. albicans* share some common features, they also exhibit many significant differences since they diverged from a common ancestor more than 500 million years ago. The shared evolutionary fate of *C. albicans* and its hosts might have allowed it to evolve several developmental programs that would be activated by environmental cues within the host and that could aid in the colonization of different niches. These colonization sites represent different environments in terms of cohabitant microbiota, pH, nutrients, and $O_2$ or

$CO_2$ levels. Thus, this distinct lifestyle might have been a driving force in rewiring *C. albicans* signaling networks (Li and Johnson 2010).

## 5.2 Evolution of Phosphoregulation

Protein phosphorylation is a ubiquitous and reversible modification that is crucial for the regulation of cellular events (Seet et al. 2006). Comparative studies of the phosphoproteome of three yeast species (*C. albicans*, *S. cerevisiae*, and *Schizosaccharomyces pombe*) suggest that protein kinases probably contribute to a substantial extent to the evolution and generation of phenotypic diversity (Beltrao et al. 2009).

An important feature of *C. albicans* is that the CUG codon is decoded as Ser instead of Leu (Santos and Tuite 1995), with an average frequency of 1–6 CUGs per gene (Massey et al. 2003; Butler et al. 2009). Since Ser is a substrate of protein kinases, this change in codon usage might increase the number of potential phosphorylation sites per protein, adding new possible layers of phosphoregulation to protein networks. In fact, the ratio of phosphoserines/total proteins determined by mass spectrometry (MS) analysis in *C. albicans* is higher (0.54) than in *S. cerevisiae* (0.39) and *S. pombe* (0.32) (Beltrao et al. 2009).

Protein kinases regulate the function of their target proteins by adding a phosphate group to specific sites, which can change the activity of the protein through two different mechanisms (Holt et al. 2009). First, phosphorylation could drive a precise conformational change in the structure of the protein because the phosphate modifies the network of hydrogen bonds of several neighboring amino acids. This type of regulation, common in metabolic enzymes, is highly context-dependent and exhibits strong evolutionary conservation. Alternatively, the addition of phosphates to disordered regions (either the N- or C-termini or internal loops) of substrates can modify their interaction with other proteins (Serber and Ferrell 2007; Strickfaden et al. 2007) or can create new interactions through the phosphopeptide-binding modules present in other molecules, such as the SH2, 14-3-3, or WW domains (Bhattacharyya et al. 2006; Seet et al. 2006; Morrison 2009). In these cases, the position of the phosphoacceptor residue(s) is less context-dependent and therefore it can undergo a higher rate of change and a greater potential to generate functional diversity (Beltrao and Serrano 2007).

Cyclin-dependent protein kinases (Cdks) control progression along the eukaryotic cell cycle. These proteins are proline-directed kinases that preferentially phosphorylate substrates with the full consensus sequence S/T-P-X-K/R (where X is any amino acid), although they can also phosphorylate the minimal consensus sequence S/T-P (Songyang et al. 1994; Echalier et al. 2010). In this chapter, based on recent findings concerning the evolution of Cdk phosphorylation sites in the Ascomycete linage (Holt et al. 2009), we analyze the density of Cdk motifs in the *C. albicans* proteome using the $S_{LR}$ algorithm (Moses et al. 2007a) and focusing on proteins that are components of regulatory modules that play important roles in environmental sensing. Since Cdks are also involved in morphogenesis and environmental signaling (Huang et al. 2007; Moseley and Nurse 2009; Wang 2009), this search could help us to speculate about how Cdk signaling might be integrated in the regulatory networks that control morphopathogenic determinants in *C. albicans*.

## 5.3 Cyclin-Dependent Kinases in *C. albicans*

In yeast, cell cycle progression is driven by a single Cdk1 (Cdc28 in *S. cerevisiae* and *C. albicans*, Cdc2 in *S. pombe*). The combinatorial association of Cdk1 with G1 or G2 cyclins is thought to generate Cdk complexes with different substrate specificities that regulate different cell cycle transitions (Loog and Morgan 2005; Bloom and Cross 2007). However, phylogenetic studies of yeast B-type cyclins and experimental yeast models are consistent with the idea of an ancestral eukaryote with a single Cdk/cyclin module driving the cell cycle (Fisher and Nurse 1996; Archambault et al. 2005; Coudreuse and Nurse 2010). It is likely that the appearance of multiple cyclins in most eukaryotic lineages would have introduced new regulatory layers to fine-tune the single core Cdk module, providing more flexibility in the control of the cell cycle in response to different inputs (Loog and Morgan 2005; Bloom and Cross 2007; Coudreuse and Nurse 2010).

In addition to driving the cell cycle, Cdks coordinate cell morphology switches (Moseley and Nurse 2009; Wang 2009). In *C. albicans*, modifications of cyclin levels produce dramatic morphological changes. This fungus contains three G1 (Ccn1, Hgc1, and Cln3) and two G2 (Clb2 and Clb4) cyclins. Depletion of Cln3, Clb2, or Clb4 in yeast cells results in hyperpolarized growth in the absence of hypha-inducing conditions (Bachewich and Whiteway 2005; Bensen et al. 2005; Chapa y Lazo et al. 2005). Ccn1 is a nonessential G1 cyclin that is expressed during the G1/S transition and is required for the maintenance, but not the initiation, of hyphal growth under certain conditions (Loeb et al. 1999; Sinha et al. 2007). Hgc1 is a hypha-specific G1 cyclin-like protein that preferentially localizes to the dividing apical cell of the hyphae (Zheng and Wang 2004; Wang et al. 2007). Transcription from the *HGC1* promoter is essential for this asymmetric cell localization, since Hgc1 no longer exhibits the preferential apical accumulation when expressed under the control of the *MAL2* promoter (Wang et al. 2007). Deletion of *HGC1* prevents hyphal growth under all hypha-inducing conditions and results in reduced virulence in mouse models (Zheng and Wang 2004). Unlike the cell cycle-regulated transcription of other cyclin genes, *HGC1* expression is activated by hypha-inducing signals through the cAMP/PKA signaling pathway. The evolution of such control would probably have been crucial for ensuring the cell cycle-independent polarized growth of hyphae (Zheng and Wang 2004; Wang 2009). In agreement with this hypothesis, the expression of one allele of *CCN1* under the control of the *HGC1* promoter rescues the *hgc1*Δ mutant (P. Gutiérrez-Escribano and J. Correa-Bordes, unpublished results).

In recent years, insight into the links between Cdks and cell polarity proteins has been obtained in *C. albicans*, highlighting the importance of Cdk1 complexes in the control of cell morphogenesis during yeast and hyphal growth (Zheng and Wang 2004; Sinha et al. 2007; Zheng et al. 2007; González-Novo et al. 2008; Wang et al. 2009). An excellent summary of the role of Cdks in the yeast–hyphal transition of *C. albicans* has been published recently (Wang 2009).

## 5.4 Predicting Cdk Targets in *C. albicans* Proteins

Past efforts aimed at the identification of HSGs have underlined the importance of the cAMP-PKA and MAPK pathways in the transcriptional activation required for hyphal growth. However, recent findings have suggested that post-translational modifications mediated by Cdks are also important mechanisms in the regulation of polarized growth immediately after hyphal induction, independently of the cAMP-PKA and MAPK pathways (Sinha et al. 2007). These results suggest the existence of an additional signaling pathway(s) that plays a major role in the control of hyphal development, which is mediated by the Cdk-phosphorylation of key regulatory proteins in response to hypha-inducing signals.

### *5.4.1 Lessons from* S. cerevisiae

Global analysis of the Cdk1-dependent *S. cerevisiae* phosphoproteome identified the position of 547 phosphorylation sites on 308 proteins, based on the specific chemical inhibition of Cdk1 and quantitative MS (Holt et al. 2009). Study of the structural context of the Cdk1 sites revealed some interesting features. First, more than 90% of the sites were located in loops and disordered regions. Second, Cdk1 substrates tended to be phosphorylated at multiple sites. Finally, Cdk1 phosphorylation sites tended to cluster in the primary sequence, suggesting that multiple phosphorylation events would modulate the same protein surface. Notably, comparisons of the substrates with their orthologs in another 32 Ascomycetes showed that the position of most of the phosphorylation sites was not highly conserved; instead, they shifted position inside the rapidly evolving disordered regions (Holt et al. 2009). In sum, although the minimal Cdk1 phosphorylation motif (S/T-P) is conserved over long evolutionary timescales, the regions containing them show rapid evolution. Thus, these features allow Cdk1 control mechanisms to adapt rapidly to developmental challenges that have arisen or may arise during the course of evolution. This flexibility in phosphorylation site positioning might have important biological implications, since the appearance of a linage-specific Cdk cluster in a protein could give rise to new regulatory controls. For example, it has been suggested that the Cdk-driven regulation of nuclear localization of the pre-Replicative Complex component Mcm3 could have appeared in the *S. cerevisiae* linage after its divergence from *C. albicans* through the acquisition of Cdk sites clustered at the ScMcm3 C-terminus (Moses et al. 2007b).

A new computational strategy aimed at identifying proteins that contain high densities of strong (S/T-P-X-R/K) and weak (S/T-P) Cdk consensus sites closely spaced in the amino acid sequence has recently been developed (Moses et al. 2007a). This method allows the identification of proteins in which Cdk clusters deviate from random expectation by calculating the likelihood ratio statistic ($S_{LR}$). This cluster-based method measures the enrichment of motifs in a sequence and their spatial clustering. In order to define an $S_{LR}$ cut-off value to use in the prediction of Cdk substrates, a comparison of the distribution of $S_{LR}$ scores using either the real Cdk consensus motifs or scrambled versions (P-R/K-X-S/T and P-S/T) was performed and a score threshold of 3.5 was defined. Therefore, cluster-based methods used in combination with other evidence, such as structural properties (Iakoucheva et al. 2004) or evolutionary conservation (Budovskaya et al. 2005), could be exploited to predict Cdk targets (Moses et al. 2007a).

### *5.4.2 Prediction of Cdk Targets in the* C. albicans *Proteome*

Given the rapid evolution of Cdk phosphorylation site positioning in the disordered regions of proteins, we propose a speculative model whereby the commensal lifestyle of *C. albicans* might have led Cdk evolution to be connected to a much broader range of signaling pathways than in *S. cerevisiae*. This would have allowed the integration of Cdk-control mechanisms with the different developmental programs triggered by environmental cues found in the host. This hypothesis would imply the existence of *Candida* linage-specific Cdk clusters in proteins involved in the response to environment signals, such as cell signaling proteins and transcriptional regulators. To test this hypothesis, we searched the *C. albicans* proteome for proteins containing putative regulatory Cdk clusters using the $S_{LR}$ algorithm. In order to reduce the number of false positives, a second criterion was used; this was that Cdk clusters had to be located in disordered regions, as determined by the PONDR algorithm (http://www.pondr.com). Finally, to test whether the identified putative regulatory clusters were linage-specific, we compared such regions with their ortholog proteins of other *Hemiascomycota* species.

To identify putative Cdk targets in *C. albicans*, we analyzed the proteome using the $S_{LR}$ algorithm for the presence of clusters of strong and weak Cdk motifs and used a threshold $S_{LR}$ score of 3.5 (named $S_{LRF}$ analysis). This analysis identified 91 proteins with an $S_{LR}$ value above the cut-off, which represents the 1.46% of the total proteins (Fig. 5.1, inset). Of these 91 predictions, 52 of them had orthologs in *S. cerevisiae*. Gene ontology (GO) analysis of the putative substrates revealed a strong enrichment for cell cycle-related functional categories (20/52; 38%). In addition, 34 of them (63%) showed Cdk1-dependent phosphorylation in *S. cerevisiae* or *C. albicans* (Ubersax et al. 2003; Beltrao et al. 2009;

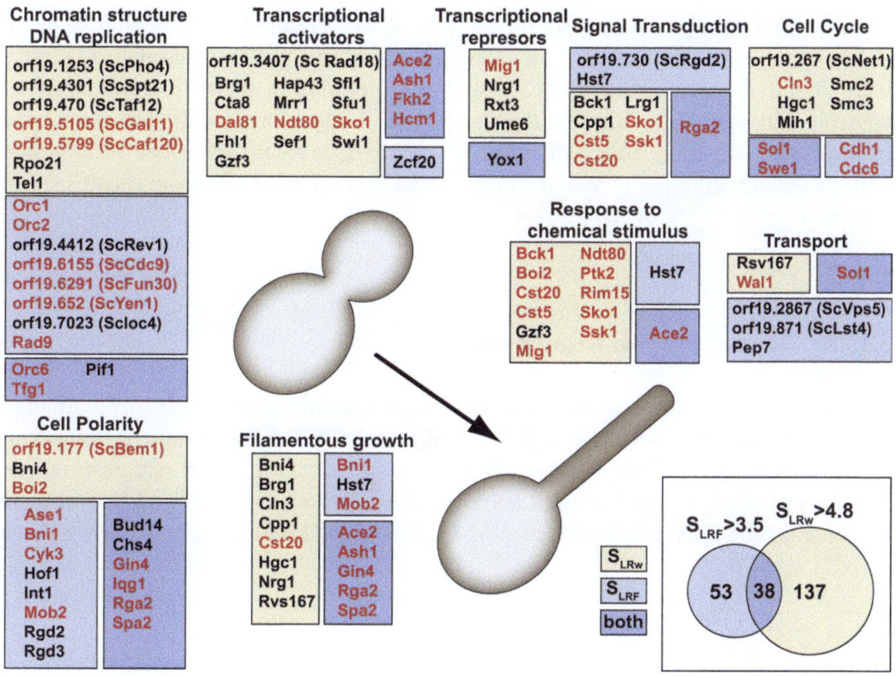


**Fig. 5.1** Selected proteins with Cdk1 regulatory clusters grouped by GO cellular process. The color of the box indicates whether a protein was identified in the strong phosphorylation site search ($S_{LRF}$), in the weak phosphorylation site analysis ($S_{LRW}$), or in both. Proteins in *red* show Cdk-dependent phosphorylation in *S. cerevisiae* and/or *C. albicans* (Ubersax et al. 2003; Beltrao et al. 2009; Holt et al. 2009). The *inset* shows a graphic representation of the results obtained in the SLRF or SLRW analysis

Holt et al. 2009). Accordingly, the value of 3.5 seemed to be a good threshold when strong and weak Cdk motifs were used. However, given the existence of Cdk substrates lacking strong consensus sites regulated by Cdk phosphorylation at weak sites (Nash et al. 2001; Strickfaden et al. 2007), we performed a second analysis searching for clustering of weak motifs only (named $S_{LRW}$ analysis). A total of 267 proteins (4.3% of the proteome) with a score above 3.5 were identified. To reduce the number of false positives, the threshold was increased to 4.8, a score at which the enrichment for cell cycle-related proteins was 24%; similar to that obtained in the identification of Cdk1 targets in *S. cerevisiae* (Holt et al. 2009). This reduced the set to 175 positive proteins (2.68%), which included 42% of the proteins identified in the $S_{LRF}$ analysis (Fig. 5.1). In sum, the combination of $S_{LRF}$ and $S_{LRW}$ allowed the identification of 228 proteins, representing the 3.6% of the *C. albicans* proteome. GO analysis revealed that in addition to cell cycle-related processes there was an enrichment in proteins belonging to other cellular processes, such as filamentous growth (33/228, 14.5%) or transcription regulation (27/228, 11.8%). A selection of putative Cdk1 substrates grouped by GO cellular component is shown in Fig. 5.1. Notably, we found that proteins involved in environmental sensing were also present in the predicted Cdk substrates. Environmental sensing is a complex process that includes several steps, and putative regulatory targets were found in proteins involved in cell wall regulation (CWR), transcriptional control, and cell signaling. In the following sections, we shall describe some examples to illustrate how Cdk-control mechanisms could have been integrated with other signaling pathways in *Candida* and that are different from those seen in *S. cerevisiae*.

### 5.4.2.1 Transcriptional Regulation

Transcription factors constitute one of the gene families enriched in pathogenic *Candida* species (Butler et al. 2009). In our analysis, we found several transcriptional regulators involved in environmental responses (Fig. 5.2a), suggesting that Cdk signaling could regulate their activity. These factors are: Sfu1 and Hap43 (iron response); Dal81 and Hsf1 (heat response); Mrr1 and Ndt80 (drug resistance); Ace2

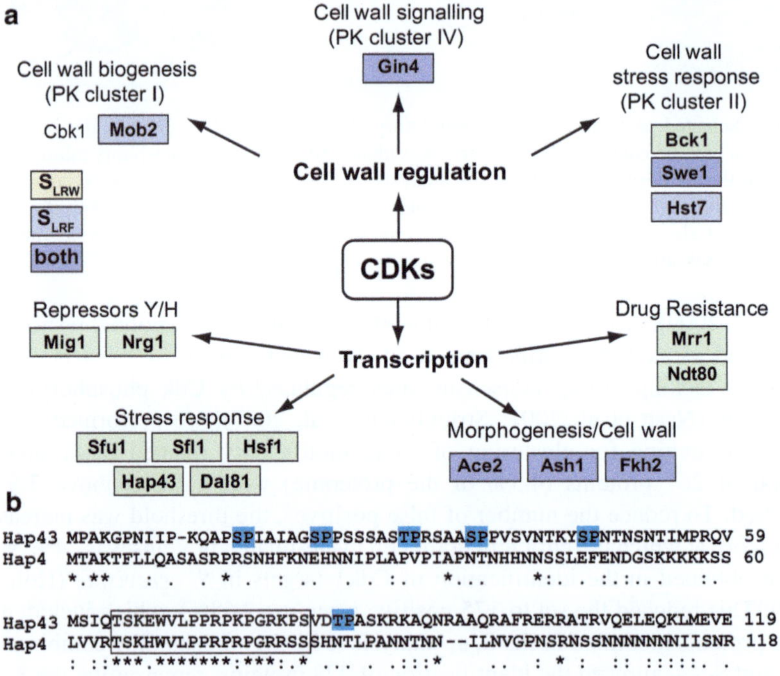


**Fig. 5.2** (**a**) Predicted Cdk1 substrates in cell wall and transcriptional regulation. In the CWR pathway, many protein kinases involved in CWR or the regulatory subunit of the Cbk1–Mob2 complex contain putative Cdk1 regulatory clusters. Cdks can also control transcription activation or repression through the phosphorylation of transcription factors in different pathways. (**b**) Alignment of the N-terminal region of *C. albicans* Hap43 and *S. cerevisiae* Hap4. Weak Cdk sites are indicated in *blue*, while the *rectangle* shows the conserved CBC domain

(biofilm); Sfl1 (caspofungin response); and Mig1 and Nrg1 (yeast-to-hypha transition). Comparison with their *S. cerevisiae* orthologs showed that five of them – Hap43, Ndt80, Hsf1, Ace2, and Nrg1 – have at least one cluster with five or more Cdk sites in regions that are not present in *S. cerevisiae*. This suggests that their regulation could have additional layers of complexity in *Candida*. As an example, an alignment of the N-terminal region of *C. albicans* Hap43 with *S. cerevisiae* Hap4 is shown in Fig. 5.2b. This region contains the CBC domain that is required for the interaction with the Hap2/3/5 complex (Bourgarel et al. 1999), which is conserved in both proteins. Interestingly, in *C. albicans* the CBC domain is located in a disordered region predicted by PONDR (1–128), which also contains six weak Cdk sites not present in *S. cerevisiae*, suggesting that Cdk1 might modulate the interaction of Hap43 with the complex. Iron homeostasis is essential for microorganisms, such as *C. albicans*, that compete for iron in a mammalian host. Hap43 is essential for iron-responsive transcriptional regulation and virulence in *C. albicans* (Hsu et al. 2011). Iron-regulatory networks have undergone a differential evolution since *S. cerevisiae* and *C. albicans* diverged, regulation in each yeast species being adapted to their specific growth conditions (Homann et al. 2009).

#### 5.4.2.2 Cell Wall Regulation

In addition to its protective role, the cell wall plays an important function in interaction with the environment, both in sensing external cues and in interacting with other cells. There is a significant expansion of cell wall gene families in pathogenic species, suggesting that CWR could be important for virulence (Butler et al. 2009). Recently, phenotypic analysis regarding sensitivity to cell wall stresses of a collection of protein kinase mutants has shown that cell wall signaling networks in *C. albicans* are expanded in comparison to those of *S. cerevisiae*, and that some signaling pathways have been rewired and integrated in the cell wall integrity response in *Candida* (Blankenship et al. 2010). Five kinases related to the CWR were identified in our analysis (Fig. 5.2a): three proteins (Bck1, Gin4, and Mob2, this latter being the regulatory subunit of the Cbk1–Mob2 complex) with conserved roles in CWR in *C. albicans* and *S. cerevisiae*, and two (Swe1 and Hst7) with apparent *C. albicans* CWR-specific functions (Blankenship et al. 2010).

These kinases have high $S_{LR}$ scores, suggesting that Cdks might modulate their kinase activity under specific circumstances. All of them contain clusters of Cdk sites of different lengths located at disordered regions that are either absent in their *S. cerevisiae* orthologs or that are at different regions of the protein. The most striking example is Gin4 ($S_{LRF}$ 18.34), which has an amino acid sequence with six contiguous strong Cdk sites ($S^{443}$, $S^{447}$, $S^{451}$, $S^{455}$, $S^{459}$, and $S^{463}$) in a disordered region. The alignment of Gin4 with its orthologs shows some interesting aspects. First, this cluster is absent in the *Saccharomyces* clade (Fig. 5.3), although other flanking Cdk sites are more or less conserved in position. Second, in the *Candida* clade, an increase in the number of Cdk sites is observed from *C. lusitaniae* to *C. albicans*. Thus, this progressive accumulation of strong Cdk sites at the same

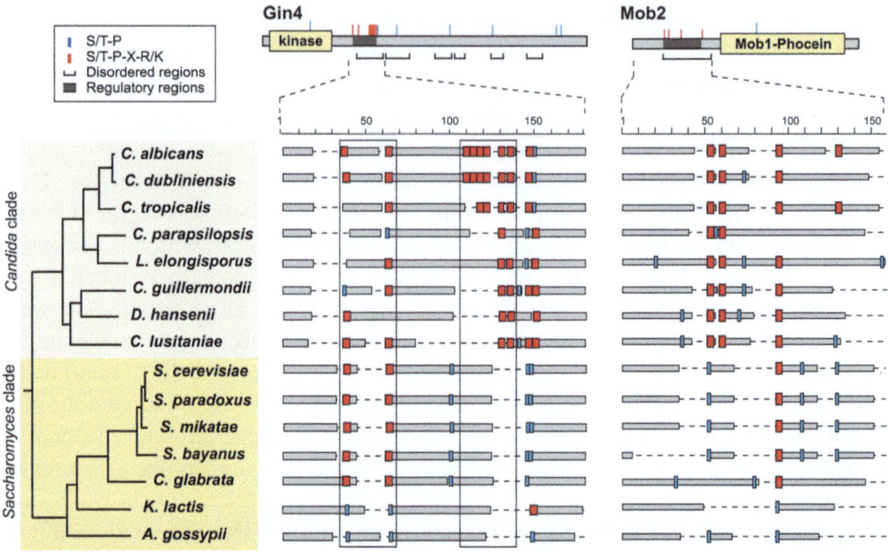


**Fig. 5.3** Evolution of clusters of Cdk regulatory sites in Hemiascomycota: examples of the appearance of clusters of regulatory sites in the *Candida* clade. At the top, a schematic representation of Mob2 and Gin4 with their different domains. Strong (*red*) and weak (*blue*) Cdk phosphorylation sites, regulatory regions predicted by SLR (*dark gray*) and disordered regions (*brackets* below the sequence) are also indicated. The sequence of the regions containing the predicted regulatory clusters (indicated with *dashed lines*) is aligned below with their orthologs from the *Candida* and *Saccharomyces* clades. The *dashed line* indicates gaps introduced to maximize the alignment

disordered region suggests a linage-specific regulation of Gin4 by Cdks. The Gin4 kinase is involved in septin organization in both yeasts (Longtine et al. 1998; Wightman et al. 2004). In *S. cerevisiae*, Gin4 is activated during mitosis in a Cdc28/Clb2-dependent manner (Altman and Kellogg 1997). This regulation is also probably conserved in *C. albicans*, since Gin4 phosphorylates the septin Cdc11 at Ser395 at the end of the cell cycle (Sinha et al. 2007). It is likely that in both yeasts cell cycle regulation could be exerted through the conserved Cdk sites, while the cluster of Cdk sites in CaGin4 might create new interactions through other phosphopeptide-binding proteins. Indeed, it has been shown that a *gin4* mutant is hypersensitive to oxidative or osmotic stresses in *C. albicans* but not in *S. cerevisiae* (Blankenship et al. 2010).

#### 5.4.2.3 Cell Signaling

MAP kinase-mediated pathways are important stress-signaling modules essential for the development of appropriate acute and adaptative responses to environmental cues. In pathogenic fungi, these pathways are important for virulence (Román et al. 2007). Our analysis suggests that the Cek1-mediated pathway could receive inputs

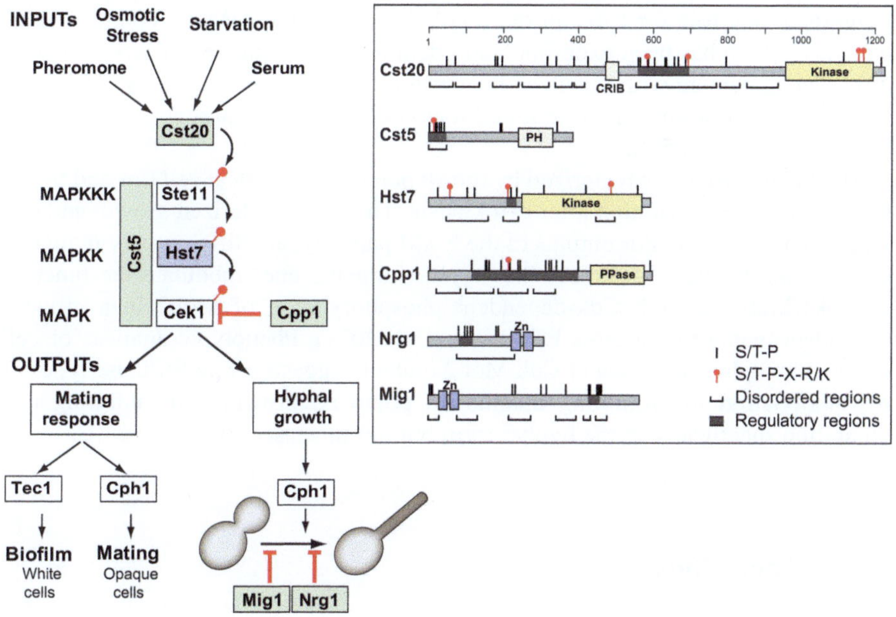


**Fig. 5.4** The Cph1-mediated MAPK pathway can be regulated by Cdk1 at different levels. Putative Cdk1 regulatory clusters are present in different components of the Cst20-Cek1 pathway and in two transcriptional repressors. To the right, a schematic representation of the six putative targets of the pathway, indicating the different domains in each protein, the position of the Cdk phosphorylation sites, the putative regulatory regions, and disordered regions

from Cdk signaling at different levels, since several components of the network were identified (Fig. 5.4). In *C. albicans*, this pathway is involved in mating and hyphal growth (Liu et al. 1994; Kohler and Fink 1996; Leberer et al. 1996; Chen et al. 2002, Cote et al. 2011; Yi et al. 2011). This network is composed of Cst20, the Ste11-Hst7-Cek1 module, the scaffolding protein Cst5, and the transcription factor Cph1. Finally, the Cpp1 phosphatase inhibits the pathway, probably by dephosphorylating Cek1 (Csank et al. 1997). During mating, this MAPK pathway activates different transcription factors depending on the cell type (Fig. 5.4) (Soll 2011). Recently, it has been shown that the exposure of *C. albicans* cells to β-glucan is controlled by the Cek1-mediated pathway (Galán-Díez et al. 2010), suggesting that this pathway might modulate innate immunoresponses triggered through dectin-1.

Our computational analysis also suggests that Cdks could regulate the RAM signaling pathway (Nelson et al. 2003) through the phosphorylation of Mob2 (Figs. 5.2 and 5.3). In *S. cerevisiae*, this pathway controls cell separation and polarized growth through the activity of the NDR kinase Cbk1 (Weiss et al. 2002), which requires interaction with Mob2 for its function. Whereas CaMob2 showed an $S_{LRF}$ value of 6.75 (the top 38), its *S. cerevisiae* ortholog had an $S_{LRF}$ value of 0.77 (the top 887 from the *S. cerevisiae* proteome). This divergence suggests a new role for Cdks in the regulation of the Cbk1/Mob2 complex in *C. albicans*, which has been experimentally demonstrated (Gutiérrez-Escribano et al. 2011). In agreement with this idea, the cluster of four full Cdk sites ($S^{44}$; $S^{51}$, $S^{67}$, and $S^{97}$) present in the amino terminal region of CaMob2 is absent in the *Saccharomyces* clade (Fig. 5.3).

Hyphal growth is characterized by robust polarized growth at cell tips and by the inhibition of cell separation after cytokinesis. Therefore, polarized growth and cell separation, the two major outputs of the RAM pathway, are differentially regulated in yeast and hyphae. We found that hyphal-inducing cues modulate the function of Cbk1/Mob2 through Cdk-dependent phosphorylation of Mob2 in a growth-dependent manner (Gutiérrez-Escribano et al. 2011). Phenotypic analysis of cell expressing a phosphodeficient Cdk Mob2 mutant suggests a role for these types of phosphorylation in promoting maintained polarized growth and inhibiting cell separation specifically in the hyphal form but not in yeast cells.

## 5.5 Conclusions

Several studies have shown the existence of rewiring in *C. albicans* transcriptional regulatory pathways (Li and Johnson 2010) and protein kinases (Blankenship et al. 2010). Recently, we have shown that Cdk is essential for the differential modification of the outputs of a core signaling system, the Cbk1/Mob2 complex, depending on the environmental signals that activate different cell fate programs (yeast or hypha) (Gutiérrez-Escribano et al. 2011). In this chapter, we have speculated that Cdks might be connected to broader range of signaling pathways than in *S. cerevisiae* through the acquisition of *Candida* linage-specific Cdk sites clustered in proteins involved in environmental responses. Our Cdk cluster-based search in combination with other evidences, such as localization in disordered regions and the evolution of such clusters in other Hemiascomycetes, could be used to predict new layers of Cdk regulation in *C. albicans* environmental sensing networks.

**Acknowledgments** This work was supported by grants from the Spanish Ministry of Science and Innovation to JCB (BFU2009-11251) and to CRV (BFU2010-15884) and the Regional Government of Extremadura (PRI08A017 and GRU09001) to JCB. The IBFG is institutionally supported by Fundación Ramón Areces.

## References

Altman R, Kellogg D (1997) Control of mitotic events by Nap1 and the Gin4 kinase. J Cell Biol 138:119–130

Archambault V, Buchler NE, Wilmes GM, Jacobson MD, Cross FR (2005) Two-faced cyclins with eyes on the targets. Cell Cycle 4:125–130. doi:10.1016/j.molcel.2004.05.025

Bachewich C, Whiteway M (2005) Cyclin Cln3p links G1 progression to hyphal and pseudohyphal development in *C. albicans*. Eukaryot Cell 4:95–102. doi:10.1128/EC.4.1.95-102.2005

Beltrao P, Serrano L (2007) Specificity and evolvability in eukaryotic protein interaction networks. PLoS Comput Biol 3:e25. doi:10.1371/journal.pcbi.0030025

Beltrao P, Trinidad JC, Fiedler D, Roguev A, Lim WA, Shokat KM, Burlingame AL, Krogan NJ (2009) Evolution of phosphoregulation: comparison of phosphorylation patterns across yeast species. PLoS Biol 7:e1000134. doi:10.1371/journal.pbio.1000134

Bensen ES, Clemente-Blanco A, Finley KR, Correa-Bordes J, Berman J (2005) The mitotic cyclins Clb2p and Clb4p affect morphogenesis in *Candida albicans*. Mol Biol Cell 16:3387–3400. doi:10.1091/mbc.E04-12-1081

Bhattacharyya RP, Remenyi A, Yeh BJ, Lim WA (2006) Domains, motifs, and scaffolds: the role of modular interactions in the evolution and wiring of cell signaling circuits. Annu Rev Biochem 75:655–680. doi:10.1146/annurev.biochem.75.103004.142710

Biswas S, Van Dijck P, Datta A (2007) Environmental sensing and signal transduction pathways regulating morphopathogenic determinants of *Candida albicans*. Microbiol Mol Biol Rev 71:348–376. doi:10.1128/MMBR.00009-06

Blankenship JR, Fanning S, Hamaker JJ, Mitchell AP (2010) An extensive circuitry for cell wall regulation in *Candida albicans*. PLoS Pathog 6:e1000752. doi:10.1371/journal.ppat.1000752

Bloom J, Cross FR (2007) Multiple levels of cyclin specificity in cell-cycle control. Nat Rev Mol Cell Biol 8:149–160. doi:10.1038/nrm2105

Bourgarel D, Nguyen CC, Bolotin-Fukuhara M (1999) HAP4, the glucose-repressed regulated subunit of the HAP transcriptional complex involved in the fermentation-respiration shift, has a functional homologue in the respiratory yeast *Kluyveromyces lactis*. Mol Microbiol 31:1205–1215

Braun BR, Johnson AD (1997) Control of filament formation in *Candida albicans* by the transcriptional repressor *TUP1*. Science 277:105–109

Braun BR, Head WS, Wang MX, Johnson AD (2000) Identification and characterization of *TUP1*-regulated genes in *Candida albicans*. Genetics 156:31–44

Budovskaya YV, Stephan JS, Deminoff SJ, Herman PK (2005) An evolutionary proteomics approach identifies substrates of the cAMP-dependent protein kinase. Proc Natl Acad Sci USA 102:13933–13938. doi:10.1073/pnas.0501046102

Butler G, Rasmussen MD, Lin MF et al (2009) Evolution of pathogenicity and sexual reproduction in eight *Candida* genomes. Nature 459:657–662. doi:10.1038/nature08064

Calderone RA, Fonzi WA (2001) Virulence factors of *Candida albicans*. Trends Microbiol 9:327–335. doi:S0966-842X(01)02094-7

Chapa y Lazo B, Bates S, Sudbery P (2005) The G1 cyclin Cln3 regulates morphogenesis in *Candida albicans*. Eukaryot Cell 4:90–94. doi:10.1128/EC.4.1.90-94.2005

Chen J, Lane S, Liu H (2002) A conserved mitogen-activated protein kinase pathway is required for mating in *Candida albicans*. Mol Microbiol 46:1335–1344. doi:10.1046/j.1365-2958.2002.03249.x

Cote P, Sulea T, Dignard D, Wu C, Whiteway M (2011) Evolutionary reshaping of fungal mating pathway scaffold proteins. MBio 2. doi:10.1128/mBio.00230-10

Coudreuse D, Nurse P (2010) Driving the cell cycle with a minimal CDK control network. Nature 468:1074–1079. doi:10.1038/nature09543

Csank C, Makris C, Meloche S, Schroppel K, Rollinghoff M, Dignard D, Thomas D, Whiteway M (1997) Derepressed hyphal growth and reduced virulence in a VH1 family-related protein phosphatase mutant of the human pathogen *Candida albicans*. Mol Biol Cell 8:2539–2551

Echalier A, Endicott JA, Noble ME (2010) Recent developments in cyclin-dependent kinase biochemical and structural studies. Biochim Biophys Acta 1804:511–519. doi:10.1016/j.bbapap. 2009.10.002

Eggimann P, Garbino J, Pittet D (2003) Epidemiology of *Candida* species infections in critically ill non-immunosuppressed patients. Lancet Infect Dis 3:685–702. doi:10.1016/S1473-3099(03)00801-6

Fisher DL, Nurse P (1996) A single fission yeast mitotic cyclin B p34cdc2 kinase promotes both S-phase and mitosis in the absence of G1 cyclins. EMBO J 15:850–860

Galán-Díez M, Arana DM, Serrano-Gómez D, Kremer L, Casasnovas JM, Ortega M, Cuesta-Domínguez A, Corbi AL, Plá J, Fernández-Ruiz E (2010) *Candida albicans* β-glucan exposure is controlled by the fungal CEK1-mediated mitogen-activated protein kinase pathway that modulates immune responses triggered through dectin-1. Infect Immun 78:1426–1436. doi:10.1128/IAI.00989-09

González-Novo A, Correa-Bordes J, Labrador L, Sánchez M, Vázquez de Aldana CR, Jiménez J (2008) Sep7 is essential to modify septin ring dynamics and inhibit cell separation during *C. albicans* hyphal growth. Mol Biol Cell 19:1509–1518. doi:10.1091/mbc.E07-09-0876

Gutiérrez-Escribano P, González-Novo A, Suárez MB, Li C-R, Wang Y, Vázquez de Aldana CR, Correa-Bordes J (2011) Cdk-dependent phosphorylation of Mob2 is essential for hyphal development in *Candida albicans*. Mol Biol Cell 22(14):2458–2469

Holt LJ, Tuch BB, Villen J, Johnson AD, Gygi SP, Morgan DO (2009) Global analysis of Cdk1 substrate phosphorylation sites provides insights into evolution. Science 325:1682–1686. doi:10.1126/science.1172867

Homann OR, Dea J, Noble SM, Johnson AD (2009) A phenotypic profile of the *Candida albicans* regulatory network. PLoS Genet 5:e1000783. doi:10.1371/journal.pgen.1000783

Hsu PC, Yang CY, Lan CY (2011) *Candida albicans* Hap43 is a repressor induced under low-iron conditions and is essential for iron-responsive transcriptional regulation and virulence. Eukaryot Cell 10:207–225. doi:10.1128/EC.00158-10

Huang D, Friesen H, Andrews B (2007) Pho85, a multifunctional cyclin-dependent protein kinase in budding yeast. Mol Microbiol 66:303–314. doi:10.1111/j.1365-2958.2007.05914.x

Iakoucheva LM, Radivojac P, Brown CJ, O'Connor TR, Sikes JG, Obradovic Z, Dunker AK (2004) The importance of intrinsic disorder for protein phosphorylation. Nucleic Acids Res 32:1037–1049. doi:10.1093/nar/gkh253

Kadosh D, Johnson AD (2005) Induction of the *Candida albicans* filamentous growth program by relief of transcriptional repression: a genome-wide analysis. Mol Biol Cell 16:2903–2912. doi:10.1091/mbc.E05-01-0073

Kohler JR, Fink GR (1996) *Candida albicans* strains heterozygous and homozygous for mutations in mitogen-activated protein kinase signaling components have defects in hyphal development. Proc Natl Acad Sci USA 93:13223–13228

Leberer E, Harcus D, Broadbent ID, Clark KL, Dignard D, Ziegelbauer K, Schmidt A, Gow NA, Brown AJ, Thomas DY (1996) Signal transduction through homologs of the Ste20p and Ste7p protein kinases can trigger hyphal formation in the pathogenic fungus *Candida albicans*. Proc Natl Acad Sci USA 93:13217–13222

Li H, Johnson AD (2010) Evolution of transcription networks–lessons from yeasts. Curr Biol 20: R746–R753. doi:10.1016/j.cub.2010.06.056

Liu H, Kohler J, Fink GR (1994) Suppression of hyphal formation in *Candida albicans* by mutation of a *STE12* homolog. Science 266:1723–1726

Lo HJ, Kohler JR, DiDomenico B, Loebenberg D, Cacciapuoti A, Fink GR (1997) Nonfilamentous *C. albicans* mutants are avirulent. Cell 90:939–949. doi:S0092-8674(00)80358-X

Loeb JD, Sepulveda-Becerra M, Hazan I, Liu H (1999) A G1 cyclin is necessary for maintenance of filamentous growth in *Candida albicans*. Mol Cell Biol 19:4019–4027

Lohse MB, Johnson AD (2009) White-opaque switching in *Candida albicans*. Curr Opin Microbiol 12:650–654. doi:10.1016/j.mib.2009.09.010

Longtine MS, Fares H, Pringle JR (1998) Role of the yeast Gin4p protein kinase in septin assembly and the relationship between septin assembly and septin function. J Cell Biol 143:719–736

Loog M, Morgan DO (2005) Cyclin specificity in the phosphorylation of cyclin-dependent kinase substrates. Nature 434:104–108. doi:10.1038/nature03329

Massey SE, Moura G, Beltrao P, Almeida R, Garey JR, Tuite MF, Santos MA (2003) Comparative evolutionary genomics unveils the molecular mechanism of reassignment of the CTG codon in *Candida spp.*. Genome Res 13:544–557. doi:10.1101/gr.811003

Morrison DK (2009) The 14-3-3 proteins: integrators of diverse signaling cues that impact cell fate and cancer development. Trends Cell Biol 19:16–23. doi:10.1016/j.tcb.2008.10.003

Moseley JB, Nurse P (2009) Cdk1 and cell morphology: connections and directions. Curr Opin Cell Biol 21:82–88. doi:10.1016/j.ceb.2008.12.005

Moses AM, Heriche JK, Durbin R (2007a) Clustering of phosphorylation site recognition motifs can be exploited to predict the targets of cyclin-dependent kinase. Genome Biol 8:R23. doi:10.1186/gb-2007-8-2-r23

Moses AM, Liku ME, Li JJ, Durbin R (2007b) Regulatory evolution in proteins by turnover and lineage-specific changes of cyclin-dependent kinase consensus sites. Proc Natl Acad Sci USA 104:17713–17718. doi:10.1073/pnas.0700997104

Murad AM, d'Enfert C, Gaillardin C, Tournu H, Tekaia F, Talibi D, Marechal D, Marchais V, Cottin J, Brown AJ (2001a) Transcript profiling in *Candida albicans* reveals new cellular functions for the transcriptional repressors CaTup1, CaMig1 and CaNrg1. Mol Microbiol 42:981–993

Murad AM, Leng P, Straffon M, Wishart J, Macaskill S, MacCallum D, Schnell N, Talibi D, Marechal D, Tekaia F, d'Enfert C, Gaillardin C, Odds FC, Brown AJ (2001b) NRG1 represses yeast-hypha morphogenesis and hypha-specific gene expression in *Candida albicans*. EMBO J 20:4742–4752. doi:10.1093/emboj/20.17.4742

Nash P, Tang X, Orlicky S, Chen Q, Gertler FB, Mendenhall MD, Sicheri F, Pawson T, Tyers M (2001) Multisite phosphorylation of a CDK inhibitor sets a threshold for the onset of DNA replication. Nature 414:514–521. doi:10.1038/35107009

Nelson B, Kurischko C, Horecka J, Mody M, Nair P, Pratt L, Zougman A, McBroom LD, Hughes TR, Boone C, Luca FC (2003) RAM: a conserved signaling network that regulates Ace2p transcriptional activity and polarized morphogenesis. Mol Biol Cell 14:3782–3803. doi:10.1091/mbc.E03-01-0018

Román E, Arana DM, Nombela C, Alonso-Monge R, Plá J (2007) MAP kinase pathways as regulators of fungal virulence. Trends Microbiol 15:181–190. doi:10.1016/j.tim.2007.02.001

Santos MA, Tuite MF (1995) The CUG codon is decoded in vivo as serine and not leucine in *Candida albicans*. Nucleic Acids Res 23:1481–1486

Seet BT, Dikic I, Zhou MM, Pawson T (2006) Reading protein modifications with interaction domains. Nat Rev Mol Cell Biol 7:473–483. doi:10.1038/nrm1960

Serber Z, Ferrell JE Jr (2007) Tuning bulk electrostatics to regulate protein function. Cell 128:441–444. doi:10.1016/j.cell.2007.01.018

Sinha I, Wang YM, Philp R, Li CR, Yap WH, Wang Y (2007) Cyclin-dependent kinases control septin phosphorylation in *Candida albicans* hyphal development. Dev Cell 13:421–432. doi:10.1016/j.devcel.2007.06.011

Slutsky B, Staebell M, Anderson J, Risen L, Pfaller M, Soll D (1987) "White-opaque transition": a second high-frequency switching system in *Candida albicans*. J Bacteriol 169:189–197

Soll DR (2009) Why does *Candida albicans* switch? FEMS Yeast Res 9:973–989. doi:10.1111/j.1567-1364.2009.00562.x

Soll DR (2011) Evolution of a new signal transduction pathway in *Candida albicans*. Trends Microbiol 19:8–13. doi:10.1016/j.tim.2010.10.001

Songyang Z, Blechner S, Hoagland N, Hoekstra MF, Piwnica-Worms H, Cantley LC (1994) Use of an oriented peptide library to determine the optimal substrates of protein kinases. Curr Biol 4:973–982. doi:S0960-9822(00)00221-9

Stoldt VR, Sonneborn A, Leuker CE, Ernst JF (1997) Efg1p, an essential regulator of morphogenesis of the human pathogen *Candida albicans*, is a member of a conserved class of bHLH proteins regulating morphogenetic processes in fungi. EMBO J 16:1982–1991. doi:10.1093/emboj/16.8.1982

Strickfaden SC, Winters MJ, Ben-Ari G, Lamson RE, Tyers M, Pryciak PM (2007) A mechanism for cell-cycle regulation of MAP kinase signaling in a yeast differentiation pathway. Cell 128:519–531. doi:10.1016/j.cell.2006.12.032

Sudbery P, Gow N, Berman J (2004) The distinct morphogenic states of *Candida albicans*. Trends Microbiol 12:317–324

Ubersax JA, Woodbury EL, Quang PN, Paraz M, Blethrow JD, Shah K, Shokat KM, Morgan DO (2003) Targets of the cyclin-dependent kinase Cdk1. Nature 425:859–864. doi:10.1038/nature02062

Wang Y (2009) CDKs and the yeast-hyphal decision. Curr Opin Microbiol 12:644–649. doi:10.1016/j.mib.2009.09.002

Wang A, Lane S, Tian Z, Sharon A, Hazan I, Liu H (2007) Temporal and spatial control of *HGC1* expression results in Hgc1 localization to the apical cells of hyphae in *Candida albicans*. Eukaryot Cell 6:253–261. doi:10.1128/EC.00380-06

Wang A, Raniga PP, Lane S, Lu Y, Liu H (2009) Hyphal chain formation in *Candida albicans*: Cdc28-Hgc1 phosphorylation of Efg1 represses cell separation genes. Mol Cell Biol 29:4406–4416. doi:10.1128/MCB.01502-08

Weiss EL, Kurischko C, Zhang C, Shokat K, Drubin DG, Luca FC (2002) The *Saccharomyces cerevisiae* Mob2p-Cbk1p kinase complex promotes polarized growth and acts with the mitotic exit network to facilitate daughter cell-specific localization of Ace2p transcription factor. J Cell Biol 158:885–900. doi:10.1083/jcb.200203094

Whiteway M, Bachewich C (2007) Morphogenesis in *Candida albicans*. Annu Rev Microbiol 61:529–553. doi:10.1146/annurev.micro.61.080706.093341

Wightman R, Bates S, Amornrrattanapan P, Sudbery P (2004) In *Candida albicans*, the Nim1 kinases Gin4 and Hsl1 negatively regulate pseudohypha formation and Gin4 also controls septin organization. J Cell Biol 164:581–591. doi:10.1083/jcb.200307176

Yi S, Sahni N, Daniels KJ, Lu KL, Huang G, Garnaas AM, Pujol C, Srikantha T, Soll DR (2011) Utilization of the mating scaffold protein in the evolution of a new signal transduction pathway for biofilm development. MBio 2:e00237-10. doi:10.1128/mBio.00237-10

Zheng X, Wang Y (2004) Hgc1, a novel hypha-specific G1 cyclin-related protein regulates *Candida albicans* hyphal morphogenesis. EMBO J 23:1845–1856. doi:10.1038/sj.emboj.7600195

Zheng XD, Lee RT, Wang YM, Lin QS, Wang Y (2007) Phosphorylation of Rga2, a Cdc42 GAP, by CDK/Hgc1 is crucial for *Candida albicans* hyphal growth. EMBO J 26:3760–3769. doi:10.1038/sj.emboj.7601814

# Chapter 6
# Cell Cycle and Morphogenesis Connections During the Formation of the Infective Filament in *Ustilago maydis*

**José Pérez-Martín**

**Abstract** *Ustilago maydis* is the causal agent of smut disease on corn plants. The infective process depends on the formation of a specific structure called infective filament consisting on a dikaryotic hyphae, which is required to penetrate the plant tissue. The formation of the infective filament in *U. maydis* is alike to a germination process, although it requires an intermediate mating step that links sexual development and virulence. This way, the induction of the pathogenic program implies strong morphological changes (bud to hypha transition) as well as genetic changes (haploid to dikaryotic transition). As a consequence, an accurate control of the cell cycle as well as morphogenesis is predicted during these transitions: the induction of the infective filament relies on a dual process that involves by one side a specific cell cycle arrest and in other side the specific activation of a hyperpolarization growth. Impairment of any of these processes will have as an outcome the inhibition of the virulence. This review has been framed in three major points: (1) Which transcriptional program is responsible for the induction of the infective filament formation, (2) How polar growth is regulated during the induction of the infective filament, and (3) Which mechanisms are responsible for cell cycle arrest during the infective filament formation.

## 6.1 Introduction: *Ustilago maydis*, a Useful Model to Understand Virulence, Cell Cycle, and Morphogenesis Connections

Many of the most important plant diseases are caused by fungal pathogens, which rely on the formation of specialized cell structures to breach the leaf surface as well as to proliferate inside the plant. In multicellular eukaryotes, the control of cell

J. Pérez-Martín (✉)
Centro Nacional de Biotecnología – CSIC, Madrid, Spain
e-mail: jperez@cnb.csic.es

J. Pérez-Martín and A. Di Pietro (eds.), *Morphogenesis and Pathogenicity in Fungi*,
Topics in Current Genetics 22, DOI 10.1007/978-3-642-22916-9_6,

cycle and morphogenesis is pivotal to cellular differentiation, which must synchronize cell division to form specific tissues and organs effectively. In a similar way, cell cycle and morphogenesis regulation would be likely to provide control points for infection development by fungal pathogens. To initiate pathogenic development, fungi respond to a set of inductive cues. Some of them are of extracellular nature (environmental signals) while others respond to intracellular conditions (developmental signals). How each of these signals is perceived and how all signals are integrated into a single response are not clear. We believe that cell cycle and morphogenetic responses have to be adjusted in response to both environmental and developmental signals, and that the integration of both classes of signals by the cell machinery will result in an outcome that define the fungal fate: in the case of pathogenic fungi, whether they enter or not the virulence program. In this review, we seek to explore the relationship between cell cycle and morphogenesis regulation and pathogenesis development in the causal agent of corn smut, *Ustilago maydis*. This fungus is an excellent system to address the relationships between cell cycle, morphogenesis, and virulence (Perez-Martin et al. 2006; Steinberg and Perez-Martin 2008). In this model organism, virulence, and sexual development are intricately interconnected. A prerequisite for generating the infectious stage is the mating of two compatible budding haploid cells to generate, after cell fusion, an infective dikaryotic filament. Therefore, the induction of the pathogenic program implies not only strong morphological changes (bud to hypha transition) but also genetic changes (haploid to dikaryotic transition). Subsequently, an accurate control of the cell cycle as well as morphogenesis is predicted during these transitions.

## 6.2 Overview of the Life Cycle of *U. maydis*: The Importance of the Infective Filament During the Virulence Process

*U. maydis* is the causal agent of smut disease on corn plants. This basidiomycete fungus belongs to an important group of plant pathogens, the smut fungi, which can cause considerable grain yield loss and economic damage. In the field, corn smut infections are dispersed by airborne diploid teliospores (Christensen 1963; Brown and Hovmoller 2002). Germination of the teliospore on the plant surface is the first step in the infection process (Fig. 6.1a, b). Upon germination, meiosis takes place and pairs of compatible haploid cells are generated (Fig. 6.1c). Pathogenic development is mediated by two independent loci: the *a*-locus, encoding a pheromone-receptor system, and the *b*-locus, encoding a pair of homeoproteins (bW and bE). The process initiates with the recognition of mating pheromone secreted by haploid cells of the opposite mating type on the plant surface (Bolker et al. 1992). This induces a cell cycle arrest (Garcia-Muse et al. 2003; Perez-Martin et al. 2006) and triggers cells to the formation of long conjugation tubes. These filaments grow toward each other and fuse at their tips (Snetselaar et al. 1996). Cytoplasmic fusion is not followed by kariogamy, resulting in a dikaryotic cell, a hallmark of many

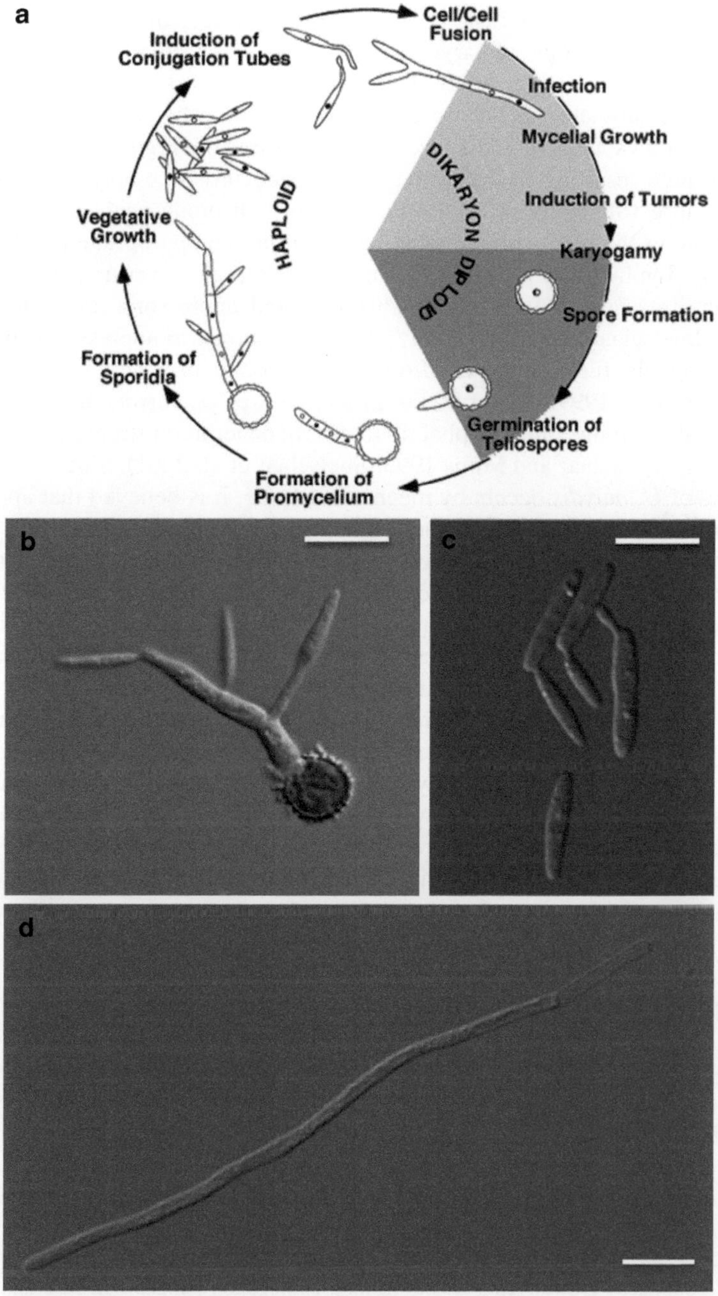


**Fig. 6.1** (**a**) Life cycle of *Ustilago maydis*. Diploid teliospore forms a promycelium that undergoes meiosis resulting in haploid sporidia. These yeast-like cells are saprofitic and grow by polar budding. If the teliospore germinates on the surface of a plant, the pathogenic program is initiated by the exchange of pheromone between haploid siblings and a switch to filamentous

basidiomycete. After cell fusion, on the plant surface, the cell cycle arrest is sustained and the single dikaryotic cell grows in a polar manner producing the infective filament (Fig. 6.1d). This hypha expands at the apical tip, and the cytoplasm accumulates in the tip cell compartment, whereas older parts of the hypha become vacuolated and are sealed off by inserting regularly spaced septa at the distal pole resulting in the formation of characteristic empty sections, which often collapse (Steinberg et al. 1998). This growth mode enables the fungus to progress along the plant surface most likely to find an appropriate point of entry. Eventually, hyphae stop polar growth in response to an as yet unidentified signal, and their tips swell to form poorly differentiated appressoria and penetrate the cuticule (Snetselaar and Mims 1992, 1993). In contrast to appressoria from other phytopathogenic fungi such as *Magnaporthe grisea* or *Colletotrichum* species (Bechinger et al. 1999; Talbot 2003), appressoria of *U. maydis* are unmelanized, rather small swellings of the hyphal tip that form penetration structures that are less constricted (Snetselaar and Mims 1993; Snetselaar et al. 2001). Since it is unlikely that entry of *U. maydis* occurs by mechanical force, it is believed that appressoria simply mark the point at which the growth direction changes. Once the filament enters the plant, cell cycle is reactivated. The formation of empty sections ceases, and mitotic divisions take place, concomitant with the development of clamp-like structures that allow the correct sorting of nuclei to maintain the dikaryotic status. This way, the fungal cells proliferate to a network of filaments with septated cell compartments each containing a pair of nuclei (Snetselaar and Mims 1992; Banuett and Herskowitz 1996). At the initial stage of infection, the hyphae traverse plant cells without an apparent host defense response. At later stages, plant tumors are induced by the fungus, followed by massive proliferation of fungal hyphae inside these tumors. Proliferation is followed by sporogenesis, a poorly understood process that includes karyogamy to produce diploid nuclei, hyphal sections fragmentation, and differentiation into heavily melanized diploid teliospores. Eventually the tumors dry up and rupture, releasing the diploid spores, which are dispersed by air, closing the life cycle of this fungus.

In many diseases caused by foliar fungal pathogens, the infection is initiated when spores attach to host surface and germinate. The process of germination implies the activation of a polarity axis and the emergence of a germ tube. For some fungi, the resulting germ tube has evolved to locate natural openings such as stomata, or alternatively has also evolved the ability to elaborate specialized infection structures, such as appressoria, that enable direct penetration of plant cuticule (Tucker and Talbot 2001). The formation of the infective filament in *U. maydis* is alike to this germination process, although it requires an intermediate

---

**Fig. 6.1** (continued) growth. This results in long conjugation hyphae that fuse and form the infective filament that contains two nuclei. This filament enters the plant and, after a short period of spreading within the host, starts to induce plant tumors, within which the fungus proliferates. After nuclear fusion and fragmentation of the hyphae, diploid black thick-walled resting spores (teliospores) are formed. (**b**) Teliospore during germination/meiosis process. (**c**) Haploid sporidia growing in liquid medium. (**d**) Infective filament. *Bar*: 15 μm mating step that links sexual development and virulence. In any way, the final result is the formation of an exploratory hypha that is required to penetrate the plant tissue. It is clear that the ability to form polarized hyphae may represent an "Achiles Heel" that can be exploited to limit fungal invasion of the plant tissue (Harris 2006). Therefore to understand how this infective filament formation is regulated is primordial to attack the infection at its early stages.

## 6.3 Formation of the Infective Filament: Transcriptional Regulation

The formation of the infective filament in *U. maydis* depends on an intricate transcriptional program at top of which is located a transcriptional regulator determined by the *b* locus (Feldbrugge et al. 2004). The *b* locus of *U. maydis* encodes the two unrelated homeodomain transcription factors bE and bW. These two proteins can form a heterodimeric complex, but only when the proteins are derived from different alleles (Gillissen et al. 1992; Kamper et al. 1995). This way, the production of this master regulator is linked to the mating process that, after cell fusion, leads to the interaction of the two subunits composing the b-factor, bW and bE, each one provided by each compatible mating partner. The current model proposes that dimerization is achieved via a limited number of hydrophobic and polar interactions within the variable N-terminal regions of the compatible bE and bW proteins (Kamper et al. 1995; Yee and Kronstad 1998). This heterodimerization to form active transcription factors is the general principle for interaction between the b proteins of smut fungi. The formation of the heterodimeric bE/bW complex is the sole determinant for the initiation of pathogenic development in *U. maydis*. This has been shown by the construction of haploid strains carrying compatible *b* alleles. Ectopic expression of these compatible b subunits in these haploid cells induces the formation of a monokaryotic filament that mimics its dikaryotic counterpart in all aspects of filamentous growth as well as apparent cell cycle arrest (Brachmann et al. 2001).

The central role of the bE/bW transcription factor in triggering pathogenic development as well as the availability of these *U. maydis* strain that expresses an inducible combination of compatible *b* genes was instrumental to obtain a comprehensive view of the genes that are regulated by the bE/bW complex because it made it possible to obtain a time-resolved view. Early attempts used differential techniques in which two stages with active and inactive bE/bW combinations were compared. This resulted in the identification of several *b*-regulated genes as well as the characterization of the binding sites for the bE/bW (*bbs*) (Brachmann et al. 2001). The consensus sequence for bE/bW transcription factor spans approximately 20 nucleotides, containing the core sequence TGA-$N_9$-TGA. With the availability of the genomic sequence (Kamper et al. 2006), it became possible to develop genome-wide DNA arrays for *U. maydis*, which represent about 93% of the predicted 6,522 genes. Using this technology, a set of 345 *b*-regulated genes were defined, of which 206 were upregulated, and 139 were repressed after induction of the bE/bW heterodimer (Heimel et al. 2010). When correlated with presence or absence of bE/bW binding sites, the *b*-regulated genes could be classified into direct and indirect targets, being the majority classified as indirect targets. This indicated that bE/bW heterodimers trigger a transcriptional cascade. To identify elements of this transcriptional cascade, putative transcription factors were analyzed and a putative C2H2 Zinc finger transcription factor, Rbf1, turned out to be the main transcriptional regulator located downstream of b-factor: 90% of the genes that show altered expression upon bE/bW-activation require the zinc finger transcription factor Rbf1. Moreover, the expression of *rbf1* in *b*-deletion strains is sufficient to promote the normally *b*-induced cell cycle arrest and initiates filamentous growth. In addition, *rbf1* contains identifiable *bbs* sequences in its promoter, suggesting that *rbf1* is a direct target of bE/bW (Heimel et al. 2010).

Other transcriptional factors such as *biz1*, *hdp1*, and *hdp2* were analyzed. The zinc-finger Biz1, that was characterized previously as a negative regulator of cyclin B1 (Flor-Parra et al. 2006), and the homeodomain Hdp1 are thought to be targets of Rbf1 while a second homeodomain Hdp2 seems to be a direct b-factor target (Heimel et al. 2010). Overall, the transcriptional regulators identified and characterized to date comprise cascade in which early signals through bE/bW are amplified to effect dikaryotic growth (Fig. 6.2).

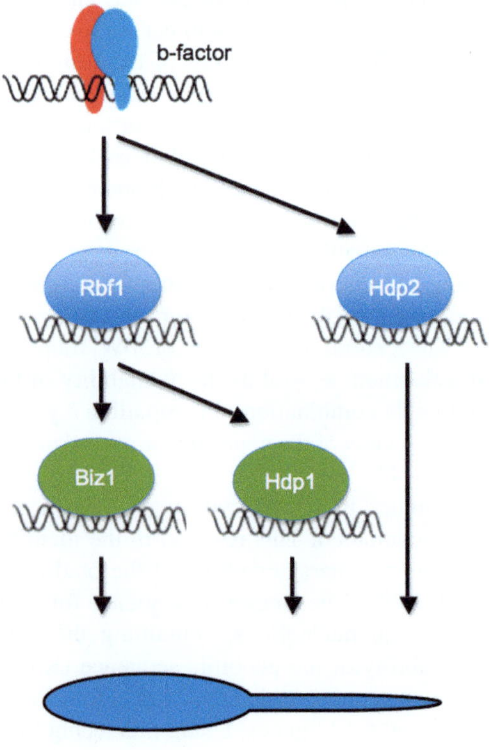


**Fig. 6.2** Scheme of the transcription factor cascade responsible for the *b*-dependent transcriptional program. The bE/bW complex operates at the top of a transcriptional cascade responsible for establishing the infective filament. It acts directly to induce a number of regulators, including *rbf1* and *hdp2*. The transcription factor Rbf1 then induces the expression of *biz1* and *hdp1*. Only in the case of *rbf1*, the *b*-dependent regulation was proved to be direct. In the case of *hdp2*, this regulation has yet to be verified as direct

Functional classification of the genes showed that cellular processes such as the restructuring of the cell wall or alterations in lipid metabolism are controlled by the *b*-mating type locus. A large number of genes affect the regulation of the cell cycle, mitosis and DNA replication, which is consistent with the observation that, after *b*-induction, cell division is stalled (see below). The other intriguing characteristic of the b-filament is the secretion of various potential effector proteins. Such effectors are thought to be involved in suppression of host defense responses and redirection of nutrient flow during biotrophic growth.

## 6.4 Formation of the Infective Filament: Activation of Hyperpolarized Growth

One of the most obvious features of the infective filament from *U. maydis* is its ability to undergo a strong polar growth. This hyperpolarized growth is characterized by apical tip expansion followed by nuclear migration. Since the cell cycle is arrested and there is a limit of cytoplasm the dikaryotic nuclear information is able to achieve, at the distal end of the tip compartment regularly spaced septa are laid down, which delimit empty sections of collapsed hyphal segments (Snetselaar and Mims 1992; Steinberg et al. 1998). By this mechanism, the length of the cytoplasm-filled tip compartment is kept constant at about 150 μm (Steinberg et al. 1998).

The sustained polar growth is not different from the polar growth described in filamentous fungi and therefore relies on several coordinated processes. Notably, the mating type-induced polarized growth depends on the Rho family GTP-binding protein Rac1 but not the closely related GTPase Cdc42, which plays a central role for cell polarization in yeast (Mahlert et al. 2006). The expression of Rac1 is not upregulated during *b* mating type-dependent dimorphic switching (Mahlert et al. 2006) suggesting that bE/bW expression results in stimulation of Rac1 activity. This activity is regulated by the GEF factor Cdc24 as well as the scaffold protein Bem1 (Alvarez-Tabares and Perez-Martin 2008). Among the proposed downstream effectors of Rac1 is the PAK kinase Cla4, which is required to sustain polar growth in *U. maydis* (Leveleki et al. 2004). In response to positional information, the morphogenetic machinery is locally reorganized such that cell surface expansion and cell wall deposition are directed to the specified site. Three elements are involved in the maintenance of the localized polar growth. The first one depends on septin-based structures that localize just behind the tip where amongst multiple functions they likely serve as diffusion barriers that help to maintain the polarity axis (Alvarez-Tabares and Perez-Martin 2010). An additional element that helps to restrict the growth at hyphal tips is the presence of specific membrane domains, which are referred to as lipid rafts or sterol-rich domains (SRDs) since they are typically enriched for sterols and sphingolipids when compared to the rest of the plasma membrane. The use of filipin to localize sterols has revealed that in

*U. maydis* SRDs concentrate at the infective filament tips (Canovas and Perez-Martin 2009). Moreover, functional studies show that disruption of SRD formation leads to loss of hyphal polarity. For example, perturbation of sphingolipid synthesis in *U. maydis* (i.e., using compound aureobasidin A) dramatically affects localization of the morphogenetic machinery at hyphal tips (Canovas and Perez-Martin 2009). Finally a third element involved in the maintenance of polarity is a balanced transport of vesicules to and from hyphal tip. A key element involved in this transport equilibrium is the so-called Spitzenkörper that is thought to function as a vesicle supply center containing exocytotic as well as endocytotic vesicles (Virag and Harris 2006). The integrity of the Spitzenkörper depends on the polarisome, a protein complex that polarizes the actin-associated secretion machinery (Harris et al. 2005). Molecular components of the polarisome such the protein Spa2 (Carbo and Perez-Martin 2008) and supposedly formins, which nucleate actin cables, as well as components of the exocyst have been proposed to communicate with the Spitzenkörper (Harris 2009). The polarisome might participate in F-actin polarization, which might in turn guide Spitzenkörper vesicles to the hyphal apex. Spitzenkörper also participates in membrane recycling processes. Indeed endocytic recycling via early endosomes is essential for proper hyphal morphology and pathogenicity in *U. maydis* (Fuchs et al. 2006).

Polarized growth of hyphae requires both microtubules and actin. As a consequence, pharmacological disruption of actin cables or microtubules leads to defects in development of dikaryotic hyphae (Fuchs et al. 2005). Actin plays a dual role. By one side, actin filaments provide "tracks" for localized delivery of exocytic vesicles to sites of cell wall deposition at the hyphal tip. As stated above, the assembly of these filaments is likely regulated in part by components of the polarisome. Myo5, a class V myosin, is the motor responsible for the transport of membranous vesicles along these actin filaments. As expected, Myo5 is required for hyphal growth as well as pathogenicity (Weber et al. 2003). Other actin-based elements, the actin patches, associate with and drive internalization of endocytic vesicles, accumulated in a sub-apical "belt" (Castillo-Lluva et al. 2007). Microtubules, on the other hand, are specifically required for long-distance transport in all *U. maydis* hyphae that extend beyond 50–60 μm. Cytoplasmic microtubules mediate long-range transport to and from the tip region, which is dependent on molecular transport along microtubules mediated by molecular motors such as dynein and kinesins (Fuchs et al. 2005; Schuchardt et al. 2005). Cell biological analyses revealed that microtubule-dependent transport processes are essential to support polar growth. For instance, deletion of *kin1* encoding the conventional kinesin, a microtubule-dependent motor, resulted in increasing numbers of bipolarly growing cells and an absence of empty sections (Lehmler et al. 1997; Steinberg et al. 1998; Schuchardt et al. 2005). In addition to vesicle transport, microtubule-based long-distance transport processes involving RNA-binding proteins are also needed for polar growth of hyphae (Vollmeister and Feldbrugge 2010). The current idea is that clusters of polysomes are closely associated with the Spitzenkörper in tip cells suggesting local mRNA translation. Potential proteins that need to be localized by mRNA transport in *U. maydis* include enzymes involved in cell wall synthesis, such as chitin synthases, components of the endocytosis machinery, regulatory proteins such as small G proteins or cell-shape determinants such as septins (Becht et al. 2006; Konig et al. 2009). The latter example is of particular interest, because recently, a third cytoskeleton element was described consisting of septin fibers running along the infective filament of the cell near the cortex. Two major defects were observed during the formation of the infective filaments in strains defective in these septins: the filament elongation was retarded in comparison to wild-type strains, and a higher proportion of the cells grew in a bipolar manner (Alvarez-Tabares and Perez-Martin 2010).

In summary, since polarized growth of infective hypha shows similar steps as polar growth in filamentous fungi, it is not surprising that cytoskeleton regulators, like Rac1, or molecular motors such as myosin V, are required for pathogenic development in *U. maydis* (Weber et al. 2003; Mahlert et al. 2006). However, how these housekeeping elements are differentially controlled during the pathogenic development is not currently understood. Transcriptomic studies showed in a few cases alterations in mRNA levels for genes encoding several cytoskeleton-related proteins, such as a diaphanous-related formin, or septins (Heimel et al. 2010). However, the clue to understand the *b*-dependent induced polar growth of the infective hypha seems to be related to a recently described kinase, Cdk5, which is required for sustained polar growth in *U. maydis* (Castillo-Lluva et al. 2007). Cdk5 belongs to a family of cyclin-dependent kinases (CDK) implicated in the regulation of morphogenesis in organisms ranging from yeast to human (Xie et al. 2006). *U. maydis* cells carrying a *cdk5* conditional mutation showed drastically reduced virulence (Castillo-Lluva et al. 2007). However, because of the essential role of Cdk5 for growth, it is not clear whether this requirement for virulence reflects specific roles of Cdk5 during the pathogenic development or whether it is an indirect effect of the various cellular abnormalities associated with the *cdk5* conditional mutation. CDK activity requires the interaction with proteins known as cyclins (Morgan 1997), which target the catalytic subunit to correct substrates. This idea is supported by the observation that a single catalytic subunit, in complex with different cyclins, can phosphorylate a different set of substrates. In *U. maydis* Cdk5 is able to associate with at least seven cyclins. Interestingly the transcription one of these cyclins, Pcl12, is dependent on *b*-induction and most likely it will be a direct target of the *b*-dependent transcriptional factor Rbf1 (Flor-Parra et al. 2007; Heimel et al. 2010). Consistently, Pcl12 is required for a proper development of the infective filament (Perez-Martin and Castillo-Lluva 2008).

How the Cdk5/Pcl12 complex controls the polar growth is poorly understood. It seems that as it was reported for human Cdk5, the *U. maydis* counterpart also affects cytoskeleton remodeling (Xie et al. 2006). This action is mediated by affecting actin-based as well as microtubule-based processes (Fig. 6.3a). We have described that one of the roles of Cdk5 was to locate the Cdc24 GEF, essential for the activation of Rac1, at the cell pole (Castillo-Lluva et al. 2007). This dependence on Cdk5 for Cdc24 localization relies on the scaffold protein Bem1 (Alvarez-Tabares and Perez-Martin 2008). Some details are still missing in this regulation such as to determine which factor is the actual target of Cdk5. In neurons, Cdk5 also

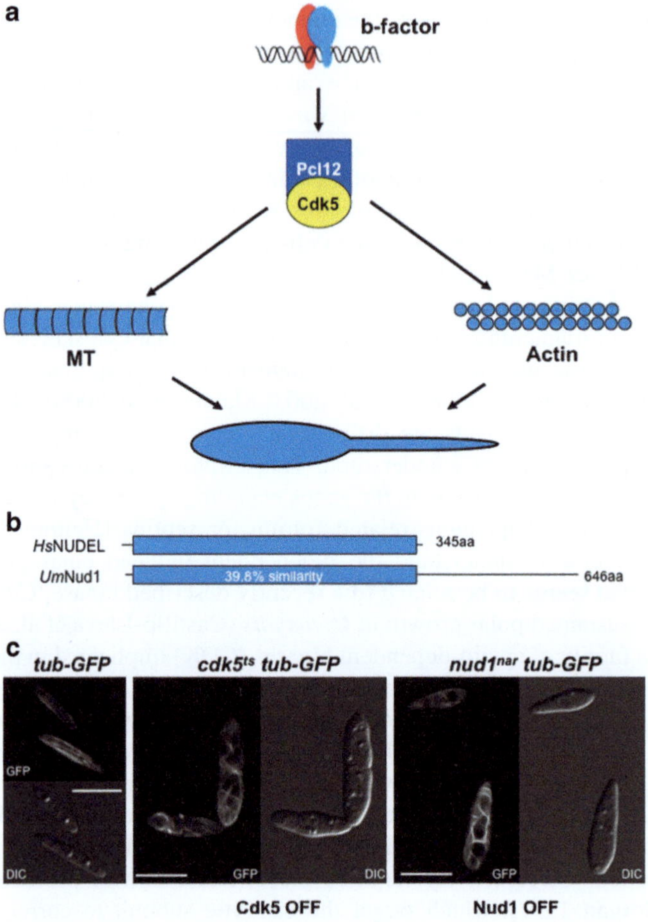


**Fig. 6.3** (**a**) Cdk5 bound to the cyclin Pcl12 is predicted to control both actin- and MT-dependent cytoskeleton, in a similar way as mammalian Cdk5 does. (**b**) Scheme of the putative NUDEL-like protein Nud1 from *U. maydis*. (**c**) Images showing the similarity of effects in MT cytoskeleton between *cdk5* and *nud1* conditional mutants. Note that mutants showed much longer, curved, and disorganized MTs. *Bar*: 10 m

affects microtubule cytoskeleton, via the regulation of dynein activity. This role is related to the ability to stabilize the microtubule cytoskeleton, and to that end it has been proposed that the protein phosphorylates NUDEL, which interact with the protein Lis1 that regulates the activity of dynein and this in turn the stability of the microtubule (Niethammer et al. 2000). This regulatory scheme is of great interest given that Lis1, NUDEL, and Cdk5 have been associated with a number of neurodegenerative diseases such as Alzheimer's disease, amyotrophic lateral sclerosis, Parkinson's disease, or Niemann-Pick (Cruz and Tsai 2004). Interestingly, in *U. maydis* there is a counterpart of Lis1, which is also involved in the morphogenesis of the hyphae (Lenz et al. 2006) as well as a gene with the capacity to encode a protein with high sequence similarity to NUDEL, that we called Nud1. Nud1 is an essential and conditional mutant that shows similar phenotype as *cdk5*$^{ts}$ mutant growing at restrictive temperature with respect to MT organization, leading to much longer, curved, and disorganized MTs (Fig. 6.3b). These defects in MT organization resembled that of dynein mutants (Straube et al. 2001), suggesting that in *U. maydis* a scheme similar to the one described in humans is plausible for the connections between Cdk5 and MT.

## 6.5 Role of Cell Cycle Arrest During the Formation of the Infective Filament

The infective filament is composed of a single cell carrying two nuclei. The average length for *U. maydis* cell is 17 μm, while the length of filament can be up to six times the length of haploid cells. In fungal cells, size and cell division are connected in a way that cell divides after it reaches a certain critical size (Rupes 2002). However, the infective filament is unable to divide in spite of its size, indicating that cell cycle is actively arrested during this process. The reasons for this apparent cell cycle arrest on the plant surface are unknown. It has been hypothesized that such a cell cycle adjustment would be required for a precise execution of the virulence program (Perez-Martin et al. 2006). In fact, the apparent cell cycle arrest of the infectious hypha on plant surface observed in *U. maydis* seems to be more general, and it is also present in rust fungi like *Uromyces phaseoli* (Heath and Heath 1979). Most likely, the cell cycle arrest may have a mechanistic reason. Cell division requires a large quantity of cytoskeletal elements for forming the mitotic spindle. By other way, sustained polar growth depends on the coordinated use of both actin- and microtubule-based cytoskeletons. It is thought that these two developmental events are incompatible, because they compete for the cytoskeletal components (Mata et al. 2000). Therefore, these two events cannot take place simultaneously. Examples of this incompatibility are well reported in metazoan development (Duncan and Su 2004). An interesting feature of *b*-induced cell cycle arrest in *U. maydis* is that it happens during G2 phase: infectious hyphae produced after the expression of compatible b proteins showed a single nucleus with 2C DNA content surrounded by an intact nuclear envelope and a defined cytoplasmic array of microtubules (Mielnichuk et al. 2009). In *U. maydis*, during the G2 phase, the cytoskeletal growth machinery is set up to support polar growth (Steinberg et al. 2001), and then a prolonged G2 phase is best suited to support tip growth during infective hyphae formation.

This finding had lead considerable efforts to define networks of regulatory genes that promote G2/M progression in *U. maydis* (Castillo-Lluva et al. 2004; Garcia-Muse et al. 2004; Castillo-Lluva and Perez-Martin 2005; Sgarlata and Perez-Martin 2005a, b; Mielnichuk and Perez-Martin 2008). As in other eukaryotic organisms, in *U. maydis*, cyclin-dependent protein kinases (Cdks) are key regulators of the cell division cycle. Two distinct Cdk-cyclin complexes are responsible for G2/M transition in *U. maydis*. While Cdk1–Clb1 is required for the G1/S and the G2/M transitions, the Cdk1–Clb2 complex is specific for the G2/M transition (Garcia-Muse et al. 2004). The G2/M transition is mainly controlled by inhibitory phosphorylation/activating dephosphorylation of the catalytic subunit Cdk1 associated with cyclin Clb2. These activities were mediated by the kinase Wee1 and the phosphatase Cdc25, respectively (Sgarlata and Perez-Martin 2005a, b). The *b*-dependent cell cycle arrest was mediated by the accumulation of phosphorylated inactive forms of Cdk1. Inability to phosphorylate this protein – either by downregulation of the kinase in charge of this, Wee1, or by the expression of a Cdk1 allele that was refractory to inhibitory phosphorylation – resulted in filaments that were not cell cycle-arrested (Mielnichuk et al. 2009). The accumulation of phosphorylated Cdk1 upon formation of b heterodimer is a consequence of the inactivation of the Cdc25 phosphatase that results from the binding of the phosphatase to Bmh1, a 14-3-3 protein that sequester it in the cytoplasm (Mielnichuk and Perez-Martin 2008). To enable the Cdc25-Bmh1 inhibitory interaction, Cdc25 has to be phosphorylated in the cognate 14-3-3 binding sites. A surprising result was the realization that Chk1, a well-known regulatory kinase involved in DNA damage responses, was the kinase responsible of this phosphorylation (Mielnichuk et al. 2009; Perez-Martin 2009). Furthermore, Chk1-dependent cell cycle arrest upon *b*-inductions is also dependent on Atr1, the cognate Chk1-activating kinase in response to DNA damage (Fig. 6.4) (de Sena-Tomás et al. 2011). It is no clear how the b-complex, a transcriptional factor that induces the virulence program, activates the Atr1-Chk1 cascade, which in normal conditions responds to DNA damage. Attempts to analyze whether activation of Atr1-Chk1 cascade during *b*-induction was concomitant to massive DNA damage gave negative results: using the formation of Rad51 foci as reporter for active DNA repair, there was no evidence for presence of massive DNA damage associated with the induction of the infective filament (Mielnichuk et al. 2009). It is worth noting that b proteins activate the transcription of a gene, *polX*, encoding a putative DNA polymerase (Brachmann et al. 2001), which belongs to the family X of DNA polymerases that are described to be involved in a number of DNA repair processes (Ramadan et al. 2004). The induction of *polX* suggests that there is DNA damage to be repaired after activation by the b regulator, although no Rad51 foci were detected. One possibility could be that the putative DNA damage is different from double-strand break damage, so alternative DNA repair pathways such as base excision repair (BER) are recruited, and therefore no need for Rad51. In other eukaryotic systems, BER-mediated signaling is independent on ATR-Chk1, but perhaps in *U. maydis* is more simplified than in higher eukaryotes for instance, and involves ATR/ATM instead of the DNA PK version. Another appealing possibility is that a limited DNA damage (for instance, a single double-strand break, not detectable using the Rad51-GFP reporter), induced by gene products regulated by b-complex, was responsible for the developmental activation of the DNA damage cascade during the induction of the virulence program in *U. maydis*. This explanation was inspired in the role of HO endonuclease during mating-type switching in *Saccharomyces cerevisiae*

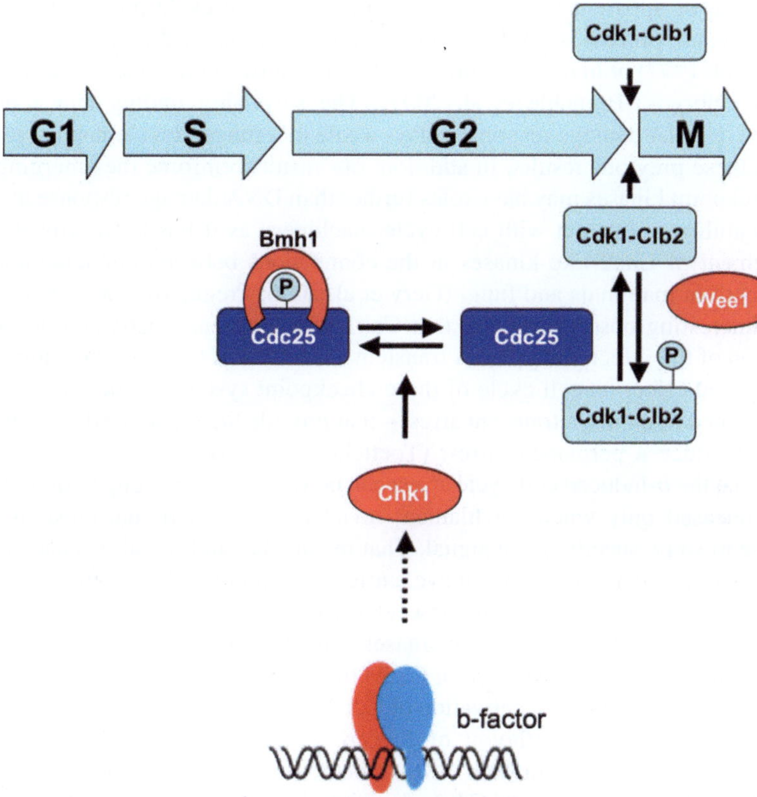


**Fig. 6.4** Scheme showing the current model for *b*-induced cell cycle arrest in *U. maydis*. G2/M transition depends on the activity of two cyclin-Cdk1 complexes, Cdk1–Clb1 and Cdk1–Clb2. The latter is subject to Wee1-dependent inhibitory phosphorylation of Cdk1. This phosphorylation is reversed by the Cdc25 phosphatase. Upon b-factor production, the checkpoint kinase Chk1 is activated by a yet uncharacterized way. Chk1 phosphorylates Cdc25, promoting thereby the interaction of the phosphatase with the 14-3-3 inhibitory protein Bmh1. As a result, an accumulation of inactive Cdk1–Clb2 complexes produces the observed G2 cell cycle arrest

(Nasmyth 1993). In opposition to these explanations suggesting coupling between DNA damage and *b* induction is worth to say that no defect in the ability to arrest cell cycle or to infect plants was apparent in cells lacking Brh2, a BRCA2-like protein that is required for DNA repair by homologous recombination (Kojic et al. 2002). Moreover, preliminary research indicated that cells lacking the *polX* polymerase were able to arrest cell cycle at levels comparable to wild-type cells (de Sena-Tomás et al. 2011). Two recent reports showed that activation of DNA damage response cascade can be triggered in the absence of DNA damage by stable association of elements of the cascade with chromatin (Bonilla et al. 2008; Soutoglou and Misteli 2008). Whether a similar mechanism could explain our observations in *U. maydis* will need additional research. It has become increasingly clear that elements from the DNA damage response cascade can be utilized even in the absence of apparent DNA damage to modulate cell cycle progression during developmental processes such a midblastula transition in *Drosophila* embryos (Sibon et al. 1997) or in the asynchronous division at two-cell-stage *Caenorhabditis elegans* embryos (Brauchle et al. 2003). The surprising finding that a protein involved in DNA damage responses plays a role in a fungal developmental process mirrors these previous results. In addition, our results reinforce the emerging idea that checkpoint kinases may have roles further than DNA damage response in virtue to their ability to interact with cell cycle machinery as it has been proved in the involvement of Chk2-like kinases in the connections between circadian and cell cycles both in mammals and fungi (Gery et al. 2006; Pregueiro et al. 2006).

An interesting observation was that Chk1 was transiently activated during the formation of the infective hypha. A transient activation of Chk1 is compatible with the proposed effect on cell cycle of these checkpoint systems, which appear to be more devoted to induce a transient arrest – that provide time to solve the problems – than to produce a permanent arrest (Toettcher et al. 2009). However, it is well known that the *b*-induced cell cycle arrest can be sustained for a long period of time, and is released only when the filament invaded the plant tissue, most likely in response to some specific plant signal. That means that additional mechanisms are required to sustain a long-term cell cycle arrest, and that Chk1 activation is just the trigger. Currently, the nature of these elements is unknown although there are several candidates. Two of them are kinases, which transcriptional levels decreased upon b-induction and are required for G2/M transition in *U. maydis*. One is the Polo kinase, which is required for activation of Cdc25 phosphatase, while the second one encodes the *U. maydis* ortholog of *S. cerevisiae* Hsl1, which is required for downregulation of Wee1. In both cases, a decrease in the levels of these proteins correlate with accumulation of Cdk1 phosphorylated forms (unpublished results). However, it is still unknown whether the decrease in the levels of these regulators is a cause or a consequence of the cell cycle arrest.

## 6.6 Conclusions: Connections Polar Growth and Cell Cycle Arrest

In this review we discussed that the induction of the infective filament in *U. maydis*, the first step in the pathogenic process, relies on a dual process that involves by one side a specific G2 cell cycle arrest and in other side the specific activation of a hyperpolarization growth. The impairment of any of these processes will have as an outcome the inhibition of the virulence. However, both polar growth and G2 arrest are interdependent, in such a way that they are two sides of the same coin: G2 cell cycle arrest has a consequence of the activation of polar growth, but induction of polar growth generates a cell cycle delay/arrest in G2 phase. This explanation is supported by different results that indicated that arresting the cell cycle in G2, for instance by downregulation of crucial elements involved in G2/M transition such as cyclin b or the phosphatase Cdc25, resulted in a sustained polar growth (Garcia-Muse et al. 2004; Sgarlata and Perez-Martin 2005a). In the same way, induction of a strong polar growth forces the cell cycle to remain in G2. For instance, we found that overexpression of *rac1*, a Rho-like GTPase that induces a strong polar growth in *U. maydis* (Mahlert et al. 2006) generates a G2 delay (Perez-Martin and Castillo-Lluva 2008). This is also the case when the *cln1* cyclin is overexpressed: the induction of a strong polar growth curses with a G2 delay (Castillo-Lluva and Perez-Martin 2005). Further research efforts will be needed to define the nature of these putative connections as well as their roles during the induction of the virulence program in phytopathogenic fungi.

**Acknowledgements** Our research was financially supported by the SIGNALPATH Marie Curie Research Training Network (MRTN-CT-2005-019277) and by grant BIO2008-04054 from the Spanish Ministerio de Ciencia e Innovación (MICINN).

## References

Alvarez-Tabares I, Perez-Martin J (2008) Cdk5 kinase regulates the association between adaptor protein Bem1 and GEF Cdc24 in the fungus *Ustilago maydis*. J Cell Sci 121:2824–2832

Alvarez-Tabares I, Perez-Martin J (2010) Septins from the phytopathogenic fungus *Ustilago maydis* are required for proper morphogenesis but dispensable for virulence. PLoS One 5:e12933

Banuett F, Herskowitz I (1996) Discrete developmental stages during teliospore formation in the corn smut fungus, *Ustilago maydis*. Development 122:2965–2976

Bechinger C, Giebel KF, Schnell M, Leiderer P, Deising HB, Bastmeyer M (1999) Optical measurements of invasive forces exerted by appressoria of a plant pathogenic fungus. Science 285:1896–1899

Becht P, Konig J, Feldbrugge M (2006) The RNA-binding protein Rrm4 is essential for polarity in *Ustilago maydis* and shuttles along microtubules. J Cell Sci 119:4964–4973

Bolker M, Urban M, Kahmann R (1992) The a mating type locus of *U. maydis* specifies cell signaling components. Cell 68:441–450

Bonilla CY, Melo JA, Toczyski DP (2008) Colocalization of sensors is sufficient to activate the DNA damage checkpoint in the absence of damage. Mol Cell 30:267–276

Brachmann A, Weinzierl G, Kamper J, Kahmann R (2001) Identification of genes in the bW/bE regulatory cascade in *Ustilago maydis*. Mol Microbiol 42:1047–1063

Brauchle M, Baumer K, Gonczy P (2003) Differential activation of the DNA replication checkpoint contributes to asynchrony of cell division in *C. elegans* embryos. Curr Biol 13:819–827

Brown JK, Hovmoller MS (2002) Aerial dispersal of pathogens on the global and continental scales and its impact on plant disease. Science 297:537–541

Canovas D, Perez-Martin J (2009) Sphingolipid biosynthesis is required for polar growth in the dimorphic phytopathogen *Ustilago maydis*. Fungal Genet Biol 46:190–200

Carbo N, Perez-Martin J (2008) Spa2 is required for morphogenesis but it is dispensable for pathogenicity in the phytopathogenic fungus *Ustilago maydis*. Fungal Genet Biol 45:1315–1327

Castillo-Lluva S, Perez-Martin J (2005) The induction of the mating program in the phytopathogen *Ustilago maydis* is controlled by a G1 cyclin. Plant Cell 17:3544–3560

Castillo-Lluva S, Garcia-Muse T, Perez-Martin J (2004) A member of the Fizzy-related family of APC activators is regulated by cAMP and is required at different stages of plant infection by *Ustilago maydis*. J Cell Sci 117:4143–4156
Castillo-Lluva S, Alvarez-Tabares I, Weber I, Steinberg G, Perez-Martin J (2007) Sustained cell polarity and virulence in the phytopathogenic fungus *Ustilago maydis* depends on an essential cyclin-dependent kinase from the Cdk5/Pho85 family. J Cell Sci 120:1584–1595
Christensen JJ (1963) Corn smut caused by *Ustilago maydis*. Am Phytopathol Soc Monogr 2:1–141
Cruz JC, Tsai LH (2004) A Jekyll and Hyde kinase: roles for Cdk5 in brain development and disease. Curr Opin Neurobiol 14:390–394
de Sena-Tomás C, Fernandez-Alvarez A, Holloman WK, Pérez-Martín J (2011) The DNA damage response signaling cascade regulates proliferation of the phytopathogenic fungus *Ustilago maydis* in planta. Plant Cell 23:1654–1665
Duncan T, Su TT (2004) Embryogenesis: coordinating cell division with gastrulation. Curr Biol 14:R305–R307
Feldbrugge M, Kamper J, Steinberg G, Kahmann R (2004) Regulation of mating and pathogenic development in *Ustilago maydis*. Curr Opin Microbiol 7:666–672
Flor-Parra I, Vranes M, Kamper J, Perez-Martin J (2006) Biz1, a zinc finger protein required for plant invasion by *Ustilago maydis*, regulates the levels of a mitotic cyclin. Plant Cell 18:2369–2387
Flor-Parra I, Castillo-Lluva S, Perez-Martin J (2007) Polar growth in the infectious hyphae of the phytopathogen *Ustilago maydis* depends on a virulence-specific cyclin. Plant Cell 19:3280–3296
Fuchs U, Manns I, Steinberg G (2005) Microtubules are dispensable for the initial pathogenic development but required for long-distance hyphal growth in the corn smut fungus *Ustilago maydis*. Mol Biol Cell 16:2746–2758
Fuchs U, Hause G, Schuchardt I, Steinberg G (2006) Endocytosis is essential for pathogenic development in the corn smut fungus *Ustilago maydis*. Plant Cell 18:2066–2081
Garcia-Muse T, Steinberg G, Perez-Martin J (2003) Pheromone-induced G2 arrest in the phytopathogenic fungus *Ustilago maydis*. Eukaryot Cell 2:494–500
Garcia-Muse T, Steinberg G, Perez-Martin J (2004) Characterization of B-type cyclins in the smut fungus *Ustilago maydis*: roles in morphogenesis and pathogenicity. J Cell Sci 117:487–506
Gery S, Komatsu N, Baldjyan L, Yu A, Koo D, Koeffler HP (2006) The circadian gene per1 plays an important role in cell growth and DNA damage control in human cancer cells. Mol Cell 22:375–382
Gillissen B, Bergemann J, Sandmann C, Schroeer B, Bolker M, Kahmann R (1992) A two-component regulatory system for self/non-self recognition in *Ustilago maydis*. Cell 68:647–657
Harris SD (2006) Cell polarity in filamentous fungi: shaping the mold. Int Rev Cytol 251:41–77
Harris SD (2009) The Spitzenkorper: a signalling hub for the control of fungal development? Mol Microbiol 73:733–736
Harris SD, Read ND, Roberson RW, Shaw B, Seiler S, Plamann M, Momany M (2005) Polarisome meets spitzenkorper: microscopy, genetics, and genomics converge. Eukaryot Cell 4:225–229
Heath IB, Heath MC (1979) Structural studies of the development of infection structures of cowpea rust, *Uromyces phaseoli* var. vignae. II. Vacuoles. Can J Bot 57:1830–1837
Heimel K, Scherer M, Vranes M et al (2010) The transcription factor Rbf1 is the master regulator for b-mating type controlled pathogenic development in *Ustilago maydis*. PLoS Pathog 6: e1001035
Kamper J, Reichmann M, Romeis T, Bolker M, Kahmann R (1995) Multiallelic recognition: nonself-dependent dimerization of the bE and bW homeodomain proteins in *Ustilago maydis*. Cell 81:73–83
Kamper J, Kahmann R, Bolker M et al (2006) Insights from the genome of the biotrophic fungal plant pathogen *Ustilago maydis*. Nature 444:97–101

Kojic M, Kostrub CF, Buchman AR, Holloman WK (2002) BRCA2 homolog required for proficiency in DNA repair, recombination, and genome stability in *Ustilago maydis*. Mol Cell 10:683–691

Konig J, Baumann S, Koepke J, Pohlmann T, Zarnack K, Feldbrugge M (2009) The fungal RNA-binding protein Rrm4 mediates long-distance transport of ubi1 and rho3 mRNAs. EMBO J 28:1855–1866

Lehmler C, Steinberg G, Snetselaar KM, Schliwa M, Kahmann R, Bolker M (1997) Identification of a motor protein required for filamentous growth in *Ustilago maydis*. EMBO J 16:3464–3473

Lenz JH, Schuchardt I, Straube A, Steinberg G (2006) A dynein loading zone for retrograde endosome motility at microtubule plus-ends. EMBO J 25:2275–2286

Leveleki L, Mahlert M, Sandrock B, Bolker M (2004) The PAK family kinase Cla4 is required for budding and morphogenesis in *Ustilago maydis*. Mol Microbiol 54:396–406

Mahlert M, Leveleki L, Hlubek A, Sandrock B, Bolker M (2006) Rac1 and Cdc42 regulate hyphal growth and cytokinesis in the dimorphic fungus *Ustilago maydis*. Mol Microbiol 59:567–578

Mata J, Curado S, Ephrussi A, Rorth P (2000) Tribbles coordinates mitosis and morphogenesis in Drosophila by regulating string/CDC25 proteolysis. Cell 101:511–522

Mielnichuk N, Perez-Martin J (2008) 14-3-3 regulates the G2/M transition in the basidiomycete *Ustilago maydis*. Fungal Genet Biol 45:1206–1215

Mielnichuk N, Sgarlata C, Perez-Martin J (2009) A role for the DNA-damage checkpoint kinase Chk1 in the virulence program of the fungus *Ustilago maydis*. J Cell Sci 122:4130–4140

Morgan DO (1997) Cyclin-dependent kinases: engines, clocks, and microprocessors. Annu Rev Cell Dev Biol 13:261–291

Nasmyth K (1993) Regulating the HO endonuclease in yeast. Curr Opin Genet Dev 3:286–294

Niethammer M, Smith DS, Ayala R, Peng J, Ko J, Lee MS, Morabito M, Tsai LH (2000) NUDEL is a novel Cdk5 substrate that associates with LIS1 and cytoplasmic dynein. Neuron 28:697–711

Perez-Martin J (2009) DNA-damage response in the basidiomycete fungus *Ustilago maydis* relies in a sole Chk1-like kinase. DNA Repair (Amst) 8:720–731

Perez-Martin J, Castillo-Lluva S (2008) Connections between polar growth and cell cycle arrest during the induction of the virulence program in the phytopathogenic fungus *Ustilago maydis*. Plant Signal Behav 3:480–481

Perez-Martin J, Castillo-Lluva S, Sgarlata C et al (2006) Pathocycles: *Ustilago maydis* as a model to study the relationships between cell cycle and virulence in pathogenic fungi. Mol Genet Genomics 276:211–229

Pregueiro AM, Liu Q, Baker CL, Dunlap JC, Loros JJ (2006) The Neurospora checkpoint kinase 2: a regulatory link between the circadian and cell cycles. Science 313:644–649

Ramadan K, Shevelev I, Hubscher U (2004) The DNA-polymerase-X family: controllers of DNA quality? Nat Rev Mol Cell Biol 5:1038–1043

Rupes I (2002) Checking cell size in yeast. TIG 18:479–485

Schuchardt I, Assmann D, Thines E, Schuberth C, Steinberg G (2005) Myosin-V, Kinesin-1, and Kinesin-3 cooperate in hyphal growth of the fungus *Ustilago maydis*. Mol Biol Cell 16:5191–5201

Sgarlata C, Perez-Martin J (2005a) The Cdc25 phosphatase is essential for the G2/M phase transition in the basidiomycete yeast *Ustilago maydis*. Mol Microbiol 58:1482–1496

Sgarlata C, Perez-Martin J (2005b) Inhibitory phosphorylation of a mitotic cyclin-dependent kinase regulates the morphogenesis, cell size and virulence of the smut fungus *Ustilago maydis*. J Cell Sci 118:3607–3622

Sibon OC, Stevenson VA, Theurkauf WE (1997) DNA-replication checkpoint control at the Drosophila midblastula transition. Nature 388:93–97

Snetselaar KM, Mims CW (1992) Sporidial fusion and infection of maize seedlings by the smut fungus *Ustilago maydis*. Mycologia 84:193–203

Snetselaar KM, Mims CW (1993) Infection of maize stigmas by *Ustilago maydis*: light and electron microscopy. Phytopathology 83:843–850

Snetselaar KM, Bolker M, Kahmann R (1996) Ustilago maydis mating hyphae orient their growth toward pheromone sources. Fungal Genet Biol 20:299–312

Snetselaar KM, Carfioli MA, Cordisco KM (2001) Pollination can protect maize ovaries from infection by *Ustilago maydis*, the corn smut fungus. Can J Bot 79:1390–1399

Soutoglou E, Misteli T (2008) Activation of the cellular DNA damage response in the absence of DNA lesions. Science 320:1507–1510

Steinberg G, Perez-Martin J (2008) Ustilago maydis, a new fungal model system for cell biology. Trends Cell Biol 18:61–67

Steinberg G, Schliwa M, Lehmler C, Bolker M, Kahmann R, McIntosh JR (1998) Kinesin from the plant pathogenic fungus *Ustilago maydis* is involved in vacuole formation and cytoplasmic migration. J Cell Sci 111:2235–2246

Steinberg G, Wedlich-Soldner R, Brill M, Schulz I (2001) Microtubules in the fungal pathogen *Ustilago maydis* are highly dynamic and determine cell polarity. J Cell Sci 114:609–622

Straube A, Enard W, Berner A, Wedlich-Soldner R, Kahmann R, Steinberg G (2001) A split motor domain in a cytoplasmic dynein. EMBO J 20:5091–5100

Talbot NJ (2003) On the trail of a cereal killer: exploring the biology of *Magnaporthe grisea*. Annu Rev Microbiol 57:177–202

Toettcher JE, Loewer A, Ostheimer GJ, Yaffe MB, Tidor B, Lahav G (2009) Distinct mechanisms act in concert to mediate cell cycle arrest. Proc Natl Acad Sci USA 106:785–790

Tucker SL, Talbot NJ (2001) Surface attachment and pre-penetration stage development by plant pathogenic fungi. Annu Rev Phytopathol 39:385–417

Virag A, Harris SD (2006) The Spitzenkorper: a molecular perspective. Mycol Res 110:4–13

Vollmeister E, Feldbrugge M (2010) Posttranscriptional control of growth and development in *Ustilago maydis*. Current Opin Microbiol 13:693–699

Weber I, Gruber C, Steinberg G (2003) A class-V myosin required for mating, hyphal growth, and pathogenicity in the dimorphic plant pathogen *Ustilago maydis*. Plant Cell 15:2826–2842

Xie Z, Samuels BA, Tsai LH (2006) Cyclin-dependent kinase 5 permits efficient cytoskeletal remodeling – a hypothesis on neuronal migration. Cereb Cortex 16(Suppl 1):i64–i68

Yee AR, Kronstad JW (1998) Dual sets of chimeric alleles identify specificity sequences for the bE and bW mating and pathogenicity genes of *Ustilago maydis*. Mol Cell Biol 18:221–232

# Chapter 7
# Appressorium Function in *Colletotrichum orbiculare* and Prospect for Genome Based Analysis

**Yasuyuki Kubo**

**Abstract** *Colletotrichum* species cause disease on cereals, grasses, legumes, ornamentals, vegetables, and fruit trees. To penetrate the host plant, they form specialized infection structures, named appressoria, that are melanized single cells developed from conidial germ tubes. Appressorium formation and host invasion by *Colletotrichum* constitutes a model to study environmental signal reception and cellular development of fungal pathogens. In this chapter, I will discuss the findings and future perspectives in four key aspects (1) morphological and functional development of infection structures; (2) host–parasite interactions and infection-related morphogenesis; (3) pathogen associated molecular patterns (PAMPs) and basal resistance; and (4) genome project of *Colletotrichum* species.

## 7.1 Colletotrichum

*Colletotrichum* species include a wide range of plant pathogens that cause serious diseases to various cereals, grasses, legumes, ornamentals, vegetables, and fruit trees. They usually form well-developed infection structures named appressoria as a host invasion structure. They are generally melanized single cells developed from germ tubes from conidia. Factors involved in appressorium development and its function were extensively analyzed by forward genetics. Studies on appressorium formation constitute a model study on environment signal reception and cellular development of fungal pathogens and also the basic study for the fungal pathogenesis, which affords information on potential targets of control agent to the

Y. Kubo (✉)
Laboratory of Plant Pathology, Graduate School of Life and Environmental Sciences, Kyoto Prefectural University, Kyoto 606-8522, Japan
e-mail: y_kubo@kpu.ac.jp

J. Pérez-Martín and A. Di Pietro (eds.), *Morphogenesis and Pathogenicity in Fungi*, Topics in Current Genetics 22, DOI 10.1007/978-3-642-22916-9_7,

pathogens. Actually *Colletotrichum* species not only include agriculturally serious disease pathogens, but also serve as model pathogens to other disease such as rice blast disease caused by *Magnaporthe oryzae* which forms similar well-developed melanized appressoria and makes hemibiotrophic infection similar to *Colletotrichum* (Prusky et al. 2000).

Among *Colletotrichum* species, well-studied species for appressorium biology include *C. orbiculare* (syn. *C. lagenarium*), *C. lindemuthianum*, *C. gloeosporioides*, *C. trifoli*, *C. graminicola*, and *C. higginsianum*. *Colletotrichum* species provide excellent models for studying fungal–plant interactions (Perfect et al. 1999) and extensive studies have been applied for those species by cytological, physiological, biochemical, and genetics approach (O'Connell and Panstruga 2006). Among them consistent and intensive researches have been accomplished on *C. orbiculare* strain 104-T (MAFF240422), an anthracnose fungus which infects cucurbitaceae group which was isolated from a cucumber plant in 1952 in Japan and infects major cucumis plants, including cucumber, melon, and water melon, and form a well-developed melanized appressoria synchronously and makes a hemibiotrophic infection to host plants (Fig. 7.1).

In this chapter, I will discuss the findings and future perspective in four sections comprising (1) morphological and functional development of infection structure, (2) host–parasite interactions and infection-related morphogenesis, (3) pathogen associated molecular patterns (PAMPs) and basal resistance, (4) genome project of *Colletotrichum* species.

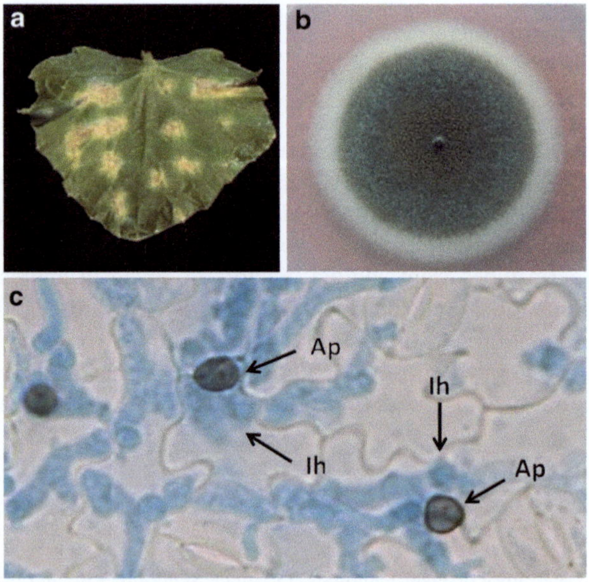


**Fig. 7.1** Morphogenesis and infection by *Colletotrichum orbiculare* strain 104-T (MAFF240422). (**a**) A cucumber leaf infected with *C. orbiculare* showing anthracnose legion. (**b**) Colony grown on potato dextrose medium forming conidia. (**c**) Cuticle infection to cucumber leaf. *Ap* appressorium, *Ih* infection hyphae

## 7.2 Morphological and Functional Development of Infection Structure

### *7.2.1 Signal Transduction*

Several factors involved in infection-related morphogenesis have been identified in *C. orbiculare* (Fig. 7.2). Signal transduction pathways, such as cyclic AMP (cAMP) dependent pathway and MAP kinase pathway are essential for germination, appressorium development, infection hyphae formation, and invasive growth. Several signal transduction-related genes associated with these morphological changes have been characterized in *C. orbiculare* and it has been shown that the MAPK and cAMP signaling pathways are linked to infection-related morphological changes in this fungus. Three well-characterized MAPK pathways in *Saccharomyces cerevisiae* were evaluated in *C. orbiculare*. A pheromone response MAPK gene *FUS3/KSS1* homolog, *CMK1* is essential for conidial germination, appressorium formation, and invasive growth (Takano et al. 2000). Upstream *MAPKKK* gene *CoMEKK1*, a *S. cerevisiae STE11* ortholog was recently identified in the disrupted gene of an AtMT (*Agrobacterium tumefaciens* mediated transformation) mutant and the gene controls the nuclear localization of Cmk1 protein in response to environmental stresses (Sakaguchi et al. 2010b). The transcription factor gene

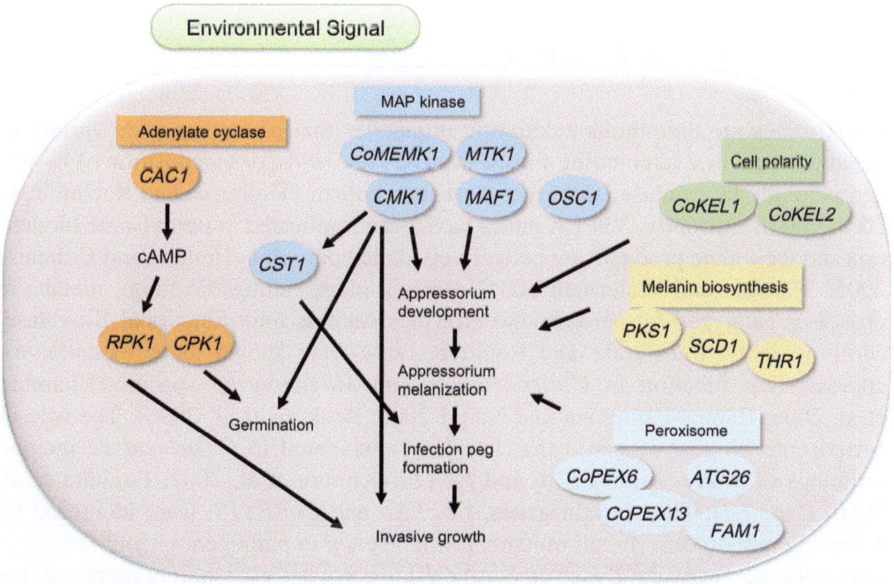


**Fig. 7.2** Genes involved in development of infection-related morphogenesis of *Colletotrichum orbiculare*. Those genes are categorized into a group of genes involved in signal transduction pathway, cellular polarity, peroxisome function, and melanin biosynthesis

*CST1*, a homolog of *S. cerevisiae* transcriptional factor *STE12*, a downstream target of *FUS3/KSS1*, is essential for formation of infection pegs from the appressoria (Tsuji et al. 2003). The osmotic stress responsive MAPK gene *SLT2* controls cell wall integrity in *S. cerevisiae*. The homolog of *SLT2*, *MAF1* is essential for appressorium development and pathogenesis (Kojima et al. 2002). The upstream *MAPKKK* gene C*TKl* was recently identified and the *mtkl* mutant showed similar phenotype to *mafl* mutant (Hiruma et al. 2010). *S. cerevisiae* osmotic responsive MAPK *HOG1* ortholog *Oscl* of *C. orbiculare* is phosphorylated under high osmotic conditions and by a phenylpyrrol fungicide fludioxonil treatment. However, the MAKP is not essential for appressorium development and pathogenesis (Kojima et al. 2004).

The cAMP signaling pathway also plays pivotal roles in transducing environmental cues for cell development and plays a critical role in regulating conidial germination and pathogenicity in *C. orbiculare* (Takano et al. 2001; Yamauchi et al. 2004). Although mutants of the *RPK1* gene encoding the PKA regulatory subunit form normal melanized appressoria, they are defective in generating penetration hyphae (Takano et al. 2001). Mutants of adenylate cyclase and PKA catalytic subunit genes have also been shown to be defective in conidial germination and pathogenicity (Yamauchi et al. 2004). Furthermore, the *cpkl* and *cacl* mutants show a defect in infectious growth in plant similar to *cmkl* mutants, indicating that cAMP signaling controls multiple steps of fungal infection in cooperative regulation with *CMKl* MAPK in *C. orbiculare*.

### *7.2.2 Peroxisome Function*

Peroxisomes are ubiquitous eukaryotic organelles that perform a wide variety of metabolic processes, including β-oxidation of fatty acids, decomposition of hydrogen peroxide by catalase, and glyoxylate metabolism (Titorenko and Rachubinski 2001, 2004). Currently, 32 PEX genes have been implicated in peroxisome biogenesis and their gene products are collectively called peroxins (Heiland and Erdmann 2005; Wanders and Waterman 2004). Among plant pathogenic fungi, metabolic processes catalyzed by peroxisomal enzymes such as multifunctional β-oxidase, carnitine acetyl transferase and isocitrate lyase were shown to be essential for appressorium function in *Colletotrichum* and *Magnaporthe* species (Bhambra et al. 2006; Ramos-Pamplona and Naqvi 2006; Asakura et al. 2006). The role of peroxisome biogenesis genes that has been elucidated in *C. orbiculare* include orthologs of *S. cerevisiae PEX6* and *PEX13* (Kimura et al. 2001; Fujihara et al. 2010) (Fig. 7.2). The peroxin genes, *CoPEX6* and *CoPEX13*, were identified by screening random insertional mutants for deficiency in pathogenesis and fatty acid utilization, respectively. Based on *S. cerevisiae* peroxin function, it is supposed that *CoPEX6* codes for AAA-peroxin that heterodimerizes with AAA-peroxin Pex1p and participates in the recycling of peroxisomal signal receptor Pex5 from the peroxisomal membrane to the cystosol, and *CoPEX13* gene codes for a docking protein for the receptor Pex5 and PTS (Peroxisome Targeting Signal) protein complex that functions for the import of PTS proteins into peroxisome. The phenotypes of both peroxin gene disruptants are similar. They are unable to utilize fatty acids as a carbon source. And expectedly, PTS1 or PTS2 fused green fluorescent protein (GFP) are not imported into peroxisome, thus import machinery for peroxisomal matrix proteins is impaired. Appressoria of both mutants are defective in melanization, due to the inability to produce acetyl-CoA for the polyketide melanin biosynthesis starter metabolite through the defect of β-oxidation of fatty acids. Moreover, the concentration of intracellular glycerol is lower in *copex13* mutant appressoria than those of the wild-type (Fujihara et al. 2010). Thus, those appressoria are defective in penetration ability on host plants. These findings indicate that fatty acid oxidation in peroxisomes is required not only for appressorium melanization but also for cell wall biogenesis and metabolic processes involved in turgor generation, all of which are essential for appressorium penetration ability. Further *PEX* genes are to be elucidated and forward genetics screen for fatty acid utilization deficiency will further reveal peroxisome function. Actually, we have identified a novel type of peroxin that shuttles between peroxisome and Woronin body, a peroxisome derived cellular organelle that function for sealing of septal pore and stops the leakage of cytoplasm when the fungal spore damaged. And recently it was also reported that autophagic degradation of peroxisomes is essential for infection-related morphogenesis in *C. orbiculare* (Asakura et al. 2009). Thus, cellular organelle integrity control would be an essential aspect for the appressorium development.

### 7.2.3 *Melanin Biosynthesis*

*C. orbiculare* forms melanized appressoria that mediate the initial direct penetration of host epidermal cells by penetrating the plant cuticle and cell wall layers. Extensive studies on appressorial melanization have been accumulated in *C. orbiculare*. Melanization of appressoria is crucial for appressorium function (Kubo and Furusawa 1991), and three melanin biosynthesis enzyme genes, *PKS1*, *SCD1*, and *THR1*, have been characterized as essential genes for turgor-mediated melanized appressorium penetration (Takano et al. 1995, 1997; Kubo et al. 1996; Perpetua et al. 1996) (Fig. 7.2). The essential significance of appressorium melanization for cuticle penetration is widely studied and quite evident; however, melanized appressoria-independent mode of penetration, hyphal tip-based entry (HTE) was recently reported as alternative form of penetration in *C. orbiculare* (Hiruma et al. 2010). Cuticle penetration through well-developed melanized appressoria is not general mode of entry among plant pathogenic fungi and confined in particular species, such as *Colletotrichum* and *Magnaporthe* species. Evolutional approach for the development of turgor-mediated melanized appressoria would be an important theme and is dealt with further below in this chapter.

### 7.2.4 Cellular Polarity

The appressorium development from conidial germination could be recognized as the process of cellular differentiation under cellular growth polarity control. In general, the ability to generate cell polarity in eukaryotic organisms is important to regulate processes such as cell division, cell differentiation, and cell migration. In budding and fission yeast, the onset of polarized growth is controlled by the determination of a growth origin that is usually marked by the deposition of a landmark protein (Chang and Peter 2003). In fission yeasts *Schizosaccharomyces pombe*, Tea1p, a kelch-repeat protein could be a candidate landmark protein localized at microtubule plus ends (Behrens and Nurse 2002). *Tea2* encodes a kinesin-like protein (Browning et al. 2000), and the Tea2p protein is loaded onto microtubule plus ends using Tea2p's intrinsic motor activity (Browning et al. 2003). The polarity of hyphae is a defining feature of filamentous fungi, allowing them to efficiently colonize and exploit new substrates. In fungal hyphae, which grow at the tip, apical extension is based on the intracellular transport of vesicles along the cytoskeleton. KipA, related to Tea2p in *Aspergillus nidulans*, is also required for polarized growth (Konzack et al. 2005). Kelch motif containing regulatory genes *CoKEL1* and *CoKEL2* of *C. orbiculare* are identified as genes involved in appressorium development (Sakaguchi et al. 2008, 2010a) (Fig. 7.2). Both Kelch motif containing proteins show their cellular location by a microtubule-dependent fashion. And the gene disrupted mutants form aberrant form of appressoria which is also defective in further development of infection hyphae, indicating that proper cellular polarity control is essential for infection-related morphogenesis. One surprising finding is that *CoKEL1* gene is essential for proper appressorium development only in vitro condition, such as on glass slides, but dispensable in vivo on plants condition, forming normal appressorium and infection hyphae into host plant epidermal cells. This indicates an existence of bypass pathway responding to plant specific signals in addition to physical signals. Genetics evidence in plant pathogens that indicates distinct responses to physical and plant signals is unique and novel. This point is further described below.

## 7.3 Host–Parasite Interactions and Infection-Related Morphogenesis

Many studies have been done for plant signal reception and appressorium formation in *Colletotrichum* species (Kolattukudy et al. 1995). Physical signals such as substrate hydrophobicity or hardness (Liu and Kolattukudy 1998) and plant derived chemical compounds such as cuticle wax derivatives and ethylene are essential factor for the induction of appressorium development in some *Colletotrichum* species (Flaishman et al. 1995). Although such signals are key for the infection structure development, genetics analysis for the signal reception mechanisms is obscure. A recent report on cellular polarity mutant analysis of *C. orbiculare* indicated the signal transduction pathway is branching into one specific for plant signals and physical signals (Sakaguchi et al. 2008). The *cokel2* mutant defective in a gene containing Kelch repeat motif, a homolog of *TEA1* gene of *S. pombe* has a deficiency in appressorium development, showing aberrant form of appressoria on an artificial surface, such as glass slides, due to the disturbance of cellular polarity as in *S. pombe*. But surprisingly, the mutant forms normal shape of appressoria on the host leaf surface and effectively proceeds development of infection hyphae. This phenomenon provides a hypothesis that there could be a plant derived signals independent of physical signal perception pathway. Supporting this, plant surface exudates restored the normal shape of appressorium formation of *cokel2* mutant on an artificial substrate. Also, calcium ion treatment, applied as a candidate second messenger of intracellular signaling involved restored appressorium formation effectively. Conclusively, *CoKEL2* is involved in physical signal perception pathway, and *C. orbiculare* has a plant signal specific bypath pathway along with physical pathway for appressorium formation. This means when we try to identify plant derived signal perception pathway leading to appressorium formation, we have to search them in condition that physical signal pathway is canceled. The *cokel2* mutant affords unique and good experimental system for that purpose. Based on this idea, we have screened about 10,000 AtMT mutants from the *cokel2* mutants and obtained several mutants specifically deficient in appressorium development in a *cokel2* mutant background but not in *CoKEL2* wild type background. This provides genetics evidence for the existence of plant derived signal specific pathway.

More recently, dual host entry manner of *C. orbiculare* was reported depending on adapted or nonadapted interactions. Hiruma et al. (2010) used *Arabidopsis thaliana – C. orbiculare* and *C. gloeosporioides* system as nonadapted combination and *C. higginsianum* as an adapted combination. They reported that nonadapted anthracnose fungi engage two alternative entry modes during pathogenesis on leaves: turgor-mediated invasion beneath melanized appressoria, and a previously undiscovered HTE that is independent of appressorium formation.

The frequency of HTE is positively regulated by carbohydrate nutrients and negatively by the fungal mitogen-activated protein kinase MAF1 cascade, a MAPK essential for appressorium formation. The *Arabidopsis pen2* mutants defective in indole glucosinolate biosynthesis or metabolism support entry of the nonadapted anthracnose fungi *C. gloeosporioides* and *C. orbiculare* when these pathogens employ HTE. Genetic evidence for appressorium-independent entry into *Arabidopsis* leaf cells was provided by the ability of nonadapted *C. orbiculare maf1* and *mtk1* (MAPPKKK upstream of *MAF1*) strains to form lesions on the *pen2* plants with the development of intracellular hyphae. These mutant strains lack a MAPK required for the early differentiation phase of appressorium formation. In vivo evidence for the HTE was provided by the infection analysis of wild type strains at wounded site which supplies exudative nutrients supporting HTE. One of the distinguished findings of this study is that they suggest that hyphal tips produced by the *Colletotrichum* germlings also retain the ability to breach the cuticle of host cells directly. Besides *Colletotrichum* and *Magnaporthe* species, many fungal pathogens invade host plants by using nonmelanized less developed appressoria, suggesting that melanization-independent entry modes represent more general forms. This study supported this idea showing that *Colletotrichum* species potentially retains appressorium-independent tip based entry fashion. Thus, this may support the idea that appressorium-independent tip based entry mode is common ancestry mode of invasion in *Colletotrichum*, and additively *Colletotrichum* species has acquired default mode of infection by turgor aided melanized appressorium. The report of Sakaguchi et al. (2008) that appressorium formation is induced by dual signals, physical, or plant specific may support this idea. HTE is plant signal-dependent entry mode and melanized turgor-dependent entry could be a default programmed development in response to physical signal. It is envisaged that during evolution *Colletotrichum* species might have developed the deposition of melanin in appressorium formation to generate high turgor and to be more accessible to plant infection.

## 7.4 PAMPs and Basal Resistance in *C. orbiculare*

The currently accepted model of plant recognition of pathogens is that plants recognize PAMPs through corresponding receptors, triggering plant immunity (Thordal-Christensen 2003; Nürnberger et al. 2004; Jones and Dangl 2006). AtMT mutant screening of *C. orbiculare* revealed a novel type of pathogenesis deficient mutant so far not reported (Tanaka et al. 2007). Characterization of the mutant indicated that this mutant enhanced induced host resistance response by its change of cellular surface nature. The mutated gene was identified as a *S. cerevisiae SSD1* gene ortholog. In *S. cerevisiae*, several different signaling pathways have been implicated in promoting cell wall integrity, and Ssd1 protein is a component of one such pathway (Kaeberlein and Guarente 2002). *C. orbiculare ssd1* mutants showed greater sensitivity to the chitin-binding dye Calcofluor White, in common with *S. cerevisiae ssd1* mutants (Kaeberlein and Guarente 2002). In addition, the mutants showed an increased and unique recognition pattern to a monoclonal antibody UB20 that specifically recognizes polysaccharides of cell surface glycoproteins in *Colletotrichum*, suggesting that *ssd1* mutants have a modified cell wall composition or architecture. Cell walls of the *S. cerevisiae ssd1* mutant are depleted in major structural polysaccharides such as beta-1,3-glucan and beta-1,6-glucan but are enriched with chitin and mannoproteins (Wheeler et al. 2003). Notably, *S. cerevisiae* mutant showed increased virulence to mice and it was suggested that the altered cell surface composition leads to misrecognition by the innate immune system and greater induction of proinflammatory cytokine, resulting in hypervirulence.

The similar increased host response and involvement of host innate immunity was verified using *Nicotiana benthamiana* as a susceptible model host of *C. orbiculare* (Kubo and Tanaka 2010; Shen et al. 2001; Tanaka et al. 2009).

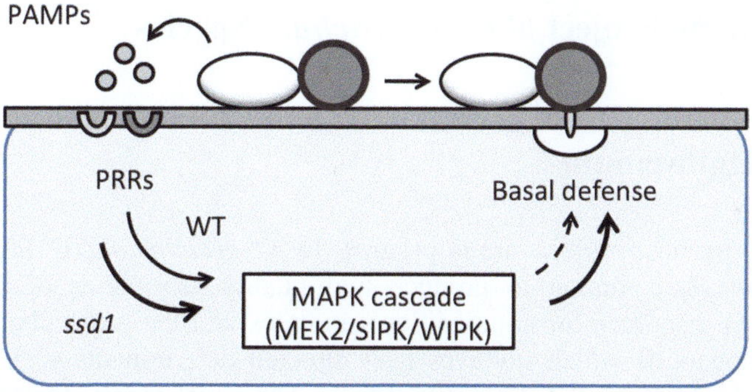


**Fig. 7.3** A model of basal resistance in *Colletotrichum orbiculare* – host plant interactions. During the course of appressorial morphogenesis on the plant surface, the host plant potentially recognizes fungal PAMPs via receptors. In the case of *ssd1* mutant the induction activity is higher than wild type causing increased activation of the MAPK cascade, MEK2-SIPK/WIPK is activated, resulting in induction of by basal defense accompanied by callose deposition

Inoculation of *N. benthamiana* leaves with conidia of the *C. orbiculare ssd1* mutant induced increased frequency in papillae formation accompanied by greater salicylic acid-induced protein kinase (SIPK) and wound induced-protein kinase (WIPK) activity than wild-type conidia. Furthermore gene silencing of SIPK/WIPK restored full infection by the *ssd1* mutant (Tanaka et al. 2009), indicating that fungal cell surface PAMPs could induce plant innate immunity via activation of the SIPK/WIPK signaling pathway, and that the altered surface composition of the *ssd1* mutant provides a stronger inducing signal than the wild-type (Fig. 7.3). Thus, a noteworthy significance of this study is in vivo evaluation of basal resistance against fungal pathogen in an adapted pathosystem, showing that alteration of fungal PAMPs could induce basal resistance to the level sufficient to suppress fungal infection.

The availability of *ssd1* mutant was further verified for the study of PAMPs receptor identification through in vivo evaluation of infectivity. There are several lines of experimental evidence that plants have receptor proteins for the recognition of bacterial and fungal PAMPs (Gómez-Gómez and Boller 2000; Zipfel et al. 2006). Chitin is also a major cell wall component in filamentous fungi and constitutes PAMPs that can be recognized by the innate immune systems of both animals and plants (Nürnberger et al. 2004). The involvement of chitin receptor in basal resistance was evidenced using *ssd1* mutant of *M. oryzae* – barley pathosystem (Tanaka et al. 2010). The *M. oryzae ssd1* mutant showed attenuated pathogenicity on barley and appressorial penetration was restricted by the increased formation of papillae at attempted entry sites. On a chitin receptor gene *HvCEBiP*-silenced barley plants, the mutant restored lesions formation accompanied by increased appressorium-mediated penetration into plant epidermal cells. Clearly, *HvCEBiP* is involved in basal resistance against appressorium-mediated infection and that basal resistance could be triggered by the recognition of chitin oligosaccharides derived from *M. oryzae*.

## 7.5 Genome Project of *Colletotrichum* Species

### *7.5.1* Colletotrichum graminicola *and* Colletotrichum higginsianum

Large-scale genome projects are in progress for *C. graminicola*, *C. higginsianum*, and *C. orbiculare* aiming to produce high-quality assemblies of the genome sequences as resources for comparative genomics and the molecular analysis of fungal pathogenesis which involves host directed determinants such as effector proteins and infection-related morphogenesis. The novel reports for powdery mildew (Spanu et al. 2010), downy mildew (Baxter et al. 2010), and smut fungus (Schirawski et al. 2010) genomics opened a new insight into the research on pathogenesis of plant pathogenic fungi/oomycete. Those studies essentially focus on genomic analysis by comparative genomics making a comparison between obligate and hemibiotrophic parasitism of fungi/oomycete in powdery and downy mildew and also revealed the pathogenesis determinants in smut fungi. Spanu et al. (2010) reported that the unique gene loss and genome expansion in *Blumeria graminis* comparing *S. cerevisiae* and ascomycete filamentous plant pathogen including the anthracnose fungus *C. higginsianum*.

*C. higginsianum* infects crucifer group including model plant *Arabidopsis thaliana* as well as *Brassica* plants providing an ideal model pathosystem in which both partners can be genetically manipulated (O'Connell et al. 2004; Narusaka et al. 2004, 2009; Ushimaru et al. 2010). *C. graminicola* is a destructive anthracnose pathogen of maize. These pathogens make a hemibiotrophic infection, but while the biotrophic phase of *C. graminicola* extends into many host cells, that of C. *higginsianum* is confined to single epidermal cells.

The recent genome project on *C. higginsianum* and *C. graminicola* has opened genome era of anthracnose fungi and the draft genome data could be publically available (Max Planck Institute for Plant Breeding Research, Cologne, Germany for *C. higginsianum* (http://www.mpiz-koeln.mpg.de/english/research/pmi-dpt/oconnell/index.html) and MIT-Broad Institute, USA, for *C. graminicola* (http://www.broadinstitute.org/annotation/genome/colletotrichum_group/MultiHome.html). Comparison of the genomes of two closely related species with contrasting pathogenic lifestyles in hemibiotrophy and host specificities enable us to study lineage-specific expansions and contractions of gene families, metabolic development and identify genes which may be involved in interactions with the host plant, for example those coding effector proteins.

The genome sequence statistics is summarized in Table 7.1. The genome of *C. graminicola* was sequenced (8X Sanger, 11X paired-end 454 pyrosequencing) resulting in an assembly comprising 1,151 contigs ($L_{50}$ contig size = 228.9 kb) in 653 scaffolds. And the genome size was estimated as 57.4 Mb in combination with optical mapping data. The genome of *C. higginsianum* was sequenced (24X 454 shot-gun, 60X Illumina paired-end, and 0.2X Sanger fosmid end reads). These data

**Table 7.1** Genome sequence statistics of *Colletotrichum*

| | *C. graminicola* | *C. higginsianum* | *C. orbiculare* | *B. graminis*[a] |
|---|---|---|---|---|
| Number of supercontigs/contigs | 653/1,151 | 367/10,235 | 549/12,179 | 6,898/15,110 |
| Total assembly size of supercontigs / contigs | 51.6/50.9 Mb | 49.3/49.1 Mb | 88.3/87.7 Mb | 119/88 Mb |
| L50[b] | 579/229 kb | 267/6.1 kb | 428/47 kb | 2.0 Mb/18 kb |
| Coverage (*x*-fold) | 19 | 83 | 88 | 140 |
| Estimated genome size[c] | 57.4 Mb | 52.5 Mb | 90.9 Mb | 120 Mb |
| Number of predicted gene | 12,006 | 15,591 | 13,382 | 5,854 |
| Number of chromosomes | 13 | 12 | 10 | |

[a]Data from Spanu et al. (2010)
[b]$L_{50}$ is the length of the smallest N50 contig, where N50 is the minimum number of contigs required to represent 50% of genome
[c]Genome size was estimated by optical mapping in *C. graminicola* and *C. higginsianum*, and by contigs depth in *C. orbiculare* and *B. graminis*

assembled to give 10,325 contigs ($L_{50}$ contig size = 6.1 kb) in 367 supercontigs. And the genome size was estimated as 52.5 Mb in combination with optical mapping data. In *C. higginsianum*, the infection stage specific cDNA libraries representing appressoria formed in vitro, appressoria penetrating leaf epidermis, biotrophic hyphae, and necrotrophic mycelium were constructed allowing identification of candidate genes involved in those morphogenetic stages.

## *7.5.2* Colletotrichum orbiculare

In addition to preceding genome project on *C. graminicola* and *C. higginsianum*, our group (Yoshitaka Takano, Kyoto University, Japan and Ken Shirasu, Riken, Japan) set up the genome project on *C. orbiculare*, an anthracnose fungus of cucumber. The aim of this project is to produce a high-quality genome assembly for *C. orbiculare* aiming for the research on (1) mechanisms of fungal pathogenicity and morphogenesis, (2) identification of factors involved in host infection such as secreted effector proteins, (3) comparative genomic analysis of the evolutionary and functional relationships with other *Colletotrichum* species such as *C. graminicola* and *C. higginsianum*.

Characteristics of *C. orbiculare* strain 104-T (MAFF240422) give us several rationales or advantages for the selection of genome sequencing project. This strain has been very stable for pathogenesis and infection-related morphogenesis for long term more than 50 years research history and genetically characterized mutant resources are available. Gene manipulation techniques such as gene targeting or random gene insertion are established by AtMT or protoplast transformation. For the host–parasite interaction study, a model plant *N. benthamiana* could be used as a susceptible host as a model pathosystem (Tanaka et al. 2009). Finally, a lot of physiological, biochemical, and genetic studies have been accomplished with this strain as described in this book chapter. Accumulation of such data has good potential and advantages for the development of further studies. From the view point of comparative genomics, *C. orbiculare* is quite related to alfalfa pathogen *C. trifoli*, and bean pathogen *C. lindemuthianum* while *C. higginsianum* and *C. graminicola* belong to independent phylogenetic clade (Latunde-Dada and Lucas 2007), thus *C. orbiculare* genome data would provide us comprehensive information and deepen the *Colletotrichum* biology.

The genome of *C. orbiculare* was sequenced by Roche 454 shot-gun reads for 15 X of expected genome size. Roche 454 paired-end reads for 7 X of expected genome size. And 66 X Illumina paired-end read for essentially miss read correction. And totally 88 X sequences were analyzed. The total number of contigs is 12,179 and total contigs length is 87.7 Mb. The total number of supercontig is 549 and the total supercontig length is 88.3 Mb. And the genome size was estimated as 90.9 Mb by contigs depth statistical analysis (Table 7.1).

The genome size of *C. orbiculare* is rather surprising, since *C. graminicola* and *C. higginsianum* genome size is 57Mb and 53Mb, respectively and more than 90 Mb was not expected. In general, most frequent size of ascomycetes fungal genome is around 30–40 Mb (Fig. 7.4). The size of 90 Mb is the highest but one in the plant pathogenic ascomycetes fungi of which genome so far analyzed. The largest is *B. graminis* with 120 Mb and surprisingly massive proliferation of transposable element accounts for 64% of genome (Spanu et al. 2010).

In combination with genomic sequence analysis, mitotic chromosomes and karyotype of *C. orbiculare* strain 104-T (MAFF240422) is being analyzed in collaboration with Masatoki Taga, Okayama University. Pulsed field gel electrophoresis (PFGE) showed that no mini-chromosomes are comprised in the genome. The smallest chromosome is estimated to be 5–6 Mb in size and other larger chromosomes could not be separated by PFGE. Cytological observation on the mitotic metaphase cells revealed that number of chromosome is $n = 10$. Furthermore, the measurements of the axial length of chromosomes gave a rough estimate of 80–100 Mb as the total genome size which is in consistent with estimated genome size of 91 Mb by genome sequencing analysis. By specific fluorescent dye staining in combination with Giemsa staining revealed that most chromosomes contained a very large, distinctive A-T-rich heterochromatin segment, locating pericentrically. The other chromosomal region seemed to be G-C-rich. From the sequencing analysis, the genome size of *C. orbiculare* is estimated almost twice as large as *C. graminicola* and *C. higginsianum*. The size could be referring to the existence of the distinctive heterochromatin region which could not be observed in

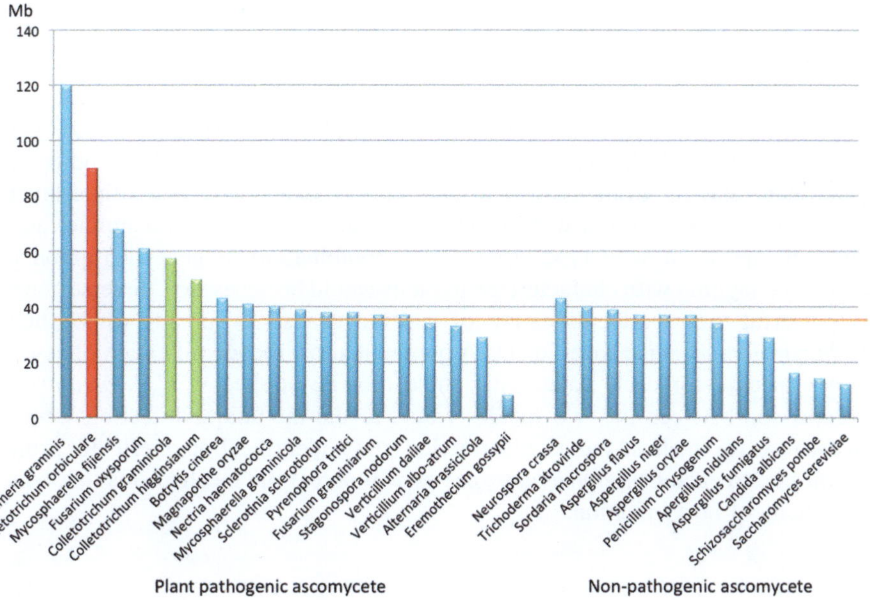


**Fig. 7.4** Genome size of distribution of ascomycetes fungi based on genome project data. *Colletotrichum orbiculare* genome size (90.8 Mb) was indicated as *red bar*, and *Colletotrichum graminicola* (57.4 Mb) and *Colletotrichum higginsianum* (52.5 Mb) are shown as *green bars*. The median genome size of euascomycetes (36.7 Mb) is shown as *orange line*

*C. graminicola* chromosomes used as control. The existence of large portion of repetitive DNA was similar to the data of *B. graminis* and it was reported that 64% of chromosome is repetitive region which frequently includes transposable element. However, the different point from the *B. graminis* chromosome is that *C. orbiculare* chromosomes constitute large distinctive heterochromatin region, but the region is not randomly dispersed in supercontigs analyzed as in *B. graminis*. It could be hypothesized that *C. orbiculare* has accumulated a large amount of repetitive sequences to constitute heterochromatin and the heterochromatin accumulation contributes to the generation of exceptionally large genome. It was reported that *B. graminis* lacks component for repeat-induced point mutations (RIPs) that are essential for meiotic and mitotic silencing, thus presumably causing genome inflation (Spanu et al. 2010). Our further study will reveal that whether such mechanisms exist in *C. orbiculare*. In our recent study, *C. higginsianum* is highly active in nonhomologous end joining (NHEJ) pathway that repairs double-strand breaks in DNA, and homologous recombination rarely occurs to the introduced DNA (Ushimaru et al. 2010). However, this genetic trait is not conspicuous in *C. orbiculare*. There might be a difference in the mechanisms for chromosome DNA repair and stability. Comparative genomics approach of *C. orbiculare* and *C. higginsianum* is expected to afford an answer.

## 7.6 Conclusions and Future Perspectives

Genome project of *Colletotrichum* species has now opened the new era for the study of pathogenesis and infection-related morphogenesis. Most simply, forward genetics will enable high throughput analysis of gene function by random and targeted gene mutagenesis. As mentioned in this book chapter, there is multiple layer of redundancy for signal transduction and gene function. Thus, not a simple random gene disruption of wild type strain but a combination of gene disruption plus activation tagging with characterized mutants would be necessary. Reverse genetics is powerful tool for identification of specific genes. As already established in *C. higginsianum*, stage specific cDNA pool will allow the identification of essential genes specific for each stage. This could also be applied to *C. orbiculare* and comparison between the two species would reveal the unique infection strategies in each. As for the host–parasite interaction study, identification of effector proteins of specific to each *Colletotrichum* species will reveal the mechanisms that control specificity of host infection.

## References

Asakura M, Okuno T, Takano Y (2006) Multiple contributions of peroxisomal metabolic function to fungal pathogenicity in *Colletotrichum lagenarium*. Appl Environ Microbiol 72:6345–6354

Asakura M, Ninomiya S, Sugimoto M, Oku M, Yamashita S, Okuno T, Sakai Y, Takano Y (2009) Atg26-mediated pexophagy is required for host invasion by the plant pathogenic fungus *Colletotrichum orbiculare*. Plant Cell 21:1291–1304

Baxter L, Tripathy S, Ishaque N, Boot N, Cabral A, Kemen E, Thines M, Ah-Fong A, Anderson R, Badejoko W, Bittner-Eddy P, Boore JL, Chibucos MC, Coates M, Dehal P, Delehaunty K, Dong S, Downton P, Dumas B, Fabro G, Fronick C, Fuerstenberg SI, Fulton L, Gaulin E, Govers F, Hughes L, Humphray S, Jiang RHY, Judelson H, Kamoun S, Kyung K, Meijer H, Minx P, Morris P, Nelson J, Phuntumart V, Qutob D, Rehmany A, Rougon-Cardoso A, Ryden P, Torto-Alalibo T, Studholme D, Wang Y, Win J, Wood J, Clifton SW, Rogers J, Van den Ackerveken G, Jones JDG, McDowell JM, Beynon J, Tyler BM (2010) Signatures of adaptation to obligate biotrophy in the *Hyaloperonospora arabidopsidis* genome. Science 10:1549–1551

Behrens R, Nurse P (2002) Roles of fission yeast tea1p in the localization of polarity factors and in organizing the microtubular cytoskeleton. J Cell Biol 157:783–793

Bhambra GK, Wang ZY, Soanes DM, Wakley GE, Talbot NJ (2006) Peroxisomal carnitine acetyl transferase is required for elaboration of penetration hyphae during plant infection by *Magnaporthe grisea*. Mol Microbiol 61:46–60

Browning H, Hayles J, Mata J, Aveline L, Nurse P, McIntosh JR (2000) Tea2p is a kinesin-like protein required to generate polarized growth in fission yeast. J Cell Biol 151:15–28

Browning H, Hackney DD, Nurse P (2003) Targeted movement of cell end factors in fission yeast. Nat Cell Biol 5:812–818

Chang F, Peter M (2003) Yeasts make their mark. Nat Cell Biol 5:294–299

Flaishman MA, Hwang CS, Kolattukudy PE (1995) Involvement of protein phosphorylation in the induction of appressorium formation in *Colletotrichum gloeosporioides* by its host surface wax and ethylene. Physiol Mol Plant Pathol 47:103–117

Fujihara N, Sakaguchi A, Tanaka S, Fujii S, Tsuji G, Shiraishi T, O'Connell R, Kubo Y (2010) Peroxisome biogenesis factor PEX13 is required for appressorium-Mediated plant infection by the anthracnose fungus, *Colletotrichum orbiculare*. Mol Plant Microbe Interact 23:436–445

Gómez-Gómez L, Boller T (2000) FLS2: an LRR receptor-like kinase involved in the perception of the bacterial elicitor flagellin in *Arabidopsis*. Mol Cell 5:1003–1011

Heiland I, Erdmann R (2005) Biogenesis of peroxisomes: topogenesis of the peroxisomeal membrane and matrix proteins. FEBS J 272:2362–2372

Hiruma K, Onozawa-Komori M, Takahashi F, Asakura M, Bednarek P, Okuno T, Schulze-Lefert P, Takano Y (2010) Entry mode–dependent function of an indole glucosinolate pathway in *Arabidopsis* for nonhost resistance against anthracnose pathogens. Plant Cell 22:2429–2443

Jones JD, Dangl JL (2006) The plant immune system. Nature 444:323–329

Kaeberlein M, Guarente L (2002) *Saccharomyces cerevisiae MPT5* and *SSD1* function in parallel pathways to promote cell wall integrity. Genetics 160:83–95

Kimura A, Takano Y, Furusawa I, Okuno T (2001) Peroxisomal metabolic function is required for appressorium-mediated plant infection by *Colletotrichum orbiculare*. Plant Cell 13:1945–1957

Kojima K, Kikuchi T, Takano Y, Oshiro E, Okuno T (2002) The mitogen-activated protein kinase gene *MAF1* is essential for the early differentiation phase of appressorium formation in *Colletotrichum lagenarium*. Mol Plant Microbe Interact 15:1268–1276

Kojima K, Takano Y, Yoshimi A, Tanaka C, Kikuchi T, Okuno T (2004) Fungicide activity through activation of a fungal signalling pathway. Mol Microbiol 53:1785–1796

Kolattukudy PE, Rogers LM, Li D, Hwang CS, Flaishman MA (1995) Surface signaling in pathogenesis. Proc Natl Acad Sci USA 9:4080–4087

Konzack S, Rischitor PE, Enke C, Fischer R (2005) The role of the kinesin motor KipA in microtubule organization and polarized growth of *Aspergillus nidulans*. Mol Biol Cell 16:497–506

Kubo Y, Furusawa I (1991) Melanin biosynthesis: prerequisite for successful invasion of the plant host by appressoria of *Colletotrichum* and *Pyricularia*. In: Cole GT, Hoch HC (eds) The fungal spore and disease initiation in plants and animals. Plenum, New York

Kubo Y, Tanaka S (2010) Pathogenesis and plant basal resistance in *Colletotrichum orbiculare* and *Magnaporthe oryzae* infection. In: Wolpert T, Shiraishi T, Allen C, Glazebrook J, Akimitsu K (eds) Genome-enabled integration of research in plant pathogen systems. APS, St. Paul, MN

Kubo Y, Takano Y, Endo N, Yasuda N, Tajima S, Furusawa I (1996) Cloning and structural analysis of the melanin biosynthesis gene *SCD1* encoding scytalone dehydratase in *Colletotrichum lagenarium*. Appl Environ Microbiol 62:4340–4344

Latunde-Dada AO, Lucas JA (2007) Localized hemibiotrophy in *Colletotrichum*: cytological and molecular taxonomic similarities among *C. destructivum*, *C. linicola* and *C. truncatum*. Plant Pathol 56:443–447

Liu ZM, Kolattukudy PE (1998) Identification of a gene product induced by hard-surface contact of *Colletotrichum gloeosporioides* conidia as a ubiquitin-conjugating enzyme by yeast complementation. J Bacteriol 180:3592–3597

Narusaka Y, Narusaka M, Park P, Kubo Y, Hirayama T, Seki M, Shiraishi T, Ishida J, Nakashima M, Enju A, Sakurai T, Satou M, Kobayashi M, Shinozaki (2004) *RCH1*, a locus in *Arabidopsis* that confers resistance to the hemibiotrophic fungal pathogen *Colletotrichum higginsianum*. Mol Plant Microbe Interact 17:749–762

Narusaka M, Shirasu K, Noutoshi Y, Kubo Y, Shiraishi T, Iwabuchi M, Narusaka Y (2009) *RRS1* and *RPS4* provide a dual resistance-gene system against fungal and bacterial pathogens. Plant J 59:672–683

Nürnberger T, Brunner F, Kemmerling B, Piate L (2004) Innate immunity in plants and animals: striking similarity and obvious differences. Immunol Rev 198:249–266

O'Connell RJ, Panstruga R (2006) Tête à tête inside a plant cell: establishing compatibility between plants and biotrophic fungi and oomycetes. New Phytol 171:699–718

O'Connell R, Herbert C, Sreenivasaprasad S, Khati M, Esquerré-Tugayé MT, Dumas B (2004) A novel *Arabidopsis-Colletotrichum* pathosystem for the molecular dissection of plant-fungal interactions. Mol Plant Microbe Interact 17:272–282

Perfect SE, Hughes HB, O'Connell RJ, Green JR (1999) *Colletotrichum*: a model genus for studies on pathology and fungal–plant interactions. Fungal Genet Biol 27:186–198

Perpetua NS, Kubo Y, Yasuda N, Takano Y, Furusawa I (1996) Cloning and characterization of a melanin biosynthetic *THR1* reductase gene essential for appressorial penetration of *Colletotrichum lagenarium*. Mol Plant Microbe Interact 9:323–329

Prusky D, Freeman S, Dickman M (2000) Host specificity, pathogenicity and host pathogen interaction of *Colletotrichum*. APS, St. Paul, MN

Ramos-Pamplona M, Naqvi NI (2006) Host invasion during rice-blast disease requires carnitine-dependent transport of peroxisomal acetyl-CoA. Mol Microbiol 61:61–75

Sakaguchi A, Miyaji T, Tsuji G, Kubo Y (2008) Kelch-repeat protein Clakel2p and calcium signaling control appressorium development in *Colletotrichum lagenarium*. Eukaryot Cell 7:102–111

Sakaguchi A, Miyaji T, Tsuji G, Kubo Y (2010a) A Kelch repeat protein Cokel1p associates with microtubules and is involved in appressorium development in *Colletotrichum orbiculare*. Mol Plant Microbe Interact 23:103–111

Sakaguchi A, Tsuji G, Kubo Y (2010b) A yeast *STE11* homologue *CoMEKK1* is essential for pathogenesis-related morphogenesis in *Colletotrichum orbiculare*. Mol Plant Microbe Interact 23:1563–1572

Schirawski J, Mannhaupt G, Münch K, Brefort T, Schipper K, Doehlemann G, Stasio MD, Rössel N, Mendoza-Mendoza A, Pester D, Müller O, Winterberg B, Meyer E, Ghareeb H, Wollenberg T, Münsterkötter M, Wong P, Walter M, Stukenbrock E, Güldener U, Kahmann R (2010) Pathogenicity determinants in smut fungi revealed by genome comparison. Science 10:1546–1548

Shen S, Goodwin PH, Hsiang T (2001) Infection of *Nicotiana* species by the anthracnose fungus, *Colletotrichum orbiculare*. Eur J Plant Pathol 107:767–773

Spanu PD, Abbott JC, Amselem J, Burgis TA, Soanes DM, Stüber K, Loren V, van Themaat E, Brown JKM, Butcher SA, Gurr SJ, Lebrun MH, Ridout CJ, Schulze-Lefert P, Talbot NJ, Ahmadinejad N, Ametz C, Barton GR, Benjdia M, Bidzinski P, Bindschedler LV, Both M, Brewer MT, Cadle-Davidson L, Cadle-Davidson MM, Collemare J, Cramer R, Frenkel O, Godfrey D, Harriman J, Hoede C, King BC, Klages S, Kleemann J, Knoll D, Koti PS, Kreplak J, López-Ruiz FJ, Lu X, Maekawa T, Mahanil S, Micali C, Milgroom MG, Montan G, Noir S, O'Connell RJ, Oberhaensli S, Parlange F, Pedersen C, Quesneville H, Reinhardt R, Rott M, Sacristán S, Schmidt SM, Schön M, Skamnioti P, Sommer H, Stephens A, Takahara H, Thordal-Christensen H, Vigouroux M, Weßling R, Wicker T, Panstruga R (2010) Genome expansion and gene loss in powdery mildew fungi reveal tradeoffs in extreme parasitism. Science 10:1543–1546

Takano Y, Kubo Y, Shimizu K, Mise K, Okuno T, Furusawa I (1995) Structural analysis of *PKS1*, a polyketide synthase gene involved in melanin biosynthesis in *Colletotrichum lagenarium*. Mol Gen Genet 249:162–167

Takano Y, Kubo Y, Kuroda I, Furusawa I (1997) Temporal transcriptional pattern of three melanin biosynthesis genes, *PKS1*, *SCD1*, and *THR1*, in appressorium-differentiating and nondifferentiating conidia of *Colletotrichum lagenarium*. Appl Environ Microbiol 63:351–354

Takano Y, Kikuchi T, Kubo Y, Hamer JE, Mise K, Furusawa I (2000) The *Colletotrichum lagenarium* MAP kinase gene *CMK*1 regulates diverse aspects of fungal pathogenesis. Mol Plant Microbe Interact 13:374–383

Takano Y, Komeda K, Kojima K, Okuno T (2001) Proper regulation of cyclic AMP-dependent protein kinase is required for growth, conidiation, and appressorium function in the anthracnose fungus *Colletotrichum lagenarium*. Mol Plant Microbe Interact 14:1149–1157

Tanaka S, Yamada K, Yabumoto K, Fujii S, Huser A, Tsuji G, Koga H, Dohi K, Mori M, Shiraishi T, O'Connell R, Kubo Y (2007) *Saccharomyces cerevisiae SSD1* orthologues are essential for host infection by the ascomycete plant pathogens *Colletotrichum lagenarium* and *Magnaporthe grisea*. Mol Microbiol 64:1332–1349

Tanaka S, Ishihama N, Yoshioka H, Huser A, O'Connell R, Tsuji G, Tsuge S, Kubo Y (2009) The *Colletotrichum orbiculare ssd1* mutant enhances *Nicotiana benthamiana* basal resistance by activating a mitogen-activated protein kinase pathway. Plant Cell 21:2517–2526

Tanaka S, Ichikawa A, Yamada K, Tsuji G, Nishiuchi T, Mori M, Koga H, Nishizawa Y, O'Connell R, Kubo Y (2010) *HvCEBiP*, a gene homologous to rice chitin receptor CEBiP, contributes to basal resistance of barley to *Magnaporthe oryzae*. BMC Plant Biol 10:288

Thordal-Christensen H (2003) Fresh insights into processes of nonhost resistance. Curr Opin Plant Biol 6:351–357

Titorenko VI, Rachubinski RA (2001) Dynamics of peroxisome assembly and function. Trends Cell Biol 11:22–29

Titorenko VI, Rachubinski RA (2004) The peroxisome: orchestrating important developmental decisions from inside the cell. J Cell Biol 164:641–645

Tsuji G, Fujii S, Fujihara N, Hirose C, Tsuge S, Shiraishi T, Kubo Y (2003) *Agrobacterium tumefaciens*-mediated transformation for random insertional mutagenesis in *Colletotrichum lagenarium*. J Gen Plant Pathol 69:230–239

Ushimaru T, Terada H, Tsuboi K, Kogou Y, Sakaguchi A, Tsuji G, Kubo Y (2010) Development of an efficient gene targeting system in *Colletotrichum higginsianum* using a non-homologous end-joining mutant and *Agrobacterium tumefaciens*-mediated gene transfer. Mol Genet Genomics 284:357–371

Wanders RJA, Waterman HR (2004) Peroxisomal disorders I: biochemistry and genetics of peroxisome biogenesis disorder. Clin Genet 67:107–133

Wheeler RT, Kupiec M, Magnelli P, Abeijon C, Fink GR (2003) A *Saccharomyces cerevisiae* mutant with increased virulence. Proc Natl Acad Sci USA 100:2766–2770

Yamauchi J, Takayanagi N, Komeda K, Takano Y, Okuno T (2004) cAMP-PKA signaling regulates multiple steps of fungal infection cooperatively with Cmk1 MAP kinase in *Colletotrichum lagenarium*. Mol Plant Microbe Interact 17:1355–1365

Zipfel C, Kunze G, Chinchilla D, Caniard A, Jones JDG, Boller T, Felix G (2006) Perception of the bacterial PAMP EF-Tu by the receptor EFR restricts *Agrobacterium*-mediated transformation. Cell 125:749–760

# Chapter 8
# Morphogenesis in *Candida albicans*: How to Stay Focused

**Martine Bassilana and Peter Follette**

**Abstract** Morphogenesis, such as the transition from budding to filamentous growth in *Candida albicans*, is a fundamental process that is of general interest both because of its cell biology and for its implication in host–pathogen interactions. Condition-specific transcriptome analyses and large-scale gene deletion or inactivation studies, together with forward and reverse genetic approaches, have uncovered a number of components involved in the regulation of the cellular reorganization that takes place in *C. albicans* during morphogenesis. This chapter will summarize the main components involved in cellular morphogenesis and provide an update of our knowledge of the regulatory steps that control the initiation and maintenance of polarized growth during *C. albicans* hyphal formation.

## 8.1 Introduction

Cell shape regulation is critical for life, from single-celled organisms to humans. In multicellular organisms, individual cells, such as neurons or red blood cells, have different shapes that are adapted to carry out specialized functions. In contrast, unicellular organisms, such as the baker's yeast *Saccharomyces cerevisiae*, can reversibly alter their shapes for different functions. For example, during mating, *S. cerevisiae* cells grow toward a partner of the opposite mating type, resulting in a specific morphology referred to as a "shmoo" (Bardwell 2005; Arkowitz 2009). Also, to forage for nutrients in nitrogen-poor conditions, *S. cerevisiae* can switch from budding growth to pseudohyphal growth, forming chains of elongated ellipsoid cells that lack cytoplasmic connections and that can invade into a semisolid

M. Bassilana (✉) • P. Follette
Centre National de la Recherche Scientifique and Université de Nice-Sophia Antipolis, Institute of Developmental Biology and Cancer, CNRS-UMR6543 Faculté des Sciences, Parc Valrose, 06108 Nice, France
e-mail: mbassila@unice.fr

J. Pérez-Martín and A. Di Pietro (eds.), *Morphogenesis and Pathogenicity in Fungi*, Topics in Current Genetics 22, DOI 10.1007/978-3-642-22916-9_8, 

support (Zaman et al. 2008). In contrast to most fungi, *S. cerevisiae* do not produce true hyphae, which are long, branching filamentous structures comprised of tubular cells with parallel walls and lacking constrictions marking septal junctions.

*S. cerevisiae* and the human opportunistic fungal pathogen *Candida albicans* both belong to the Saccharomycotina subphylum and diverged approximately 800 million years ago (Hedges 2002). *C. albicans* belongs to a clade containing organisms in which the CTG codon encodes a serine instead of a leucine, a reassignment which appears to have occurred about 170 million years ago (Massey et al. 2003; Fitzpatrick et al. 2006). In contrast to *S. cerevisiae*, *C. albicans* has never been found as haploid cells and was long thought to be asexual. Pioneering work from Magee's and Johnson's groups, however, has shown using independent approaches that mating can occur (Hull et al. 2000; Magee and Magee 2000). *C. albicans* mating has since received considerable attention, and recently homothallic mating (same sex mating) has been observed (Alby et al. 2009). Nevertheless, *C. albicans* meiosis has not been observed to date; in its absence, the return to the diploid status presumably occurs via genome reduction, through a parasexual cycle (Bennett and Johnson 2003; Forche et al. 2008).

Before mating, *C. albicans* undergoes a reversible transition between the normal yeast morphology (white) and an elongated larger cell form (opaque), which is the mating-competent cell type (Miller and Johnson 2002). White-opaque switching was initially demonstrated in the WO-1 strain by Slutsky et al. (1987), who observed that opaque cells gave rise to darker and flatter colonies than the domed white colonies typical of *C. albicans*. Work on mating, white-opaque switching, and their interconnection has been summarized in recent reviews (Soll 2009, 2011; Lohse and Johnson 2010; Morschhauser 2010) and will not be discussed further in this chapter.

In contrast to *S. cerevisiae*, *C. albicans* can exist not only as round yeasts (blastopores) and pseudohyphae, but also as true hyphae (Sudbery et al. 2004). These different cell shapes contribute to the fitness of this organism and its ability to colonize diverse environments. *C. albicans* is a commensal organism that normally lives inoffensively among the gastrointestinal tract microflora. Upon alteration of the host's physiology, however, this fungus can become a pathogen, primarily causing benign superficial mucosal infections. *C. albicans* is particularly threatening in immunocompromised patients, where it can result in fatal systemic infections. *C. albicans* can, therefore, be found in diverse physiological niches, such as the gastrointestinal tract, blood, and skin, as well as on medical devices such as catheters. Accordingly, *C. albicans* can grow as planktonic cultures (individual cells or pseudohyphae/hyphae floating in liquid media) or on solid supports as a biofilm (cells that adhere to each other and to the solid support) (Blankenship and Mitchell 2006; Nobile and Mitchell 2006; Ramage et al. 2009). Finally, *C. albicans* can form chlamydospores, a distinctive form used for diagnosis which is observed only in *C. albicans* and its closest relative *C. dubliniensis*. Chlamydospores are thick-walled, round cells that develop on suspender cells situated on hyphae or pseudohyphae in nutrient-poor and oxygen-restricted environments (Staib and Morschhauser 2007).

Understanding the determinants, signaling pathways, and cell physiological alterations involved in morphogenetic change in *C. albicans* has been the focus of considerable work, some of which has been summarized in comprehensive recent reviews (Berman 2006; Biswas et al. 2007; Whiteway and Bachewich 2007; Cottier and Muhlschlegel 2009). This chapter aims to provide an update on recent developments in the cellular and molecular biology of the reversible switch from budding to filamentous growth, and to discuss briefly its importance for *C. albicans* pathogenicity. Given the limited knowledge of the molecular mechanisms underlying filamentous growth in *C. albicans*, we will also refer to analogous work on *S. cerevisiae* and in some cases other filamentous fungi as well.

## 8.2 Morphogenesis and Virulence

Because of their capacity for invasive growth, filamentous cells (hyphae and pseudohyphae) appear to be well suited for virulence: filaments could be responsible, for example, for penetrating new tissues and thereby propagating infection. Along these lines, filaments could provide mechanical force that helps *C. albicans* penetrate epithelia and escape from phagocyte cells following internalization (Kumamoto and Vinces 2005). Consistent with this, filamentous forms can be found in samples taken from patients with Candida infections, for instance, in periodontal tissues (Jarvensivu et al. 2004), or at infection sites in different infection models, such as in vitro circulatory *C. albicans*–endothelium interaction model (Wilson and Hube 2010), ears of living mice (Mitra et al. 2010), *C. elegans* (Pukkila-Worley et al. 2009) or zebrafish (Chao et al. 2010). As an example, Fig. 8.1 illustrates *C. albicans* filamentous cells escaping macrophage phagocytosis. At the same time, pathogenesis appears to be more specifically associated with the ability of the cells to switch between nonfilamentous and filamentous morphologies (Lo et al. 1997; Gow 2002; Saville et al. 2003, 2008; Zheng and Wang 2004). Indeed, mutants locked in either growth form are non-virulent (Braun and Johnson 1997; Lo et al. 1997).

Some evidence for a relationship between filamentation and virulence has come from the observation that numerous mutations affect both processes. Many of these mutants involve transcription factors, however, which are likely to affect diverse processes, making it difficult to demonstrate a strict relationship between morphogenesis and virulence. Transcription factors regulate the expression of genes involved directly in filamentation, such as the hyphal G1 cyclin Hgc1 (Zheng and Wang 2004), as well as others involved in different aspects of virulence, such as proteases (Naglik et al. 2003, 2004) and adhesins (Zhu and Filler 2010). A recent study has examined the relationship between virulence, morphogenetic switching, and cell proliferation in *C. albicans* (Noble et al. 2010). In this study, a large number of deletion strains were examined for their effects on virulence in vivo in a mouse systemic Candidiasis assay and on filamentation and proliferation in vitro. The correlation between filamentation and virulence was far from perfect: almost

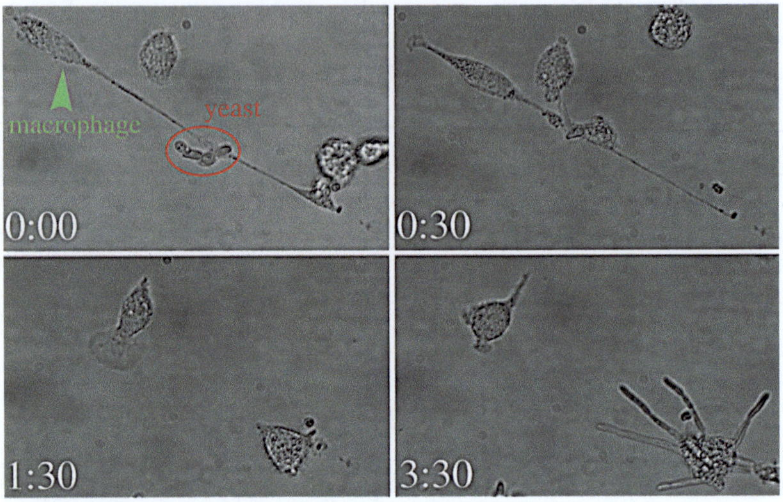


**Fig. 8.1** Filamentous *C. albicans* cells escape macrophage phagocytosis. Time-lapse images of *C. albicans* and SV40-immortalized murine macrophage (Lo et al. 1997) interactions. Images from K. Boulukos and M. Bassilana

half of the strains showing reduced infectivity displayed normal filamentation ability, and two-thirds of the strains with the strongest filamentation defects showed normal infectivity. Still, as the virulence studies were performed by directly injecting fungal inoculum in the mouse blood stream, these results do not exclude the possibility that filamentous forms play a critical role in epithelial invasion, for instance, as a prerequisite for systemic infection. Together, these results indicate that while morphogenetic switching and virulence can be genetically separated, they are clearly associated in *C. albicans*.

## 8.3 Inducers of Filamentous Growth and Transcriptional Regulation

As noted above, in its natural environment, *C. albicans* can exist in a range of morphogenetic forms, presumably in response to specific changes in the host environment. Similarly, in vitro, a number of conditions have been identified that can trigger the transition from budding to hyphal or pseudohyphal growth. These stimuli generally result in the activation of signaling pathways that ultimately act through one or more transcription factors to effect changes in gene regulation, leading to the observed morphologic changes.

### 8.3.1 Inducers of Filamentous Growth

Numerous inducers of filamentous growth exist; many of these resemble the conditions found in a host, and include neutral pH, body temperature (37°C), and serum. Other triggers include nutrient starvation (e.g., growth on Spider medium), *N*-acetylglucosamine (GlcNAc), proline, and embedded growth within semisolid media. While in some circumstances the activation of a particular pathway can trigger filamentation by itself, in many cases activation of multiple pathways can act together to bring about filamentation. For example, a common inducing condition is growth at 37°C in the presence of serum; this combination brings about a more robust induction than either of these two conditions alone. Thus, various inducing conditions activate distinct signaling pathways, including the Ras1-cAMP-PKA (Hogan and Sundstrom 2009) and MAP kinase (Monge et al. 2006) pathways, which can act alone or in concert to activate transcription, ultimately leading to filamentous growth (Liu 2001; Dhillon et al. 2003; Biswas et al. 2007; Kumamoto 2008).

### 8.3.2 Transcription Factors

Transcription factors can act negatively or positively to control the budding to hyphal growth switch under different inducing conditions (Liu 2001). A number of transcription factors act negatively to repress filamentation; accordingly, their deletion results in filamentation even in the absence of inducers. Three such well-characterized regulators are Nrg1 (Braun et al. 2001; Murad et al. 2001), Rfg1 (Kadosh and Johnson 2001; Khalaf and Zitomer 2001), and the global repressor Tup1 (Braun and Johnson 1997). In cells deleted for any of these three genes, numerous filament-specific genes are expressed even in the absence of inducing conditions and filamentation ensues, via either hyphal or pseudohyphal growth (Braun and Johnson 1997; Braun et al. 2001; Kadosh and Johnson 2001; Khalaf and Zitomer 2001). A number of target genes that are regulated by Rfg1 and Nrg1 have been identified, revealing some overlap but distinct targets as well (Kadosh and Johnson 2005). Finally, while overexpression of Nrg1 inhibits filamentation under inducing conditions (Saville et al. 2003), overexpression of Rfg1 does not (Cleary et al. 2010); interestingly, however, Rfg1 overexpression does have an effect under yeast growth conditions, as it induces pseudohyphal growth (Cleary et al. 2010).

Positive regulators of filamentation include, among others, Cph1 (Liu et al. 1994; Lo et al. 1997), Efg1 (Stoldt et al. 1997; Doedt et al. 2004), Flo8 (Cao et al. 2006), Rim101 (Davis et al. 2000; Davis 2009), Ace2 (Kelly et al. 2004; Mulhern et al. 2006), and Csr1 (Kim et al. 2008). Each of these transcription factors has been linked to the induction of filamentous growth under a particular inducing condition, and their deletion impairs this process; in the other direction, overexpression or constitutive activation of these factors generally leads to filamentous growth in the absence of inducing conditions. For example, Cph1, the *C. albicans* homolog of *S. cerevisiae* Ste12, is required for filamentous growth on solid media, but is dispensable for filamentous growth in liquid media supplemented with serum. Efg1, on the contrary, is required for filamentation on solid or liquid media in the presence of serum or GlcNAc, but not under embedded or anaerobic conditions, conditions in which it actually acts as a repressor of filamentation (Ernst 2000; Setiadi et al. 2006).

Ume6 is a recently identified positive transcription factor (Banerjee et al. 2008; Zeidler et al. 2009). *UME6* was identified both in a microarray-based screen for genes expressed in the presence of serum at 37°C (Banerjee et al. 2008), and during a characterization of *C. albicans* homologs of meiotic genes from *S. cerevisiae* (Zeidler et al. 2009). *UME6* is expressed upon induction of filamentation, and *UME6* deletion results in impaired filamentous growth and reduced expression of a number of hyphal genes. Ume6 appears to act downstream of Efg1 and Cph1, and the Nrg1–Tup1 complex seems to repress *UME6* expression in non-induced cells (Banerjee et al. 2008; Zeidler et al. 2009). Interestingly, *UME6* plays a role in dictating the specific filamentous growth state as either pseudohyphae or true hyphae. For example, while wild-type or heterozygous cells undergo hyphal growth when grown on or embedded in agar, cells homozygous for a *UME6* deletion exhibit only pseudohyphal growth (Zeidler et al. 2009). Also, while a high level of *UME6* is sufficient to drive cells toward true hyphal growth under non-inducing conditions, a lower level triggers predominantly pseudohyphal growth (Carlisle et al. 2009). These results suggest that pseudohyphae and hyphae formation is not regulated by distinct morphogenetic programs. One particularly interesting target of Ume6 is the hypha-specific G1 cyclin-encoding gene *HGC1* (Zheng and Wang 2004; Wang 2009; Carlisle and Kadosh 2010).

## 8.4 Hyphal Growth Regulation

Although budding and hyphal growth are both polarized, the cell biology of these two processes is quite different. One important difference is that hyphal elongation is regulated independently of the cell cycle in *C. albicans* (Hazan et al. 2002; Berman 2006), even though the regulation of morphogenesis during hyphal growth requires cyclins such as Cln1 (Loeb et al. 1999), Cln3 (Bachewich and Whiteway 2005; Chapa y Lazo et al. 2005), Clb2, Clb4 (Bensen et al. 2005), and in particular Hgc1 (Zheng and Wang 2004). As cell cycle regulation in *C. albicans* is treated in another chapter of this book, this will not be discussed further here.

Furthermore, in contrast to budding growth, hyphal growth requires blocking the isotropic switch to orient growth exclusively to a restricted area, stabilizing the growth axis, and preventing cell separation after cytokinesis. Indeed, hyphal growth is a characteristic of fungal development, consisting of a polarized hyphal tip extension and subsequent branching. The rate of hyphal tip elongation can differ as a function of the environment and also varies between different fungal species: for example, it is approximately 100 μm/min for *Neurospora crassa*, 3 μm/min for *Ashbya gossypii* in mature mycelia, and 0.3 μm/min for *C. albicans*. To sustain such highly polarized growth, fungi must maintain growth in one direction, toward the apex of the hypha, and rapidly transport material to this apex to generate new membrane and cell wall. The hyphal apex is thus the site of intense exocytic–endocytic cycles. Interestingly, a recent modeling study indicates that the shape of tip-growing cells, including hyphae, can be characterized by the interplay between cell wall mechanics and assembly (Campas and Mahadevan 2009). The following sections highlight our current knowledge about the major components that regulate the initiation and maintenance of this highly polarized process, and summarize the main changes that occur in cellular organization during hyphal growth.

### *8.4.1 Polarity Determinants and Small G-Proteins*

Polarized hyphal growth in response to extracellular stimuli requires the cell to restrict cytoskeletal organization to a unique site that is determined via specific spatial landmarks and cues. Components to establish polarity and components to orient and maintain polarized growth are involved in this process. As hyphal growth is cell cycle independent, initiation of the germ tube can arise from cells that are already polarized or from cells that present no apparent asymmetry. In the former case, cells do not need to define a new site of growth, but presumably need to block the isotropic switch to start cell elongation. In the latter situation, however, cells need to define a site of growth de novo, a process that could require spatial landmark proteins and/or lipids.

In diploid *S. cerevisiae* yeast cells, the transmembrane glycoproteins Bud8 and Bud9, which regulate the pattern of cell division from bipolar to unipolar during the switch from budding to pseudohyphal growth (Taheri et al. 2000), act as cortical landmarks. The role of such landmark proteins is less clear in *C. albicans*, however, as homologs for these cortical landmarks are not apparent in the genome. On the contrary, it was shown that highly polarized ergosterol-rich domains were present only during hyphal growth (Martin et al. 2004), and such lipids could play a role in the recruitment and clustering of proteins that are critical for polarized hyphal growth. Alternatively, one could imagine that the stochastic activation of a conserved small G-protein, most likely Cdc42, is sufficient to determine the incipient site of germ tube initiation. Indeed, *S. cerevisiae* deletion mutants of the Ras-like protein Rsr1/Bud1 show random budding as well as normal morphology and growth, indicating that Cdc42 can be activated stochastically (Nern and Arkowitz 2000; Wedlich-Soldner et al. 2003).

As small G-proteins play a central role in polarized growth, sustained polarized growth requires the maintenance of spatially restricted small G-protein activation. In *S. cerevisiae* as well as in *A. gossypii*, Cdc42 plays a central role during polarized growth. In addition to Cdc42, in other fungi such as *U. maydis*, another highly conserved small G-protein, Rac1, is necessary and sufficient for the budding to hyphal growth switch (Mahlert et al. 2006), while Cdc42 is required for cell separation. Indeed, in filamentous fungi, Rac homologs and Cdc42 have been implicated in different functions, some of them overlapping (Harris 2011). Although there are no Rac homologs in *S. cerevisiae* and *A. gossypii*, a Rac homolog is present in *C. albicans* (Bassilana and Arkowitz 2006). Rho1 also plays a significant role in polarized growth (Levin 2005). In *S. cerevisiae*, Rho1 is required for cell wall integrity through its interaction with the single protein kinase C Pkc1. It also regulates cell wall biogenesis through interactions with the glucan synthase complex, which produces β-(1,3)-glucan, a major component of the yeast cell wall. *C. albicans* has a Rho1 homolog which is essential and required for growth in the presence of cell wall perturbants such as calcofluor white (Smith et al. 2002b), suggesting that the primary functions of Rho1 in *S. cerevisiae* are likely to be conserved in *C. albicans*.

#### 8.4.1.1 Cdc42

*CDC42* is an essential gene that is required for cell polarity and vesicle trafficking in yeasts as well as in mammals (Johnson 1999; Park and Bi 2007; Heasman and Ridley 2008; Perez and Rincon 2010). Cdc42 is highly conserved, with approximately 80% identity between the fungal and human homologs. Not surprisingly, in *C. albicans*, Cdc42 is essential and required for hyphal growth (Michel et al. 2002; Ushinsky et al. 2002; Bassilana et al. 2003; VandenBerg et al. 2004). Consistent with this, a GFP-Cdc42 fusion protein persistently localizes as a cluster at the hyphal apex; this persistent localization requires F-actin (Hazan and Liu 2002) and the exocyst component Sec3 (Li et al. 2007).

Rho G-proteins cycle between an inactive GDP-bound and an active GTP-bound state. In *S. cerevisiae*, Cdc42 is activated by a specific guanine nucleotide exchange factor (GEF), Cdc24, which belongs to the proto-oncogene Dbl family (Rossman and Sondek 2005). *CDC24* is an essential gene, and *cdc24* mutants phenocopy *cdc42* mutants. Both *cdc42* and *cdc24* mutants fail to produce hyphae in *C. albicans* (Bassilana et al. 2003, 2005) and *A. gossypii* (Wendland and Philippsen 2001). Cdc42 is negatively regulated by three specific GTPase-activating proteins (GAPs) in *S. cerevisiae*: Rga1, Rga2, and Bem3; notably, the deletion of *RGA1*, but not of *RGA2* or *BEM3*, results in increased invasive growth (Smith et al. 2002a). Among these GAPs, only homologs for Rga2 and Bem3 have been identified in *C. albicans*; their combined deletion results in enhanced filamentation (Court and Sudbery 2007). Phosphorylation of Rga2 (Court and Sudbery 2007) via the Cdc28–Hgc1 complex (Zheng et al. 2007) is critical for Cdc42 regulation during hyphal growth: phosphorylation inhibits Rga2, allowing sustained Cdc42 activation at the growth site. An increase in Cdc42 activation could also result from Cdc24 upregulation, as a twofold transient increase in *CDC24* transcripts, dependent on the transcription factor Tec1, was observed during initiation of hyphal growth in response to serum

(Bassilana et al. 2005). It will be important to measure the level of activated Cdc42 that is present during hyphal growth to determine whether it actually increases, although this has not been reported to date. Biochemical approaches, e.g., pull-down experiments using an effector domain such as a Cdc42/Rac interacting binding (CRIB) domain, have not been particularly reproducible, perhaps due to cell-to-cell variations and/or issues related to the stability or accessibility of activated Cdc42. Alternative in vivo approaches with fluorescent reporters may prove more informative.

Upstream components are required to restrict Cdc42 activation to a specific location. In *S. cerevisiae*, the Cdc42/Cdc24 module, together with a scaffold protein, Bem1, is recruited to and activated at the site of growth via another small G-protein module consisting of the Ras-related protein Rsr1/Bud1, its activator Bud5, and its inactivator Bud2 (Park and Bi 2007). Cell polarization is, therefore, achieved via two small G-protein modules acting in tandem. In *S. cerevisiae*, deletion of *RSR1* randomizes bud position, and in *A. gossypii*, *rsr1* deletion mutants can initiate but not maintain hyphal growth in a single direction, resulting in slower growth and bulging (Bauer et al. 2004). In *C. albicans*, Bud1 also plays a critical role in filamentous growth on solid media (Yaar et al. 1997; Bassilana et al. 2003; Hausauer et al. 2005) and in hyphal guidance (Hausauer et al. 2005), as well as in hyphal tip orientation during thigmotropic and galvanotropic growth (Brand et al. 2008). Another important regulator of polarized growth is the SH3 domain-containing protein Bem1 (Chenevert et al. 1992). In *S. cerevisiae*, cells expressing a fusion protein of Bem1 that has restricted movement at the plasma membrane can generate two buds simultaneously (Howell et al. 2009). Bem1 is conserved throughout fungi, including *A. gossipii*, *Yarrowia lipolytica*, and *C. albicans*. In *C. albicans*, *BEM1* is essential (Michel et al. 2002) and is required for hyphal growth (Michel et al. 2002; Bassilana et al. 2003), although little is currently known about its regulation during this process. Interestingly, in *Ustilago maydis*, the interaction between Bem1 and GEF Cdc24 was shown to be dependent on Cdk5 (Alvarez-Tabares and Perez-Martin 2008), a cyclin-dependent kinase that is homologous to *S. cerevisiae* Pho85 and that has been implicated in morphogenesis. It would be interesting to examine whether such a link is found in other fungi.

#### 8.4.1.2 Rac1

In *C. albicans*, Rac1 is nonessential but is required for solid media invasion, including when cells are matrix embedded, yet not for hyphal formation in liquid media (Bassilana and Arkowitz 2006). In *Y. lipolytica*, *RAC1* deletion impairs hyphal growth but does not abolish the ability of the organism to polarize actin at the site of growth and to form pseudohyphae (Hurtado et al. 2000). In contrast, in *Aspergillus niger*, the Rac homolog RacA regulates polarity maintenance via actin dynamics (Kwon et al. 2011). The actin cytoskeleton does not appear to be altered in *C. albicans rac1* deletion mutants during budding growth (Bassilana and Arkowitz 2006), suggesting that Rac1 does not play a major role in cell polarity.

Dynamics studies of *C. albicans* Rac1 indicate that it can cycle in and out of the nucleus (Vauchelles et al. 2010), while such nucleo-cytoplasmic shuttling was not observed for Cdc42, raising the possibility that Rac1 is involved in gene regulation.

Overexpression studies suggest that Cdc42 and Rac1, which share about 60% sequence identity, do not have overlapping functions in *C. albicans*, indicating that they are differently regulated and/or are part of different signaling pathways. Figure 8.2 illustrates a speculative scheme of the Cdc42- and Rac1-dependent signaling pathways. In humans, the activation of Rho G-proteins such as Cdc42 and Rac1 requires GEF activity coming not only from the proto-oncogene Dbl family but also from the related Dock180 (*D*edicator *o*f *c*yto*k*inesis) family (Meller et al. 2005). Homologs of Dock180 are present in yeast as well as in filamentous fungi. In *C. albicans*, Rac1 is specifically activated by a Dock180 homolog, Dck1 (Hope et al. 2008), suggesting that the function of Cdc42 and Rac1 is dictated by their specific GEFs. Furthermore, Lmo1, a homolog of the scaffold protein *E*ngu*l*f-ment and cell *mo*tility (ELMO) which is required for optimal Dock180-dependent

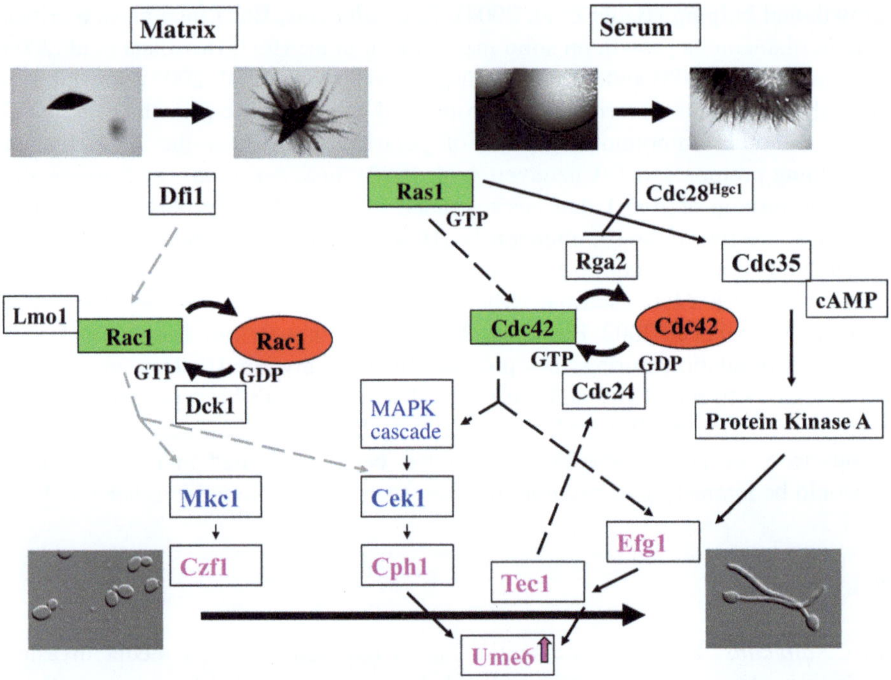


**Fig. 8.2** Different stimuli trigger distinct small G-proteins to induce *C. albicans* filamentous growth. Matrix-induced filamentous growth and serum-induced hyphal growth signaling pathways (in *gray* and *black*, respectively) are illustrated. Putative connections, direct or indirect, between components of these signaling pathways are indicated by *dashed lines*. Dfi1 is an integral plasma membrane protein that has been recently shown to be required for matrix-induced filamentous growth (Zucchi et al. 2010). For simplicity, only some of the known components are depicted in this scheme. Top images: nonfilamentous and filamentous colonies in the indicated environment. Bottom images: budding and hyphal form cells

Rac1 activation (Komander et al. 2008), was also shown to function together with Rac1 and Dck1, presumably upstream of the Cek1 and Mkc1 kinases (Hope et al. 2010). To date, no GEFs from the Dock180 family have been identified as Cdc42 or Rac1 activators in other fungi.

## *8.4.2 Polarized Growth and Maintenance*

In order to polarize growth, cells need to orient their cytoskeleton toward the site of growth, i.e., the location of the polarity establishment proteins, and to maintain cytoskeleton reorganization at the apex of the cell. Polarized growth in *S. cerevisiae* requires the interaction of Cdc42 with various critical proteins and complexes such as the polarisome via Bud6 (Kozminski et al. 2003), the actin nucleator Bni1 (Evangelista et al. 1997; Kozminski et al. 2003), and the exocyst via its subunits Exo70 and Sec3 (Wu et al. 2010).

### 8.4.2.1 Polarisome

The polarisome is a protein complex that is critical for directing the localized assembly of actin filaments at the site of polarization. In *S. cerevisiae*, the polarisome is a 12S multiprotein complex (Sheu et al. 1998) that includes the coiled-coil domain proteins Spa2 (*sp*indle *p*ole *a*ssociated) (Snyder 1989; Gehrung and Snyder 1990), Bud6 (Amberg et al. 1997), and Pea2 (Valtz and Herskowitz 1996). Bud6 is an actin monomer-binding protein, and Spa2, which also interacts with the other components of the polarisome, binds directly to the formin Bni1 (Fujiwara et al. 1998). Localized formins control growth through positioning and polarization of actin cables, and Bni1, one of the two *S. cerevisiae* formins, has been found to interact both genetically and physically with actin-related protein 2 (Arp2) (Evangelista et al. 2002b). Interestingly, *S. cerevisiae* mutants expressing only the isolated actin nucleation/assembly domain of Bnr1p or Bni1p as the sole formin can grow well, suggesting that additional mechanisms must be present to orient actin cables (Gao and Bretscher 2009).

In *A. gossypii*, orthologs for all the *S. cerevisiae* polarisome components have been identified, and Spa2 has been shown to be required for a maximal rate of polar growth (Knechtle et al. 2003). In other filamentous fungi, on the contrary, including *C. albicans*, only homologs of Spa2 and Bud6, but not of Pea2, are present (Harris and Momany 2004). Spa2 and Bud6 are both critical for normal hyphal growth in *C. albicans* (Zheng et al. 2003; Song and Kim 2006), with both *spa2* and *bud6* deletion mutants generating round cells with wide, elongated bud necks and, in hyphae-inducing conditions, thicker hyphae or hyphae with swollen tips. In *bni1* deletion mutants, both Spa2 and Bud6 still localize to the hyphal tip (Li et al. 2005). In contrast, the sustained localization of Spa2 and Bni1 to the hyphal tip, after formation of the first septin ring, requires the exocyst component Sec3 (Li et al. 2007), suggesting a connection between actin polymerization and secretion.

#### 8.4.2.2 Cytoskeleton: Actin

The fungal cytoskeleton is essentially composed of highly dynamic networks of microtubules (Horio 2007) and actin microfilaments (Moseley and Goode 2006). In *S. cerevisiae*, microfilaments are solely responsible for polarized growth (Pruyne et al. 2004). They are organized into cables, which are used as tracks to transport secretory vesicles as well as for organelles and their segregation; into cortical patches, which are dynamic structures critical for endocytosis; and into contractile rings, which are important for cytokinesis.

As in *S. cerevisiae*, the actin cytoskeleton is essential for polarized growth in both *A. gossypii* and *C. albicans*. During hyphal elongation, polarization of actin toward the *C. albicans* hyphal tip was initially observed using fluorescent phalloidin staining (Anderson and Soll 1986). This polarization of actin patches and cables persists during hyphal growth, in contrast to the dynamic rearrangement of actin structures that takes place in the same cells during budding; this suggests that the hyphal tip-associated actin polarization is regulated independently of the cell cycle (Hazan et al. 2002). Disruption of the actin cytoskeleton with latrunculin A, a drug that binds and stabilizes monomeric actin and therefore disassembles filamentous actin (F-actin), leads to swelling of hyphal tips and lysis in *A. gossypii* (Knechtle et al. 2006). Similarly, in *C. albicans*, addition of cytochalasin A, which specifically disrupts actin cables, results in swelling at the hyphal tip (Crampin et al. 2005), while the F-actin stabilizing drug jasplakinolide blocks hyphae formation altogether (Wolyniak and Sundstrom 2007).

Interestingly, several recent results suggest a role for actin dynamics in the regulation of hyphal gene expression. For example, *HWP1* gene expression is upregulated following hyphal blockade by jasplakinolide, although this upregulation is not observed with cytochalasin A or latrunculin A (Wolyniak and Sundstrom 2007); in fact, latrunculin A reduces the transcriptional induction of *HWP1* and other hyphal-specific genes such as *ECE1* and *HYR1* (Hazan and Liu 2002). This effect of actin cytoskeleton perturbants on gene expression is likely to reflect a role for F- or G-actin-dependent signal transduction pathway(s) during filamentation. Accordingly, recent evidence indicates that serum causes G-actin to enter a tripartite complex with the adenylyl cyclase Cyr1 and the cyclase-associated protein Cap1; this leads to an increase in cytosolic cAMP levels, which is required for morphogenesis (Zou et al. 2010). These results raise the possibility that the balance between G- and F-actin may provide an additional level of regulation of signaling pathways required for hyphal growth.

Microfilament turnover, which is necessary for the proper assembly of actin cables, is also important for the establishment of cell polarity. Turnover can be regulated by a number of proteins, including the primary filament nucleators Arp2/3 and formins. The Arp2/3 complex (Machesky and Gould 1999) nucleates filaments into branched networks; optimal nucleation activity depends on nucleation-promoting factors such as the Wiskott–Aldrich Syndrome protein (WASP) (Higgs and Pollard 2001). Surprisingly, in contrast to the situation in *S. cerevisiae* and *S. pombe*, the Arp2/3 complex is not essential in *C. albicans*, although deletion of *ARP2* blocks the yeast-to-hyphal switch (Epp et al. 2010b). In addition, *arp2* deletion mutants are globally defective in hypha-specific gene induction (Epp et al. 2010b). As increased expression of *UME6* in an *nrg1Δ/Δ arp2Δ/Δ* mutant is insufficient to restore hyphal formation, it is likely that Arp2/3 does not function in such signaling pathways. Arp2/3-mediated assembly of actin has been shown to be an essential part of endocytosis (Kaksonen et al. 2006); yet interestingly, endocytosis still occurs in *C. albicans arp2* deletion mutants (Epp et al. 2010b). This suggests that actin-driven polymerization can occur during endocytosis in *C. albicans* via Arp2/3-independent routes.

As with Arp2/3, the *C. albicans* WASP homolog Wal1 is dispensable for viability, but deletion mutants are unable to form hyphal filaments; they can, however, still initiate polarized growth, resulting in elongated pseudohyphal cells (Walther and Wendland 2004). The *wal1* mutant exhibits phenotypes similar to those observed in mutants deleted for the class I myosin Myo5, which is itself required for Arp2/3-dependent actin nucleation assembly (Oberholzer et al. 2002). Importantly, in mammals, N-WASP interacts with both the small GTPase Cdc42 and the phosphoinositide phosphate $PIP_2$ to direct Arp2/3 actin nucleation (Prehoda et al. 2000); the Cdc42 interacting domain is, however, not present in fungal WASP homologs, indicating that polarization of the actin network requires additional components. Strikingly, recent work with an in vitro actin assembly assay using microbeads functionalized with the *S. cerevisiae* WASP homolog Las17 shows that phosphoinositide kinase patches, including the phosphoinositide kinases Stt4 and Mss4, are recruited to the formed actin patches (Michelot et al. 2010). These results raise the important question of the role of phosphoinositides in *C. albicans* hyphal growth. Figure 8.3 shows a schematic representation of the main components required for *C. albicans* polarized hyphal growth and their potential connections.

Formins are large, multi-domain proteins that assemble polarized actin cables and the cytokinetic contractile ring, which is composed of unbranched actin filaments (Evangelista et al. 2002a; Oberholzer et al. 2002; Sagot et al. 2002; Goode and Eck 2007). Formins are thought to function as scaffolds that link Rho-type GTPases to components of the actin cytoskeleton. In *S. cerevisiae*, the formins (Bni1 and Bnr1) are individually not essential, but cell growth nevertheless requires the presence of at least one of them. Bni1 and Bnr1 are also required for cell polarity and cytokinesis, with some overlapping functions (Pruyne et al. 2004). In contrast, the *A. gossypii* Bni1 homolog is essential, while either of the two Bnr forms, Bnr1 and Bnr2, is dispensable for viability (Schmitz et al. 2006). *C. albicans*, like *S. cerevisiae*, has two formins, Bni1 and Bnr1, and requires one or the other for viability. Deletion of *BNR1* has little effect on *C. albicans* cell polarity. In contrast, *C. albicans bni1* deletion mutants produce multinucleate cells and random budding patterns as well as, under liquid hyphae-inducing conditions, germ tubes which grow into long, swollen hyphae (Li et al. 2005). Deletion of any one of the four

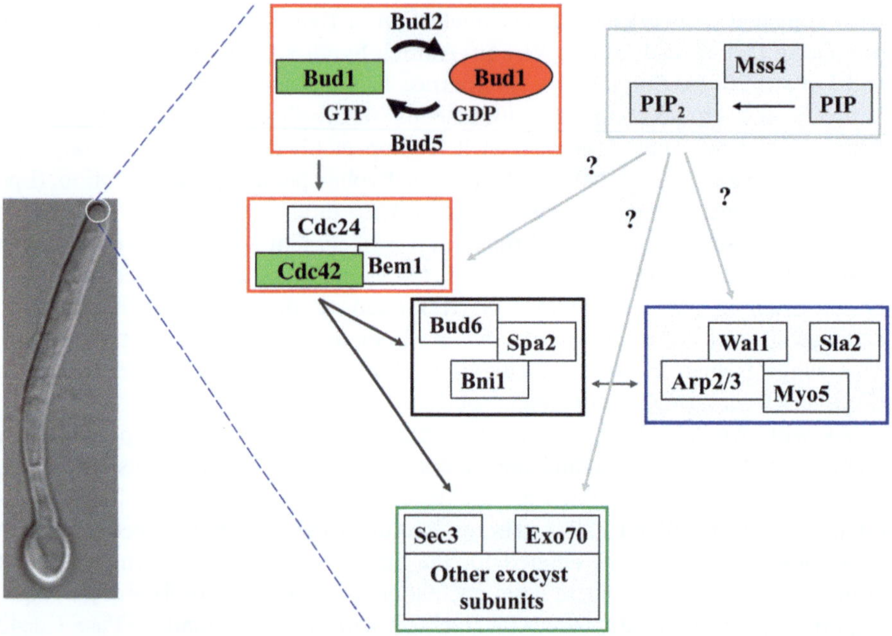


**Fig. 8.3** Components required for polarized *C. albicans* hyphal growth. *Red rectangles* indicate small G-protein modules, i.e., the Bud1/Rsr1 module and the Cdc42 module, and the *black square* denotes polarisome proteins. Proteins required for exocytosis and endocytosis are indicated in *green* and *blue rectangles*, respectively. Putative interactions between the phosphoinositide phosphate PI(4,5)$P_2$ and the relevant modules are indicated in *gray*. The number of components illustrated in this scheme is limited for clarity

coiled-coil domains of Bni1 results in the swollen hypha phenotype (Li et al. 2005), a phenotype also observed in *spa2* and *bud6* deletion mutants (Zheng et al. 2003), suggesting that the interaction of Bni1 with components of the polarisome, via coiled-coil domains, is critical for hyphal growth. At the same time, the observation that a Bnr1-GFP fusion protein localizes to the hyphal tip in a *bni1* mutant suggests that Bni1 and Bnr1 can complement each other for some functions during *C. albicans* hyphal growth (Li et al. 2005).

Finally, the conserved small G-proteins called ADP-ribosylation factors (Arfs) have also been shown to be important in actin cable and cortical patch formation in various organisms (Lambert et al. 2007); their role has not been examined in *C. albicans* or in filamentous fungi, however. Nevertheless, the ARF-GTPase activating effector protein Age3 was recently shown to be required for hyphal growth (Lettner et al. 2010) and for drug resistance and virulence (Epp et al. 2010a) in *C. albicans*.

#### 8.4.2.3 Cytoskeleton: Microtubules

The relative importance of the actin cytoskeleton and microtubules in filamentous growth varies between fungi. While microtubules are critical for the maintenance of fast hyphal growth in fungi such as *N. crassa* and *U. maydis* (Harris and Momany 2004), they are not required for hyphal extension in *A. gossypii*, as hyphae continue to elongate in the presence of nocodazole (Gladfelter 2006). In *C. albicans*, the importance of cytoplasmic microtubules for hyphal growth is less clear (Anderson and Soll 1986; Yokoyama et al. 1990; Akashi et al. 1994), and conflicting results have been obtained with different drugs to disrupt microtubules (Crampin et al. 2005; Rida et al. 2006). However, results with nocodazole suggest that microtubules do not play a prominent role in *C. albicans* hyphal elongation (Rida et al. 2006).

### *8.4.3 Septins*

In order to maintain hyphal growth, *C. albicans* cells need to block cell separation after cytokinesis. Septins, which are cytoskeletal GTP-binding proteins that can self-assemble into homo-oligomers, hetero-oligomers, and filaments, are thought to be involved in this process. They have been found throughout the eukaryotic kingdom, with the exception of plants (Pan et al. 2007; Weirich et al. 2008; Gladfelter 2010). Septins vary in number and function with cell type, and have been shown to function in cytoskeleton organization, cell division, and exocytosis. In *S. cerevisiae*, septins are cell cycle regulated via phosphorylation: they localize in G1 at the incipient bud site to form a ring, which serves as the future site of cytokinesis, and then dissociate after cytokinesis. In *A. gossypii*, septins are not essential; they initially assemble into thin filaments at growing hyphal tips, before ultimately forming stable rings at the cell cortex, a process controlled by specific kinases (DeMay et al. 2009). In *C. albicans*, homologs to all of the *S. cerevisiae* septins are present. Only *CDC3* and *CDC12* are essential; *cdc10* and *cdc11* deletion mutants are viable, but cannot maintain a unique direction necessary for proper hyphal growth (Warenda and Konopka 2002), suggesting that septins stabilize the single axis of polarity.

Septins are thus important for morphogenesis, and their localization in hyphae suggests that they play a direct role in both polarized growth and sustained hyphal growth. In *C. albicans* hyphae, septins are localized to three different sites: a diffuse band at the base of the emerging germ tube, a double ring at the septal junction, and a diffuse cap at the hyphal tip (Sudbery 2001). Elegant studies from Sinha et al. (2007) demonstrated that after hyphal induction, phosphorylation of Cdc11 by the cyclin-dependent kinase Cdc28 occurs stepwise, first in complex with Ccn1, and then with Hgc1. The Hgc1-regulated Cdc28 phosphorylation of Cdc11 thus ensures the maintenance of polarized hyphal growth by stabilizing the septin ring. Septin ring dynamics was also shown to be different in yeast and hyphal form cells. Specifically, the exchange of Cdc10 is faster in hyphae than in yeast cells (Gonzalez-Novo et al. 2008). These Cdc10 dynamics appear to be regulated by Sep7, which is itself also phosphorylated by Cdc28$^{Hgc1}$, as *sep7* hyphae have a slower Cdc10 exchange and increased cell separation in sub-apical compartments. Interestingly, deletion of the exocyst component Sec3 in either *cdc10* or *cdc11* deletion mutants essentially restores normal hyphal growth (Li et al. 2007), indicating a link between septins and polarized exocytosis.

## *8.4.4 Post-Golgi Vesicle Traffic*

### 8.4.4.1 Exocytosis

Sustained hyphal growth requires targeted exocytosis for the addition of new membrane and cell wall material to the apex. The targeting of vesicles to the site of polarized growth during filamentation is accomplished through tethering (Brown and Pfeffer 2010). Tethering requires the exocyst, a protein complex composed of eight subunits, six of which are associated with vesicles and the other two (Exo70 and Sec3) with the plasma membrane; these latter two proteins thus act as spatial landmarks for polarized secretion. In *S. cerevisiae*, both Exo70 and Sec3 interact with small Rho G-proteins: Exo70 with Cdc42 and Rho3, and Sec3 with Cdc42 and Rho1 (Wu et al. 2008). During exocytosis, while both Rho3 and Cdc42 bind the effector Exo70, recent work suggests that they have different affinities for this exocyst subunit (Wu et al. 2010). Furthermore, the respective distributions of Rho3 and Cdc42 at the membrane differ, likely resulting from posttranslational palmitoylation of Rho3 at its N-terminus (Wu and Brennwald 2010). Together, these and other results indicate that Rho3 and Cdc42 act at distinct stages during polarized growth.

In *C. albicans*, little is known about the functions of Exo70 and Rho1 in exocytosis, but those of Sec3 and Rho3 have been recently investigated, with the deletion of either gene resulting in a similar phenotype. In promoter shutdown experiments with *RHO3* under the control of the *MET3* promoter, *C. albicans* cells were able to initiate germ tube formation but could not maintain hyphal growth in the absence of *RHO3* expression: the hyphal tips rapidly began to swell, suggestive of a defect in cell polarization (Dunkler and Wendland 2007). Consistent with this result, deletion of the RhoGAP *RGD1*, which should increase the levels of GTP-bound Rho3 in vivo, results in increased filamentation (Ness et al. 2010).

The deletion of *SEC3* also results in *C. albicans* cells that are able to form germ tubes but not hyphae: after assembly of the first septin ring, cell tips swell (Li et al. 2007). Sec3 is thus dispensable for initial polarized growth, but is necessary for the maintenance of hyphal growth. Deletion of *SEC3* results in the mis-localization of other exocyst components, such as Sec15 and Sec4, as well as of Cdc42, Spa2, Bni1, and actin patches. These results suggest that although the exocyst is not required for germ tube tip localization of the polarisome, it is necessary for the persistent localization of polarisome components at the hyphal tip. Finally, the co-localization of the exocyst with septins, and the concomitant loss of polarized growth and septin ring assembly in the *sec3* mutant point toward a critical role for septins in exocyst orientation and regulation; this same scenario has also been proposed with respect to neurite outgrowth. Together, these results suggest that in elongated cells, septins play a critical role in defining or differentiating the tip or apex of the cell.

Different lines of evidence point to the potential interaction of exocyst components with phospholipids. For example, the recently determined structure of the N-terminal region of Sec3 revealed the presence of a cryptic PH domain (Baek et al. 2010; Yamashita et al. 2010), and indeed this region of Sec3 has been shown to bind the phosphatidylinositol 4,5-bisphosphate $PI(4,5)P_2$ (Zhang et al. 2008). It has also been shown that Exo70 has a phospholipid-binding site that can bind $PI(4,5)P_2$ (Ory and Gasman 2011). It would be interesting to investigate the importance of these Exo70 and Sec3 interactions with phosphoinositide phosphates such as $PIP_2$ in *C. albicans*. While the importance of phosphoinositide phosphates has been established in mammals and yeast (Strahl and Thorner 2007; Vicinanza et al. 2008; Yakir-Tamang and Gerst 2009; Saarikangas et al. 2010), little is known about their role in hyphal growth. Nevertheless, in *C. albicans*, null mutants of a PI (5)P-phosphatase, Inp51, are defective in invasive growth on solid media (Badrane et al. 2008), and the mRNA encoding the PI4P 5-kinase Mss4 was present among the 40 mRNAs found localized to the apex in a She3-dependent fashion during hyphal growth (Elson et al. 2009).

#### 8.4.4.2 Spitzenkörper

During hyphal growth, the Spitzenkörper (literally, "apical tip body"), an electron dense structure (Grove and Bracker 1970), is thought to act as a supply center of vesicles to the hyphal tip (Harold 2002; Harris et al. 2005; Virag and Harris 2006; Steinberg 2007). This structure, which is specific to filamentous fungi, can be observed by FM4-64 membrane staining as a sub-apical spot and appears to be dependent on both the actin and microtubule cytoskeletons (Crampin et al. 2005). The class V myosin Myo2 and its regulatory light chain Mlc1 both localize to this structure, as do the Rab GTPase Sec4 (Jones and Sudbery 2010) and its GEF Sec2 (Bishop et al. 2010). *F*luorescence *r*ecovery *a*fter *p*hotobleaching (FRAP) and *f*luorescence *l*oss *i*n *p*hotobleaching (FLIP) studies in *C. albicans* suggest that the Spitzenkörper is a more dynamic structure than other complexes at the hyphal tip such as the polarisome or the exocyst (Jones and Sudbery 2010), which is consistent with the notion of a vesicle supply center.

In *A. gossypii*, specific polarisome components (Spa2, Pea2, and Bni1) accumulate in the Spitzenkörper, while others (Bud6 and Cdc42) are restricted to the cortex (Kohli et al. 2008). Interestingly, the distribution of actin patches is also different between slow and fast growing hyphae. The finding that a Spitzenkörper is not present at slow growth speeds in *A. gossypii* (Kohli et al. 2008) suggests, however, that this structure is not necessary for hyphal growth per se, but rather represents a buildup of excess vesicles.

#### 8.4.4.3 Endocytosis

Endocytosis is also important for polarized growth in filamentous fungi (Penalva 2010) and in *C. albicans*, in which mutations affecting the endocytic pathway impair hyphal growth. Rvs proteins, which contain *B*in-*A*mphiphysin-*R*vs (BAR) homology domains, have been shown in other organisms to sense and promote membrane curvature. Consistent with this, Rvs161 appears to be involved in both *C. albicans* endocytosis (using FM4-64 to visualize this process) and hyphal growth in response to serum, especially on solid media (Douglas et al. 2009). Two other genes linking endocytosis and filamentation are *MYO5*, a key component of the endocytic internalization machinery, and *SLA2*, which encodes a component of the actin cortical patch. Unlike in *S. cerevisiae*, neither of these genes is essential in *C. albicans*, although they are both required for hyphal growth (Oberholzer et al. 2002). As mentioned earlier, *arp2/3* deletion mutants, while clearly delayed in endocytosis, can still endocytose, as assessed by FM4-64 or Lucifer Yellow uptake (Epp et al. 2010b), indicating that endocytosis does not strictly require the Arp2/3 complex in *C. albicans*.

The vacuole is the final destination of the endocytic pathway, and vacuolar protein sorting (Vps) factors act during the final steps of endocytosis. In *S. cerevisiae*, two distinct trafficking routes deliver proteins from the Golgi to the vacuole (1) via the late endosome, which is dependent upon the Rab GTPase Vps21p, and (2) via a distinct set of vesicles that bypass the late endosome. In *C. albicans*, both pathways contribute to polarized hyphal growth (Palmer 2010). Specifically, deletion of the endocytic components *VPS28* and *VPS32*, which are components of the ESCRT-I and ESCRT-III complexes, respectively, causes defects in alkaline-induced hyphal formation (Cornet et al. 2005). Finally, Abg1, a protein resident in vacuolar membranes, also appears to be involved in endocytosis (Veses et al. 2009). Given the large changes that take place in the plasma membrane surface during hyphal growth, it is not hard to imagine that both endocytosis and exocytosis are critical for this type of growth. Nonetheless, how these processes are regulated will require further studies.

### *8.4.5 Organelle Repositioning*

The cell morphology changes that take place during hyphal growth require the repositioning of organelles such as the nucleus and Golgi apparatus. Nuclear positioning differs greatly between fungal species (Gladfelter and Berman 2009). In *C. albicans*, nuclear positioning also varies between the different cell forms, as the site of nuclear division and septation is different between budding and hyphal cells. In hyphae, nuclei migrate to the presumptive site of septation, about 15 μm into the growing germ tube, before division (Finley and Berman 2005) and subsequent return of one nucleus to the basal cell. In contrast, during yeast and pseudohyphal growth, nuclei divide across the bud neck. Nuclear dynamics in hyphae require microtubules, and the dynein–dynactin motor complex is necessary for efficient nuclear migration and proper hyphal growth (Martin et al. 2004; Finley et al. 2008).

In *C. albicans* hyphal cells, vacuoles are also asymmetrically distributed, with larger vacuoles found in the sub-apical region (Barelle et al. 2003; Veses and Gow 2008). Vacuole dynamics is linked to hyphal growth and branch formation. Mutants with defects in vacuole biogenesis, such as the Vps mutant *vps11*, fail to form hyphae (Palmer et al. 2003), while mutants with defects in vacuole inheritance, such as *vac8*, exhibit abnormal hyphal growth and branching (Barelle et al. 2006; Veses et al. 2009). In contrast, mutants with defects in vacuole fusion, such as those generated by deletion of the SH3 domain-containing Boi2, form hyphae despite having fragmented vacuoles (Reijnst et al. 2010). Taken together, these results suggest that the cytoplasmic-to-vacuolar volume ratio is important for hyphal growth.

In addition, the majority of the Golgi complex is redistributed during *C. albicans* hyphal formation and is maintained at the distal portion of hyphae, as observed using epitope-tagged CaVrg4 (Rida et al. 2006). This redistribution of the Golgi at the distal portion of hyphae is dependent on the actin cable-nucleating formin Bni1 and is independent of microtubules (Rida et al. 2006). In contrast, both the endoplasmic reticulum, as visualized by an HDEL-tagged Kar2-GFP fusion protein, and mitochondria, as visualized by DiOC6 staining, appear to be randomly distributed in *C. albicans* (Rida et al. 2006). In certain filamentous fungi, however, this is not the case: the endoplasmic reticulum accumulates apically in *U. maydis* (Wedlich-Soldner et al. 2002), and in *N. crassa*, mitochondria are localized to the hyphal tip, although this localization does not appear to be strictly required for hyphal growth (Levina and Lew 2006). These observations could indicate that long-range transport between the ER and Golgi occurs in *C. albicans* hyphal filaments, while the clustering of Golgi at the distal part of the hyphae may be necessary for the efficient addition of new membrane and cell wall material to the apex. While the distribution of these organelles during hyphal growth is important, it will be even more critical to understand their movements and dynamics during this growth process.

## 8.5 Conclusions

Regulation of the reversible transition from budding to filamentous growth in *C. albicans*, together with the maintenance of polarized growth in hyphae, is important for host–pathogen interactions. Although a number of components required for *C. albicans* hyphal growth have now been identified, the molecular mechanisms underlying filamentous growth are still poorly understood. While polarized growth has been well studied at the molecular level in the yeast *S. cerevisiae*, hyphal growth is a different type of polarized growth that is characteristic of filamentous fungi but absent in baker's yeast. As a result, it is difficult to extrapolate directly from the mechanisms that are important for polarized growth in *S. cerevisiae*, including the molecular interactions that take place between different components and the cross-talk that occurs between different pathways.

A range of molecular tools have become available in *C. albicans* over the last 5–10 years which have allowed the identification of key components required for filamentous growth. The interactions that occur between these components and the different functional modules during the initiation and maintenance of hyphal polarized growth, however, remain largely unknown. While many of the components of signaling pathways required for the yeast-to-hyphal transition have been identified, we still lack an understanding of their spatiotemporal activities. For example, when, where, and for how long are different kinases or G-proteins activated during such morphologic changes? To obtain precise mechanistic information and enable modeling, it will be essential to develop fluorescent reporters to allow the monitoring of such processes in individual live cells and to quantify signals from these reporters during hyphal growth induced by different stimuli. In addition, improvements in fluorescent microscopy techniques should now permit in-depth analyses of the distribution of organelles during hyphal growth in live cells. Finally, the reverse transition from filamentous growth to budding yeast, although not discussed here, is equally important for understanding the regulation of filamentous growth and should also be further investigated.

In conclusion, numerous studies from a large number of laboratories have identified many proteins involved in *C. albicans* filamentous growth in response to diverse stimuli and/or virulence assay conditions. It is now critical that quantitative approaches and global analyses be carried out to further elucidate the network of molecular events that take place during these dramatic morphologic changes.

**Acknowledgments** We wish to thank Robert Arkowitz for stimulating discussions and critical reading of the manuscript. MB was supported by the Centre National de la Recherche Scientifique and PF by the Université de Nice-Sophia Antipolis. This work was supported by funding from FRM-BNP Paribas and from the French National Research Agency (Grant ANR-09-BLAN-0299-01).

## References

Akashi T, Kanbe T, Tanaka K (1994) The role of the cytoskeleton in the polarized growth of the germ tube in *Candida albicans*. Microbiology 140:271–280

Alby K, Schaefer D, Bennett RJ (2009) Homothallic and heterothallic mating in the opportunistic pathogen *Candida albicans*. Nature 460:890–893

Alvarez-Tabares I, Perez-Martin J (2008) Cdk5 kinase regulates the association between adaptor protein Bem1 and GEF Cdc24 in the fungus *Ustilago maydis*. J Cell Sci 121:2824–2832

Amberg DC, Zahner JE, Mulholland JW, Pringle JR, Botstein D (1997) Aip3p/Bud6p, a yeast actin-interacting protein that is involved in morphogenesis and the selection of bipolar budding sites. Mol Biol Cell 8:729–753

Anderson JM, Soll DR (1986) Differences in actin localization during bud and hypha formation in the yeast *Candida albicans*. J Gen Microbiol 132:2035–2047

Arkowitz RA (2009) Chemical gradients and chemotropism in yeast. Cold Spring Harb Perspect Biol 1:a001958

Bachewich C, Whiteway M (2005) Cyclin Cln3p links G1 progression to hyphal and pseudohyphal development in *Candida albicans*. Eukaryot Cell 4:95–102

Badrane H, Nguyen MH, Cheng S, Kumar V, Derendorf H, Iczkowski KA, Clancy CJ (2008) The *Candida albicans* phosphatase Inp51p interacts with the EH domain protein Irs4p, regulates phosphatidylinositol-4,5-bisphosphate levels and influences hyphal formation, the cell integrity pathway and virulence. Microbiology 154:3296–3308

Baek K, Knodler A, Lee SH, Zhang X, Orlando K, Zhang J, Foskett TJ, Guo W, Dominguez R (2010) Structure-function study of the N-terminal domain of exocyst subunit Sec3. J Biol Chem 285:10424–10433

Banerjee M, Thompson DS, Lazzell A, Carlisle PL, Pierce C, Monteagudo C, Lopez-Ribot JL, Kadosh D (2008) UME6, a novel filament-specific regulator of *Candida albicans* hyphal extension and virulence. Mol Biol Cell 19:1354–1365

Bardwell L (2005) A walk-through of the yeast mating pheromone response pathway. Peptides 26:339–350

Barelle CJ, Bohula EA, Kron SJ, Wessels D, Soll DR, Schafer A, Brown AJ, Gow NA (2003) Asynchronous cell cycle and asymmetric vacuolar inheritance in true hyphae of *Candida albicans*. Eukaryot Cell 2:398–410

Barelle CJ, Richard ML, Gaillardin C, Gow NA, Brown AJ (2006) *Candida albicans* VAC8 is required for vacuolar inheritance and normal hyphal branching. Eukaryot Cell 5:359–367

Bassilana M, Arkowitz RA (2006) Rac1 and Cdc42 have different roles in *Candida albicans* development. Eukaryot Cell 5:321–329

Bassilana M, Blyth J, Arkowitz RA (2003) Cdc24, the GDP-GTP exchange factor for Cdc42, is required for invasive hyphal growth of *Candida albicans*. Eukaryot Cell 2:9–18

Bassilana M, Hopkins J, Arkowitz RA (2005) Regulation of the Cdc42/Cdc24 GTPase module during *Candida albicans* hyphal growth. Eukaryot Cell 4:588–603

Bauer Y, Knechtle P, Wendland J, Helfer H, Philippsen P (2004) A Ras-like GTPase is involved in hyphal growth guidance in the filamentous fungus *Ashbya gossypii*. Mol Biol Cell 15:4622–4632

Bennett RJ, Johnson AD (2003) Completion of a parasexual cycle in *Candida albicans* by induced chromosome loss in tetraploid strains. EMBO J 22:2505–2515

Bensen ES, Clemente-Blanco A, Finley KR, Correa-Bordes J, Berman J (2005) The mitotic cyclins Clb2p and Clb4p affect morphogenesis in *Candida albicans*. Mol Biol Cell 16:3387–3400

Berman J (2006) Morphogenesis and cell cycle progression in *Candida albicans*. Curr Opin Microbiol 9:595–601

Bishop A, Lane R, Beniston R, Chapa-y-Lazo B, Smythe C, Sudbery P (2010) Hyphal growth in *Candida albicans* requires the phosphorylation of Sec2 by the Cdc28-Ccn1/Hgc1 kinase. EMBO J 29:2930–2942

Biswas S, Van Dijck P, Datta A (2007) Environmental sensing and signal transduction pathways regulating morphopathogenic determinants of *Candida albicans*. Microbiol Mol Biol Rev 71:348–376

Blankenship JR, Mitchell AP (2006) How to build a biofilm: a fungal perspective. Curr Opin Microbiol 9:588–594

Brand A, Vacharaksa A, Bendel C, Norton J, Haynes P, Henry-Stanley M, Wells C, Ross K, Gow NA, Gale CA (2008) An internal polarity landmark is important for externally induced hyphal behaviors in *Candida albicans*. Eukaryot Cell 7:712–720

Braun BR, Johnson AD (1997) Control of filament formation in *Candida albicans* by the transcriptional repressor TUP1. Science 277:105–109

Braun BR, Kadosh D, Johnson AD (2001) NRG1, a repressor of filamentous growth in *C. albicans*, is down-regulated during filament induction. EMBO J 20:4753–4761

Brown FC, Pfeffer SR (2010) An update on transport vesicle tethering. Mol Membr Biol 27:457–461

Campas O, Mahadevan L (2009) Shape and dynamics of tip-growing cells. Curr Biol 19:2102–2107

Cao F, Lane S, Raniga PP, Lu Y, Zhou Z, Ramon K, Chen J, Liu H (2006) The Flo8 transcription factor is essential for hyphal development and virulence in *Candida albicans*. Mol Biol Cell 17:295–307

Carlisle PL, Banerjee M, Lazzell A, Monteagudo C, López-Ribot JL, Kadosh D (2009) Expression levels of a filament-specific transcriptional regulator are sufficient to determine *Candida albicans* morphology and virulence. Proc Natl Acad Sci USA 106:599–604

Carlisle PL, Kadosh D (2010) *Candida albicans* Ume6, a filament-specific transcriptional regulator, directs hyphal growth via a pathway involving Hgc1 cyclin-related protein. Eukaryot Cell 9:1320–1328

Chao CC, Hsu PC, Jen CF, Chen IH, Wang CH, Chan HC, Tsai PW, Tung KC, Lan CY, Chuang YJ (2010) Zebrafish as a model host for *Candida albicans* infection. Infect Immun 78:2512–2521

Chapa Y, Lazo B, Bates S, Sudbery P (2005) The G1 cyclin Cln3 regulates morphogenesis in *Candida albicans*. Eukaryot Cell 4:90–94

Chenevert J, Corrado K, Bender A, Pringle J, Herskowitz I (1992) A yeast gene (BEM1) necessary for cell polarization whose product contains two SH3 domains. Nature 356:77–79

Cleary IA, Mulabagal P, Reinhard SM, Yadev NP, Murdoch C, Thornhill MH, Lazzell AL, Monteagudo C, Thomas DP, Saville SP (2010) Pseudohyphal regulation by the transcription factor Rfg1p in *Candida albicans*. Eukaryot Cell 9:1363–1373

Cornet M, Bidard F, Schwarz P, Da Costa G, Blanchin-Roland S, Dromer F, Gaillardin C (2005) Deletions of endocytic components VPS28 and VPS32 affect growth at alkaline pH and virulence through both RIM101-dependent and RIM101-independent pathways in *Candida albicans*. Infect Immun 73:7977–7987

Cottier F, Muhlschlegel FA (2009) Sensing the environment: response of *Candida albicans* to the X factor. FEMS Microbiol Lett 295:1–9

Court H, P Sudbery (2007) Regulation of Cdc42 GTPase activity in the formation of hyphae in *Candida albicans*. Mol Biol Cell 18:265–281

Crampin H, Finley K, Gerami-Nejad M, Court H, Gale C, Berman J, Sudbery P (2005) *Candida albicans* hyphae have a Spitzenkorper that is distinct from the polarisome found in yeast and pseudohyphae. J Cell Sci 118:2935–2947

Davis DA (2009) How human pathogenic fungi sense and adapt to pH: the link to virulence. Curr Opin Microbiol 12:365–370

Davis D, Wilson RB, Mitchell AP (2000) RIM101-dependent and-independent pathways govern pH responses in *Candida albicans*. Mol Cell Biol 20:971–978

DeMay BS, Meseroll RA, Occhipinti P, Gladfelter AS (2009) Regulation of distinct septin rings in a single cell by Elm1p and Gin4p kinases. Mol Biol Cell 20:2311–2326

Dhillon NK, Sharma S, Khuller GK (2003) Signaling through protein kinases and transcriptional regulators in *Candida albicans*. Crit Rev Microbiol 29:259–275

Doedt T, Krishnamurthy S, Bockmuhl DP, Tebarth B, Stempel C, Russell CL, Brown AJ, Ernst JF (2004) APSES proteins regulate morphogenesis and metabolism in *Candida albicans*. Mol Biol Cell 15:3167–3180

Douglas LM, Martin SW, Konopka JB (2009) BAR domain proteins Rvs161 and Rvs167 contribute to *Candida albicans* endocytosis, morphogenesis, and virulence. Infect Immun 77:4150–4160

Dunkler A, Wendland J (2007) *Candida albicans* Rho-type GTPase-encoding genes required for polarized cell growth and cell separation. Eukaryot Cell 6:844–854

Elson SL, Noble SM, Solis NV, Filler SG, Johnson AD (2009) An RNA transport system in *Candida albicans* regulates hyphal morphology and invasive growth. PLoS Genet 5:e1000664

Epp E, Vanier G, Harcus D, Lee AY, Jansen G, Hallett M, Sheppard DC, Thomas DY, Munro CA, Mullick A, Whiteway M (2010a) Reverse genetics in *Candida albicans* predicts ARF cycling is essential for drug resistance and virulence. PLoS Pathog 6:e1000753

Epp E, Walther A, Lepine G, Leon Z, Mullick A, Raymond M, Wendland J, Whiteway M (2010b) Forward genetics in *Candida albicans* that reveals the Arp2/3 complex is required for hyphal formation, but not endocytosis. Mol Microbiol 75:1182–1198

Ernst JF (2000) Transcription factors in *Candida albicans* – environmental control of morphogenesis. Microbiology 146:1763–1774

Evangelista M, Blundell K, Longtine MS, Chow CJ, Adames N, Pringle JR, Peter M, Boone C (1997) Bni1p, a yeast formin linking cdc42p and the actin cytoskeleton during polarized morphogenesis. Science 276:118–122

Evangelista M, Pruyne D, Amberg DC, Boone C, Bretscher A (2002a) Formins direct Arp2/3-independent actin filament assembly to polarize cell growth in yeast. Nat Cell Biol 4:32–41

Evangelista M, Pruyne D, Amberg DC, Boone C, Bretscher A (2002b) Formins direct Arp2/3-independent actin filament assembly to polarize cell growth in yeast. Nat Cell Biol 4:260–269

Finley KR, Berman J (2005) Microtubules in *Candida albicans* hyphae drive nuclear dynamics and connect cell cycle progression to morphogenesis. Eukaryot Cell 4:1697–1711

Finley KR, Bouchonville KJ, Quick A, Berman J (2008) Dynein-dependent nuclear dynamics affect morphogenesis in *Candida albicans* by means of the Bub2p spindle checkpoint. J Cell Sci 121:466–476

Fitzpatrick DA, Logue ME, Stajich JE, Butler G (2006) A fungal phylogeny based on 42 complete genomes derived from supertree and combined gene analysis. BMC Evol Biol 6:99

Forche A, Alby K, Schaefer D, Johnson AD, Berman J, Bennett RJ (2008) The parasexual cycle in *Candida albicans* provides an alternative pathway to meiosis for the formation of recombinant strains. PLoS Biol 6:e110

Fujiwara T, Tanaka K, Mino A, Kikyo M, Takahashi K, Shimizu K, Takai Y (1998) Rho1p-Bni1p-Spa2p interactions: implication in localization of Bni1p at the bud site and regulation of the actin cytoskeleton in *Saccharomyces cerevisiae*. Mol Biol Cell 9:1221–1233

Gao L, Bretscher A (2009) Polarized growth in budding yeast in the absence of a localized formin. Mol Biol Cell 20:2540–2548

Gehrung S, Snyder M (1990) The SPA2 gene of *Saccharomyces cerevisiae* is important for pheromone-induced morphogenesis and efficient mating. J Cell Biol 111:1451–1464

Gladfelter AS (2006) Nuclear anarchy: asynchronous mitosis in multinucleated fungal hyphae. Curr Opin Microbiol 9:547–552

Gladfelter AS (2010) Guides to the final frontier of the cytoskeleton: septins in filamentous fungi. Curr Opin Microbiol 13:720–726

Gladfelter A, Berman J (2009) Dancing genomes: fungal nuclear positioning. Nat Rev Microbiol 7:875–886

Gonzalez-Novo A, Correa-Bordes J, Labrador L, Sanchez M, Vazquez de Aldana CR, Jimenez J (2008) Sep7 is essential to modify septin ring dynamics and inhibit cell separation during *Candida albicans* hyphal growth. Mol Biol Cell 19:1509–1518

Goode BL, Eck MJ (2007) Mechanism and function of formins in the control of actin assembly. Annu Rev Biochem 76:593–627

Gow NA (2002) *Candida albicans* switches mates. Mol Cell 10:217–218

Grove SN, Bracker CE (1970) Protoplasmic organization of hyphal tips among fungi: vesicles and Spitzenkorper. J Bacteriol 104:989–1009

Harold FM (2002) Force and compliance: rethinking morphogenesis in walled cells. Fungal Genet Biol 37:271–282

Harris SD (2011) Cdc42/Rho GTPases in fungi: variations on a common theme. Mol Microbiol 79:1123–1127
Harris SD, Momany M (2004) Polarity in filamentous fungi: moving beyond the yeast paradigm. Fungal Genet Biol 41:391–400
Harris SD, Read ND, Roberson RW, Shaw B, Seiler S, Plamann M, Momany M (2005) Polarisome meets spitzenkorper: microscopy, genetics, and genomics converge. Eukaryot Cell 4:225–229
Hausauer DL, Gerami-Nejad M, Kistler-Anderson C, Gale CA (2005) Hyphal guidance and invasive growth in *Candida albicans* require the Ras-like GTPase Rsr1p and its GTPase-activating protein Bud2p. Eukaryot Cell 4:1273–1286
Hazan I, Liu H (2002) Hyphal tip-associated localization of Cdc42 is F-actin dependent in *Candida albicans*. Eukaryot Cell 1:856–864
Hazan I, Sepulveda-Becerra M, Liu H (2002) Hyphal elongation is regulated independently of cell cycle in *Candida albicans*. Mol Biol Cell 13:134–145
Heasman SJ, Ridley AJ (2008) Mammalian Rho GTPases: new insights into their functions from in vivo studies. Nat Rev Mol Cell Biol 9:690–701
Hedges SB (2002) The origin and evolution of model organisms. Nat Rev Genet 3:838–849
Higgs HN, Pollard TD (2001) Regulation of actin filament network formation through ARP2/3 complex: activation by a diverse array of proteins. Annu Rev Biochem 70:649–676
Hogan DA, Sundstrom P (2009) The Ras/cAMP/PKA signaling pathway and virulence in *Candida albicans*. Future Microbiol 4:1263–1270
Hope H, Bogliolo S, Arkowitz RA, Bassilana M (2008) Activation of Rac1 by the guanine nucleotide exchange factor Dck1 is required for invasive filamentous growth in the pathogen *Candida albicans*. Mol Biol Cell 19:3638–3651
Hope H, Schmauch C, Arkowitz RA, Bassilana M (2010) The *Candida albicans* ELMO homologue functions together with Rac1 and Dck1, upstream of the MAP Kinase Cek1, in invasive filamentous growth. Mol Microbiol 76:1572–1590
Horio T (2007) Role of microtubules in tip growth of fungi. J Plant Res 120:53–60
Howell AS, Savage NS, Johnson SA, Bose I, Wagner AW, Zyla TR, Nijhout HF, Reed MC, Goryachev AB, Lew DJ (2009) Singularity in polarization: rewiring yeast cells to make two buds. Cell 139:731–743
Hull CM, Raisner RM, Johnson AD (2000) Evidence for mating of the "asexual" yeast *Candida albicans* in a mammalian host. Science 289:307–310
Hurtado CA, Beckerich JM, Gaillardin C, Rachubinski RA (2000) A rac homolog is required for induction of hyphal growth in the dimorphic yeast *Yarrowia lipolytica*. J Bacteriol 182:2376–2386
Jarvensivu A, Hietanen J, Rautemaa R, Sorsa T, Richardson M (2004) Candida yeasts in chronic periodontitis tissues and subgingival microbial biofilms in vivo. Oral Dis 10:106–112
Johnson DI (1999) Cdc42: an essential Rho-type GTPase controlling eukaryotic cell polarity. Microbiol Mol Biol Rev 63:54–105
Jones LA, Sudbery PE (2010) Spitzenkorper, exocyst, and polarisome components in *Candida albicans* hyphae show different patterns of localization and have distinct dynamic properties. Eukaryot Cell 9:1455–1465
Kadosh D, Johnson AD (2001) Rfg1, a protein related to the *Saccharomyces cerevisiae* hypoxic regulator Rox1, controls filamentous growth and virulence in *Candida albicans*. Mol Cell Biol 21:2496–2505
Kadosh D, Johnson AD (2005) Induction of the *Candida albicans* filamentous growth program by relief of transcriptional repression: a genome-wide analysis. Mol Biol Cell 16:2903–2912
Kaksonen M, Toret CP, Drubin DG (2006) Harnessing actin dynamics for clathrin-mediated endocytosis. Nat Rev Mol Cell Biol 7:404–414
Kelly MT, MacCallum DM, Clancy SD, Odds FC, Brown AJ, Butler G (2004) The *Candida albicans* CaACE2 gene affects morphogenesis, adherence and virulence. Mol Microbiol 53:969–983

Khalaf RA, Zitomer RS (2001) The DNA binding protein Rfg1 is a repressor of filamentation in *Candida albicans*. Genetics 157:1503–1512

Kim MJ, Kil M, Jung JH, Kim J (2008) Roles of Zinc-responsive transcription factor Csr1 in filamentous growth of the pathogenic Yeast *Candida albicans*. J Microbiol Biotechnol 18:242–247

Knechtle P, Dietrich F, Philippsen P (2003) Maximal polar growth potential depends on the polarisome component AgSpa2 in the filamentous fungus *Ashbya gossypii*. Mol Biol Cell 14:4140–4154

Knechtle P, Wendland J, Philippsen P (2006) The SH3/PH domain protein AgBoi1/2 collaborates with the Rho-type GTPase AgRho3 to prevent nonpolar growth at hyphal tips of *Ashbya gossypii*. Eukaryot Cell 5:1635–1647

Kohli M, Galati V, Boudier K, Roberson RW, Philippsen P (2008) Growth-speed-correlated localization of exocyst and polarisome components in growth zones of *Ashbya gossypii* hyphal tips. J Cell Sci 121:3878–3889

Komander D, Patel M, Laurin M, Fradet N, Pelletier A, Barford D, Cote JF (2008) An alpha-helical extension of the ELMO1 pleckstrin homology domain mediates direct interaction to DOCK180 and is critical in Rac signaling. Mol Biol Cell 19:4837–4851

Kozminski KG, Beven L, Angerman E, Tong AH, Boone C, Park HO (2003) Interaction between a Ras and a Rho GTPase couples selection of a growth site to the development of cell polarity in yeast. Mol Biol Cell 14:4958–4970

Kumamoto CA (2008) Niche-specific gene expression during *C. albicans* infection. Curr Opin Microbiol 11:325–330

Kumamoto CA, Vinces MD (2005) Alternative *Candida albicans* lifestyles: growth on surfaces. Annu Rev Microbiol 59:113–133

Kwon MJ, Arentshorst M, Roos ED, van den Hondel CA, Meyer V, Ram AF (2011) Functional characterization of Rho GTPases in *Aspergillus niger* uncovers conserved and diverged roles of Rho proteins within filamentous fungi. Mol Microbiol 79:1151–1167

Lambert AA, Perron MP, Lavoie E, Pallotta D (2007) The *Saccharomyces cerevisiae* Arf3 protein is involved in actin cable and cortical patch formation. FEMS Yeast Res 7:782–795

Lettner T, Zeidler U, Gimona M, Hauser M, Breitenbach M, Bito A (2010) *Candida albicans* AGE3, the ortholog of the *S. cerevisiae* ARF-GAP-encoding gene GCS1, is required for hyphal growth and drug resistance. PLoS One 5:e11993

Levin DE (2005) Cell wall integrity signaling in *Saccharomyces cerevisiae*. Microbiol Mol Biol Rev 69:262–291

Levina NN, Lew RR (2006) The role of tip-localized mitochondria in hyphal growth. Fungal Genet Biol 43:65–74

Li CR, Wang YM, De Zheng X, Liang HY, Tang JC, Wang Y (2005) The formin family protein CaBni1p has a role in cell polarity control during both yeast and hyphal growth in *Candida albicans*. J Cell Sci 118:2637–2648

Li CR, Lee RT, Wang YM, Zheng XD, Wang Y (2007) *Candida albicans* hyphal morphogenesis occurs in Sec3p-independent and Sec3p-dependent phases separated by septin ring formation. J Cell Sci 120:1898–1907

Liu H (2001) Transcriptional control of dimorphism in *Candida albicans*. Curr Opin Microbiol 4:728–735

Liu H, Kohler J, Fink GR (1994) Suppression of hyphal formation in *Candida albicans* by mutation of a STE12 homolog. Science 266:1723–1726

Lo HJ, Kohler JR, DiDomenico B, Loebenberg D, Cacciapuoti A, Fink GR (1997) Nonfilamentous *C. albicans* mutants are avirulent. Cell 90:939–949

Loeb JD, Sepulveda-Becerra M, Hazan I, Liu H (1999) A G1 cyclin is necessary for maintenance of filamentous growth in *Candida albicans*. Mol Cell Biol 19:4019–4027

Lohse MB, Johnson AD (2010) Temporal anatomy of an epigenetic switch in cell programming: the white-opaque transition of *C. albicans*. Mol Microbiol 78:331–343

Machesky LM, Gould KL (1999) The Arp2/3 complex: a multifunctional actin organizer. Curr Opin Cell Biol 11:117–121

Magee BB, Magee PT (2000) Induction of mating in *Candida albicans* by construction of MTLa and MTLalpha strains. Science 289:310–313

Mahlert M, Leveleki L, Hlubek A, Sandrock B, Bolker M (2006) Rac1 and Cdc42 regulate hyphal growth and cytokinesis in the dimorphic fungus *Ustilago maydis*. Mol Microbiol 59:567–578

Martin R, Walther A, Wendland J (2004) Deletion of the dynein heavy-chain gene DYN1 leads to aberrant nuclear positioning and defective hyphal development in *Candida albicans*. Eukaryot Cell 3:1574–1588

Massey SE, Moura G, Beltrao P, Almeida R, Garey JR, Tuite MF, Santos MA (2003) Comparative evolutionary genomics unveils the molecular mechanism of reassignment of the CTG codon in Candida spp. Genome Res 13:544–557

Meller N, Merlot S, Guda C (2005) CZH proteins: a new family of Rho-GEFs. J Cell Sci 118:4937–4946

Michel S, Ushinsky S, Klebl B, Leberer E, Thomas D, Whiteway M, Morschhauser J (2002) Generation of conditional lethal *Candida albicans* mutants by inducible deletion of essential genes. Mol Microbiol 46:269–280

Michelot A, Costanzo M, Sarkeshik A, Boone C, Yates JR III, Drubin DG (2010) Reconstitution and protein composition analysis of endocytic actin patches. Curr Biol 20:1890–1899

Miller MG, Johnson AD (2002) White-opaque switching in *Candida albicans* is controlled by mating-type locus homeodomain proteins and allows efficient mating. Cell 110:293–302

Mitra S, Dolan K, Foster TH, Wellington M (2010) Imaging morphogenesis of *Candida albicans* during infection in a live animal. J Biomed Opt 15:010504

Monge RA, Roman E, Nombela C, Pla J (2006) The MAP kinase signal transduction network in *Candida albicans*. Microbiology 152:905–912

Morschhauser J (2010) Regulation of white-opaque switching in *Candida albicans*. Med Microbiol Immunol 199:165–172

Moseley JB, Goode BL (2006) The yeast actin cytoskeleton: from cellular function to biochemical mechanism. Microbiol Mol Biol Rev 70:605–645

Mulhern SM, Logue ME, Butler G (2006) *Candida albicans* transcription factor Ace2 regulates metabolism and is required for filamentation in hypoxic conditions. Eukaryot Cell 5:2001–2013

Murad AM, Leng P, Straffon M, Wishart J, Macaskill S, MacCallum D, Schnell N, Talibi D, Marechal D, Tekaia F, d'Enfert C, Gaillardin C, Odds FC, Brown AJ (2001) NRG1 represses yeast-hypha morphogenesis and hypha-specific gene expression in *Candida albicans*. EMBO J 20:4742–4752

Naglik JR, Challacombe SJ, Hube B (2003) *Candida albicans* secreted aspartyl proteinases in virulence and pathogenesis. Microbiol Mol Biol Rev 67:400–428, table of contents

Naglik J, Albrecht A, Bader O, Hube B (2004) *Candida albicans* proteinases and host/pathogen interactions. Cell Microbiol 6:915–926

Nern A, Arkowitz RA (2000) G proteins mediate changes in cell shape by stabilizing the axis of polarity. Mol Cell 5:853–864

Ness F, Prouzet-Mauleon V, Vieillemard A, Lefebvre F, Noel T, Crouzet M, Doignon F, Thoraval D (2010) The *Candida albicans* Rgd1 is a RhoGAP protein involved in the control of filamentous growth. Fungal Genet Biol 47:1001–1011

Nobile CJ, Mitchell AP (2006) Genetics and genomics of *Candida albicans* biofilm formation. Cell Microbiol 8:1382–1391

Noble SM, French S, Kohn LA, Chen V, Johnson AD (2010) Systematic screens of a *Candida albicans* homozygous deletion library decouple morphogenetic switching and pathogenicity. Nat Genet 42:590–598

Oberholzer U, Marcil A, Leberer E, Thomas DY, Whiteway M (2002) Myosin I is required for hypha formation in *Candida albicans*. Eukaryot Cell 1:213–228

Ory S, Gasman S (2011) Rho GTPases and exocytosis: what are the molecular links? Semin Cell Dev Biol 22:27–32
Palmer GE (2010) Endosomal and AP-3-dependent vacuolar trafficking routes make additive contributions to *Candida albicans* hyphal growth and pathogenesis. Eukaryot Cell 9:1755–1765
Palmer GE, Cashmore A, Sturtevant J (2003) *Candida albicans* VPS11 is required for vacuole biogenesis and germ tube formation. Eukaryot Cell 2:411–421
Pan F, Malmberg RL, Momany M (2007) Analysis of septins across kingdoms reveals orthology and new motifs. BMC Evol Biol 7:103
Park HO, Bi E (2007) Central roles of small GTPases in the development of cell polarity in yeast and beyond. Microbiol Mol Biol Rev 71:48–96
Penalva MA (2010) Endocytosis in filamentous fungi: Cinderella gets her reward. Curr Opin Microbiol 13:684–692
Perez P, Rincon SA (2010) Rho GTPases: regulation of cell polarity and growth in yeasts. Biochem J 426:243–253
Prehoda KE, Scott JA, Mullins RD, Lim WA (2000) Integration of multiple signals through cooperative regulation of the N-WASP-Arp2/3 complex. Science 290:801–806
Pruyne D, Gao L, Bi E, Bretscher A (2004) Stable and dynamic axes of polarity use distinct formin isoforms in budding yeast. Mol Biol Cell 15:4971–4989
Pukkila-Worley R, Peleg AY, Tampakakis E, Mylonakis E (2009) *Candida albicans* hyphal formation and virulence assessed using a *Caenorhabditis elegans* infection model. Eukaryot Cell 8:1750–1758
Ramage G, Mowat E, Jones B, Williams C, Lopez-Ribot J (2009) Our current understanding of fungal biofilms. Crit Rev Microbiol 35:340–355
Reijnst P, Walther A, Wendland J (2010) Functional analysis of *Candida albicans* genes encoding SH3-domain-containing proteins. FEMS Yeast Res 10:452–461
Rida PC, Nishikawa A, Won GY, Dean N (2006) Yeast-to-hyphal transition triggers formin-dependent Golgi localization to the growing tip in *Candida albicans*. Mol Biol Cell 17:4364–4378
Rossman KL, Sondek J (2005) Larger than Dbl: new structural insights into RhoA activation. Trends Biochem Sci 30:163–165
Saarikangas J, Zhao H, Lappalainen P (2010) Regulation of the actin cytoskeleton-plasma membrane interplay by phosphoinositides. Physiol Rev 90:259–289
Sagot I, Klee SK, Pellman D (2002) Yeast formins regulate cell polarity by controlling the assembly of actin cables. Nat Cell Biol 4:42–50
Saville SP, Lazzell AL, Monteagudo C, Lopez-Ribot JL (2003) Engineered control of cell morphology in vivo reveals distinct roles for yeast and filamentous forms of *Candida albicans* during infection. Eukaryot Cell 2:1053–1060
Saville SP, Lazzell AL, Chaturvedi AK, Monteagudo C, Lopez-Ribot JL (2008) Use of a genetically engineered strain to evaluate the pathogenic potential of yeast cell and filamentous forms during *Candida albicans* systemic infection in immunodeficient mice. Infect Immun 76:97–102
Schmitz HP, Kaufmann A, Kohli M, Laissue PP, Philippsen P (2006) From function to shape: a novel role of a formin in morphogenesis of the fungus *Ashbya gossypii*. Mol Biol Cell 17:130–145
Setiadi ER, Doedt T, Cottier F, Noffz C, Ernst JF (2006) Transcriptional response of *Candida albicans* to hypoxia: linkage of oxygen sensing and Efg1p-regulatory networks. J Mol Biol 361:399–411
Sheu YJ, Santos B, Fortin N, Costigan C, Snyder M (1998) Spa2p interacts with cell polarity proteins and signaling components involved in yeast cell morphogenesis. Mol Cell Biol 18:4053–4069
Sinha I, Wang YM, Philp R, Li CR, Yap WH, Wang Y (2007) Cyclin-dependent kinases control septin phosphorylation in *Candida albicans* hyphal development. Dev Cell 13:421–432

Slutsky B, Staebell M, Anderson J, Risen L, Pfaller M, Soll DR (1987) "White-opaque transition": a second high-frequency switching system in *Candida albicans*. J Bacteriol 169:189–197
Smith GR, Givan SA, Cullen P, Sprague GF Jr (2002a) GTPase-activating proteins for Cdc42. Eukaryot Cell 1:469–480
Smith SE, Csank C, Reyes G, Ghannoum MA, Berlin V (2002b) *Candida albicans* RHO1 is required for cell viability in vitro and in vivo. FEMS Yeast Res 2:103–111
Snyder M (1989) The SPA2 protein of yeast localizes to sites of cell growth. J Cell Biol 108:1419–1429
Soll DR (2009) Why does *Candida albicans* switch? FEMS Yeast Res 9:973–989
Soll DR (2011) Evolution of a new signal transduction pathway in *Candida albicans*. Trends Microbiol 19:8–13
Song Y, Kim JY (2006) Role of CaBud6p in the polarized growth of *Candida albicans*. J Microbiol 44:311–319
Staib P, Morschhauser J (2007) Chlamydospore formation in *Candida albicans* and *Candida dubliniensis* – an enigmatic developmental programme. Mycoses 50:1–12
Steinberg G (2007) Hyphal growth: a tale of motors, lipids, and the Spitzenkorper. Eukaryot Cell 6:351–360
Stoldt VR, Sonneborn A, Leuker CE, Ernst JF (1997) Efg1p, an essential regulator of morphogenesis of the human pathogen *Candida albicans*, is a member of a conserved class of bHLH proteins regulating morphogenetic processes in fungi. EMBO J 16:1982–1991
Strahl T, Thorner J (2007) Synthesis and function of membrane phosphoinositides in budding yeast, *Saccharomyces cerevisiae*. Biochim Biophys Acta 1771:353–404
Sudbery PE (2001) The germ tubes of *Candida albicans* hyphae and pseudohyphae show different patterns of septin ring localization. Mol Microbiol 41:19–31
Sudbery P, Gow N, Berman J (2004) The distinct morphogenic states of *Candida albicans*. Trends Microbiol 12:317–324
Taheri N, Kohler T, Braus GH, Mosch HU (2000) Asymmetrically localized Bud8p and Bud9p proteins control yeast cell polarity and development. EMBO J 19:6686–6696
Ushinsky SC, Harcus D, Ash J, Dignard D, Marcil A, Morchhauser J, Thomas DY, Whiteway M, Leberer E (2002) CDC42 is required for polarized growth in human pathogen *Candida albicans*. Eukaryot Cell 1:95–104
Valtz N, Herskowitz I (1996) Pea2 protein of yeast is localized to sites of polarized growth and is required for efficient mating and bipolar budding. J Cell Biol 135:725–739
VandenBerg AL, Ibrahim AS, Edwards JE Jr, Toenjes KA, Johnson DI (2004) Cdc42p GTPase regulates the budded-to-hyphal-form transition and expression of hypha-specific transcripts in *Candida albicans*. Eukaryot Cell 3:724–734
Vauchelles R, Stalder D, Botton T, Arkowitz RA, Bassilana M (2010) Rac1 dynamics in the human opportunistic fungal pathogen *Candida albicans*. PLoS One 5:e15400
Veses V, Gow NA (2008) Vacuolar dynamics during the morphogenetic transition in *Candida albicans*. FEMS Yeast Res 8:1339–1348
Veses V, Richards A, Gow NA (2009) Vacuole inheritance regulates cell size and branching frequency of *Candida albicans* hyphae. Mol Microbiol 71:505–519
Vicinanza M, D'Angelo G, Di Campli A, De Matteis MA (2008) Phosphoinositides as regulators of membrane trafficking in health and disease. Cell Mol Life Sci 65:2833–2841
Virag A, Harris SD (2006) The Spitzenkorper: a molecular perspective. Mycol Res 110:4–13
Walther A, Wendland J (2004) Polarized hyphal growth in *Candida albicans* requires the Wiskott-Aldrich Syndrome protein homolog Wal1p. Eukaryot Cell 3:471–482
Wang Y (2009) CDKs and the yeast-hyphal decision. Curr Opin Microbiol 12:644–649
Warenda AJ, Konopka JB (2002) Septin function in *Candida albicans* morphogenesis. Mol Biol Cell 13:2732–2746
Wedlich-Soldner R, Schulz I, Straube A, Steinberg G (2002) Dynein supports motility of endoplasmic reticulum in the fungus *Ustilago maydis*. Mol Biol Cell 13:965–977

Wedlich-Soldner R, Altschuler S, Wu L, Li R (2003) Spontaneous cell polarization through actomyosin-based delivery of the Cdc42 GTPase. Science 299:1231–1235

Weirich CS, Erzberger JP, Barral Y (2008) The septin family of GTPases: architecture and dynamics. Nat Rev Mol Cell Biol 9:478–489

Wendland J, Philippsen P (2001) Cell polarity and hyphal morphogenesis are controlled by multiple rho-protein modules in the filamentous ascomycete *Ashbya gossypii*. Genetics 157:601–610

Whiteway M, Bachewich C (2007) Morphogenesis in *Candida albicans*. Annu Rev Microbiol 61:529–553

Wilson D, Hube B (2010) Hgc1 mediates dynamic *Candida albicans*-endothelium adhesion events during circulation. Eukaryot Cell 9:278–287

Wolyniak MJ, Sundstrom P (2007) Role of actin cytoskeletal dynamics in activation of the cyclic AMP pathway and HWP1 gene expression in *Candida albicans*. Eukaryot Cell 6:1824–1840

Wu H, Brennwald P (2010) The function of two Rho family GTPases is determined by distinct patterns of cell surface localization. Mol Cell Biol 30:5207–5217

Wu H, Rossi G, Brennwald P (2008) The ghost in the machine: small GTPases as spatial regulators of exocytosis. Trends Cell Biol 18:397–404

Wu H, Turner C, Gardner J, Temple B, Brennwald P (2010) The Exo70 subunit of the exocyst is an effector for both Cdc42 and Rho3 function in polarized exocytosis. Mol Biol Cell 21:430–442

Yaar L, Mevarech M, Koltin Y (1997) A *Candida albicans* RAS-related gene (CaRSR1) is involved in budding, cell morphogenesis and hypha development. Microbiology 143:3033–3044

Yakir-Tamang L, Gerst JE (2009) Phosphoinositides, exocytosis and polarity in yeast: all about actin? Trends Cell Biol 19:677–684

Yamashita M, Kurokawa K, Sato Y, Yamagata A, Mimura H, Yoshikawa A, Sato K, Nakano A, Fukai S (2010) Structural basis for the Rho- and phosphoinositide-dependent localization of the exocyst subunit Sec3. Nat Struct Mol Biol 17:180–186

Yokoyama K, Kaji H, Nishimura K, Miyaji M (1990) The role of microfilaments and microtubules in apical growth and dimorphism of *Candida albicans*. J Gen Microbiol 136:1067–1075

Zaman S, Lippman SI, Zhao X, Broach JR (2008) How Saccharomyces responds to nutrients. Annu Rev Genet 42:27–81

Zeidler U, Lettner T, Lassnig C, Muller M, Lajko R, Hintner H, Breitenbach M, Bito A (2009) UME6 is a crucial downstream target of other transcriptional regulators of true hyphal development in *Candida albicans*. FEMS Yeast Res 9:126–142

Zhang X, Orlando K, He B, Xi F, Zhang J, Zajac A, Guo W (2008) Membrane association and functional regulation of Sec3 by phospholipids and Cdc42. J Cell Biol 180:145–158

Zheng X, Wang Y (2004) Hgc1, a novel hypha-specific G1 cyclin-related protein regulates *Candida albicans* hyphal morphogenesis. EMBO J 23:1845–1856

Zheng XD, Wang YM, Wang Y (2003) CaSPA2 is important for polarity establishment and maintenance in *Candida albicans*. Mol Microbiol 49:1391–1405

Zheng XD, Lee RT, Wang YM, Lin QS, Wang Y (2007) Phosphorylation of Rga2, a Cdc42 GAP, by CDK/Hgc1 is crucial for *Candida albicans* hyphal growth. EMBO J 26:3760–3769

Zhu W, Filler SG (2010) Interactions of *Candida albicans* with epithelial cells. Cell Microbiol 12:273–282

Zou H, Fang HM, Zhu Y, Wang Y (2010) *Candida albicans* Cyr1, Cap1 and G-actin form a sensor/effector apparatus for activating cAMP synthesis in hyphal growth. Mol Microbiol 75:579–591

Zucchi PC, Davis TR, Kumamoto CA (2010) A *Candida albicans* cell wall-linked protein promotes invasive filamentation into semi-solid medium. Mol Microbiol 76:733–748

# Chapter 9
# Morphogenesis in *Paracoccidioides brasiliensis*

**Iran Malavazi and Gustavo Henrique Goldman**

**Abstract** Paracoccidioidomycosis (PCM) is a deep systemic mycosis originally described in 1908 caused by the ascomycete *Paracoccidioides brasiliensis*, a thermodimorphic fungal pathogen. The disease is autochthonous to Latin America and areas of higher incidence occur in countries such as Brazil, Argentina, Colombia, and Venezuela. Inside the mammalian host, PCM is characterized by a granulomatous inflammation that invades conjunctival tissue or viscera. *P. brasiliensis* grows as yeast in cultures incubated at 37°C or inside the host and in a filamentous saprophytic form at low temperatures (26°C) or in the environmental niches. The relative temperature-dependent simplistic mechanisms whereby *P. brasiliensis* orchestrates its developmental program for switching morphological forms have been under intense scrutiny for several decades, and although major advances have been achieved, much remains to be uncovered for the identification of new virulence determinants and therapeutical targets, and for the comprehension of the pathophysiology of this fungus. Although recent studies have begun to identify genes and overall pathways required for the thermodimorphic process and pathogenicity, master regulators mediating morphogenesis, virulence, and the mycelium-to-yeast transition are still not recognized. Transcriptional profiling studies have highlighted the importance of differential expression genes in the mycelial and yeast phases of *P. brasiliensis* and have analyzed putative genes involved in virulence and morphogenesis based on evidences from other dimorphic

I. Malavazi (✉)
Departamento de Genética e Evolução,Centro de Ciências Biológicas e da Saúde, Universidade Federal de São Carlos, São Carlos, Brazil
e-mail: imalavazi@ufscar.br

G.H. Goldman
Laboratório Nacional de Ciência e Tecnologia do Bioetanol, Caixa Postal 6170, Campinas 13083-970, Brazil

Faculdade de Ciências Farmacêuticas de Ribeirão Preto, Universidade de São Paulo, São Paulo, Brazil

J. Pérez-Martín and A. Di Pietro (eds.), *Morphogenesis and Pathogenicity in Fungi*, Topics in Current Genetics 22, DOI 10.1007/978-3-642-22916-9_9,

fungi or pathways involved in virulence traits. Here, we consider the current information obtained at the transcriptional level in *P. brasiliensis* as the basis for an update of the main pathways involved in the virulence and pathogenicity of this fungus, and the experimental possibilities generated by the newly released genomic sequences that will drive the forthcoming years of systematic research of *P. brasiliensis*.

## 9.1 *Paracoccidioides brasiliensis* and Paracoccidioidomycosis: An Overview

Among over the 100,000 environmentally described fungal species, six phylogenetically related ascomycetes species lie in the category of the so-called medically important dimorphic fungi, comprising *Blastomyces dermatitidis*, *Coccidioides immitis*, *Histoplasma capsulatum*, *Penicillium marneffei*, *Sporothrix schenckii*, and *Paracoccidioides brasiliensis*. *P. brasiliensis* is especially closest to the ascomycetes that cause pulmonary evident manifestation, forming a distinct clade with *H. capsulatum* and *B. dermatitidis* (Bialek et al. 2000). Fungal dimorphism is a very striking feature and regulated process in these ascomycetes (Phylum Ascomycota, Subphylum Pezizomycotina, Class Eurotiomycetes). This classification places *P. brasiliensis* along with the thermodimorphic *H. capsulatum*, *B. dermatitidis*, and *C. immitis* into the Order Onygenales, Family Onygenaceae (San-Blas et al. 2002; Silva et al. 2008). Inside the host, the extent of the disease elicited by this group of fungal pathogens ranges from limited infection to systemic (deep) mycoses which are not limited to the epithelial surface of the organism, invading conjunctival tissue or viscera.

PCM is a systemic mycosis originally described by Adolfo Lutz in 1908 (Schwarz and Baum 1965). PCM is autochthonous to Latin America in a region comprised from latitude 20° N (México) to 35° S (Argentina), and the highest incidence of the disease occurs in South American countries (Brazil, Argentina, Colombia, and Venezuela). The disease is endemic among rural areas, affecting predominantly male individuals in their economically productive years. Brazil has 80% of the reported cases and in the endemic areas, about 10 million people are infected (Brummer et al. 1993; Restrepo et al. 2001; Restrepo and Tobón 2005). No outbreaks have so far been reported and due to its ability to enter into prolonged periods of latency, incidence can be demonstrated outside the recognized endemic areas in patients who have moved from these areas, as has been demonstrated by PCM diagnostic markers (Restrepo et al. 2008).

PCM is characterized by a granulomatous inflammation and impairment of cell-mediated immune response, the main mechanism of defense against *P. brasiliensis* (Brummer et al. 1993). Clinically, PCM infection is observed in two distinct forms: (1) An acute (or subacute), also known as juvenile, form and (2) the adult chronic form. The acute form is more severe, presents a faster clinical development, and therefore includes the cases with worst clinical prognosis which can be illustrated by clinical conditions such as reticuloendotelial system hypertrophy, bone marrow dysfunction, lymphoproliferative disorder, or severe hematogenic dissemination (Brummer et al. 1993). This scenario contrasts with the chronic form which accounts for more than 90% of cases, but can take as long as months or years to become fully established in a clinical outline mainly characterized by pulmonary manifestation in about 90% of the adults (Brummer et al. 1993; Borges-Walmsley et al. 2002). The PCM in its chronic form is also characterized by endogenous reinfection, which consists of reactivation of a quiescent focus of infection after long periods of dormancy or after the interruption of treatment. It is thought that spots of fibrotic replacement in the target tissues colonized by the *P. brasiliensis* can function as latent reservoirs of the fungi for latter recrudescence.

## 9.2 Morphological Transition and Dimorphism in *P. brasiliensis*

In fungal pathogens, adaptation to the host environment takes place at different levels and can be facilitated in some cases by the ability of the organisms to survive on environmental hostile niches, such as under conditions of limited oxygen, high temperature, and nutritional, osmotic, and pH variations. However, in dimorphic fungi, this panorama seems to be not so obvious because the infective fungal propagule is obligatorily not the pathogenic form inside the host. In this group of fungal pathogens, apparently such a sudden environmental alteration when the fungal infectious particles reach the host implies its ability to adjust rapidly in order to survive and invade the new environment. This feature clearly links morphology to virulence because cell shape facilitates host tissue penetration, dissemination, intracellular colonization, and expression of virulence factors. Although suffering a dramatic change in cell morphology, the dimorphic process keeps unique and particular features including its reversibility and the fact that the dimorphic transition is not a vital component of the fungal life cycle. Instead, it is not only an opportunistic adaptation to the environmental conditions, but also a prerequisite for progressive infection (San-Blas et al. 2002).

Dimorphic fungi are formally characterized by their ability to convert themselves between two different morphological forms, each of them associated with specific environmental niche corresponding to entirely different lifestyles and survival conditions. In these organisms, the temperature variation between the environments and the hosts triggers the dimorphic transition (Medoff et al. 1987). This variation has been attributed as the most important factor for launching the dimorphic transition in *B. dermatitidis*, *H. capsulatum*, and *P. brasiliensis* once strains of these organisms whose ability to grow as yeast is impaired are avirulent in animal models of infection (Medoff et al. 1986; Franco 1987; Rooney and Klein 2002). The morphological shift of these species can be achieved "in vitro" by simply altering the incubation temperature. This is a very important physiological trait in this group of organisms and not a constant for all fungal species in which the pathogenic form differs from the saprophytic state. In other fungal pathogens such as *Candida albicans*, morphologic alteration to the filamentous pathogenic form not only requires temperature variation but also host serum (Whiteway and Oberholzer 2004; Whiteway and Bachewich 2007). Furthermore, the morphological alteration in *C. albicans* is not total with the presence of the three peculiar morphotypes: hyphae, pseudohyphae, and yeasts (Brown and Gow 1999; Borges-Walmsley et al. 2002). In *P. brasiliensis*, very few factors other than temperature, e.g., the sulfur source (see Sect. 9.3.1), have been identified.

*P. brasiliensis* grows as yeast in cultures incubated at 37°C or inside the host. However, at low temperatures (26°C) and in the environmental niches, it grows in a filamentous saprophytic form (San-Blas and San-Blas 1984; Brummer et al. 1993). The relative temperature-dependent simplistic mechanisms whereby *P. brasiliensis* orchestrates its developmental program for switching morphological forms have been under intense scrutiny for several decades, and although major advances have been achieved, much remains to be uncovered for the identification of new virulence determinants and therapeutical targets and for the comprehension of the pathophysiology of this fungus. Recent studies have begun to identify genes and overall pathways required for the thermodimorphic process and pathogenicity; however, master regulators mediating morphogenesis, virulence, and most importantly the mycelium-to-yeast transition in *P. brasiliensis* are still not recognized.

In the soil or related environment, *P. brasiliensis* exists as a mold which produces various types of propagules such as arthroconidia and unicellular spores (3.5–5.0 μm) when cultured under conditions of nutritional deprivation (San-Blas 1986; Franco et al. 1989; Brummer et al. 1993). The biological relevance of these structures as infectious propagules for humans has not been fully determined, and some doubts still remain about the most efficient infectious particle. These saprobic and infectant fungal elements when cultured in laboratory at 37°C or inside the mammalian host grow as yeast cells of various sizes ranging from 4 to 30 μm. One of the most outstanding characteristics of the yeast phase of *P. brasiliensis* is its unique and peculiar budding pattern. No other yeast cell in the fungal kingdom possesses such a diverse way of generating daughter cells. The yeast form of *P. brasiliensis* is characterized by a polymorphic cell growth, i.e., the existence of mother and bud cells during growth with extreme variations in cell size and shape within the same cellular population, rendering to the yeast a pilot's wheel appearance (Brummer et al. 1993; Almeida et al. 2009). This cell shape is a pathognomonic feature that is used as a clear indication of *P. brasiliensis* infection in histopathological diagnosis (Brummer et al. 1993). What makes *P. brasiliensis* yeast cell so different from other yeast-like cells? If we consider the pathogen *C. albicans* for instance, its budding pattern resembles that of the nonpathogen *S. cerevisiae*, yet the former is a real pathogen. *C. albicans* needs serum to turn its yeast morphology into hyphae, but as mentioned earlier, the influence of the host's milieu on the *P. brasiliensis* morphogenetic program is not fully known. When achieving yeast form, *P. brasiliensis* seems to change all the patterns of recognition for polar growth seen in other organisms, indicating that the morphogenetic machinery is able to recognize multiple sites for polarization in the cell membrane indicating that the yeast cell shape and the budding pattern in *P. brasiliensis* may follow, if not a different set of "rules," at least a "lax control" of the establishment and maintenance of polarity during growth (Almeida et al. 2009). Nonetheless, there are very few studies addressing these issues in *P. brasiliensis* aiming to establish a connection between pathogenicity and cell shape. If this different budding pattern is somewhat required for virulence or disease progression in this fungus has not been determined so far (Svidzinski et al. 1999; Kurokawa et al. 2005).

The positioning of cell components involved in polarized growth in eukaryotic cells is consensually reported in a hierarchically organized manner (Pringle et al. 1995; Nelson 2003). In this model, the first group of proteins called landmark proteins located on the cell membrane specifies the sites of polarized growth for the second class of proteins. These second-level proteins transduce the signal locally generated by the landmark proteins to the third class of proteins, the morphogenetic machinery (composed of the cytoskeleton and associated vesicle transport complexes) needed to deliver the components required for cell expansion at the initially specified polarization site. However, some studies have shown that *S. cerevisiae* cells are capable of polarity establishment in the absence of any obvious positional signal generated by landmark proteins (revised by Harris 2008). In this case, polar growth would occur in a stochastic process dependent upon feedback loops that amplify initially weak signals to the morphogenetic machinery (Nern and Arkowitz 2000; Wedlich-Soldner et al. 2003). Based on these observations and considering the weird pattern of polar growth in *P. brasiliensis*, it is tempting to speculate how yeast manage themselves to establish the site of polarization, and consequently what model hierarchically or stochastically fits into the *P. brasiliensis* lifestyle. In this sense, this organism seems to be a good example and an interesting model organism to study cell polarization in the future. In *P. brasiliensis*, maybe the only protein whose function has been partially studied by molecular genetics approach is the homolog of *S. cerevisiae* CDC42, which is one of the most important landmark proteins in several organisms (see Sect. 9.4), including budding yeast, filamentous fungi, and animal cells. The Rho-related GTPase CDCD42 regulates polar growth by mediating the transfer of positional information to the downstream morphogenetic machinery (Harris 2008).

The influence of temperature on the mold-to-yeast form is an intrinsic part of the process in all dimorphic fungi, but the question behind it is "how do these fungi sense the temperature variation to trigger the phase switch and regulate its adaptation to the new environment?" it is clear that temperature itself would not be enough to launch the thermodimorphic transition in the absence of signaling cascades committed with dimorphism. Nemecek et al. (2006) reported the first genetic evidence that pathogenicity in dimorphic fungi requires the temperature-dependent mold-to-yeast transition, which is dependent on the activity of a hybrid histidine kinase (HK) named dimorphism-regulating protein kinase (*DRK1*). *DRK1* was initially found in *B. dermatitidis* (Nemecek et al. 2006) and also characterized in *H. capsulatum.* This histidine kinase acts as a central sensor for host adaptation mediating signaling cascades which induce the morphological transition. *B. dermatitidis DRK1* can complement the mutant Sln1 of *S. cerevisiae* which lacks the only hybrid histidine kinase known to regulate the response to osmotic [*SLN1*-regulated *H*igh *O*smolarity *G*lycerol (HOG) pathway] and oxidative stresses. Impairment of *DRK1*, in both *B. dermatitids* and *H. capsulatum*, causes cells to be arrested in the mycelial phase. This mutation causes pleotropic effects on cell wall composition, integrity, and sporulation. *DRK1* loss-of-function strains are also avirulent in animal models (Nemecek et al. 2006; Klein and Tebbets 2007). The role of HKs may represent a parallel level of signaling transduction (along with the main pathways involving serine, threonine, or tyrosine kinases), the functions of which in *P. brasiliensis* have not even started to be unraveled.

As noted before, dimorphism is a common feature of fungal cells. Some aspects of the yeast-to-mycelium or mycelium-to-yeast transition may be similar even between taxonomically unrelated species. However, other features, especially those regarding the response to environmental factors, may be different between species (Gow 1995). Therefore, the study of the hybrid histidine kinases signal transduction elements in *P. brasiliensis* is an open avenue for investigation as soon as the homologs are identified. There are five sensor HK genes and a single histidine-containing phosphotransfer intermediate gene in all three *Paracoccidioides* genomes, and two genes encoding response regulators (http://www.broadinstitute.org/annotation/genome/paracoccidioides_brasiliensis/MultiHome.html). Also, the elements linking the external stimuli to the histidine kinase and downstream signaling cascade are still not identified.

## 9.3 Identification of Genetic Determinants Involved in the Dimorphic Transition, Morphogenesis, and Virulence in *P. brasiliensis*

The understanding of basic biological traits of *P. brasiliensis* has significantly improved in the last decades with the use of molecular biology techniques applied to the pathogen itself and also with the identification of its infection and immunological properties. However, a very limited number of candidate pathogenicity genes and components of pathogenicity have been identified in this organism. Conceivably, part of this panorama can be related to the limited conditions for genetic manipulation of this organism, such as the limited spectrum of DNA-mediated transformation protocols for gene replacement. Aside from providing new insights into the biology of this important dimorphic fungal pathogen, the identification of new genetic determinants involved in pathogenesis may represent the perspective of new therapeutic targets for antifungal chemotherapy.

In fact, global regulators that mediate morphogenesis, virulence, and host adaptation are hitherto basically unknown in this fungus. Besides this, extrinsic factors such as the physical and chemical determinants that impact on the growth of this organism in its dual morphological phases are also not fully understood. One example of this scenario is that the endemic areas of *P. brasiliensis* have yet not been completely linked to a specific environmental condition that makes it suitable for the development of the mycelial phase of the fungus. The same is not true for the related dimorphic species *C. immitis*, *H. capsulatum*, or *B. dermatitidis* in which the relationship between growth and environment is better understood. *C. immitis* endemic areas are marked by halophilic conditions and high temperature, which reduce populations of microbial competitors and contribute to fungal growth in the soil and to disease spread (Egeberg et al. 1964; Cao et al. 2007). *H. capsulatum* and *B. dermatitidis* grow in moist and acidic soil of the mid-west and southeast areas of the USA (Gauthier and Klein 2008). Likewise, in *P. marneffei*, several parameters have been assessed "in vitro" in order to draw conclusions about the preferred conditions for growth and sporulation of the saprophytic form of *P. marneffei*. Cao et al. (2007) have demonstrated that growth requirements such as temperature, pH, and salt concentration may vary in different isolates causing differences in growth patterns.

*P. brasiliensis* is rarely isolated from soil, and few successful studies have been conducted to isolate the pathogen from nature and some of them are controversial (Franco et al. 2000; Bagagli et al. 2008). Besides this, although mycelial fragments are capable of causing infection, some reports also suggest that conidia would be the more suitable as infectious propagules in nature (McEwen et al. 1987; Cock et al. 2000). Meanwhile, the most compelling route for infection continues to be the inhalation of airborne propagules in the residents of endemic areas, but the long latency period of the disease and the lack of outbreaks create difficulties in determining under which circumstances the primary infection occurs and also the predominant form of infection (Franco et al. 2000; Tercarioli et al. 2007). Few reports describe the ability of *P. brasiliensis* to grow in soil with different chemical and physical compositions. Tercarioli et al. (2007) have shown that *P. brasiliensis* cannot grow in some soil samples, especially those containing high values of exchangeable aluminum (H + Al). Interestingly, the molecular detection of *P. brasiliensis* was not possible from the surface soil of endemic areas, whereas the organism was detected only in soil samples from nine-banded armadillo (*Dasypus novemcinctus*) burrows. Understanding the adaptive response of *P. brasiliensis* to different growth conditions may reveal an investigatory link to morphogenesis and genes that can help to elucidate how the organism survives in its endemic areas and reprograms its ability to grow inside the animal hosts.

However, identifying phase-specific genes related to morphogenesis and possibly to virulence, mainly those preferentially expressed in one of the dimorphic states, can be complicated in dimorphic fungal pathogens. The conditions that induce morphogenesis "in vivo" or "in vitro" can also induce cell responses that are not necessarily connected with morphogenesis, but rather are required exclusively for physiological adaptation to the new environment or related to the growth phase "in vitro" (Berman and Sudbery 2002; Marques et al. 2006). These physiological effects, although greatly important to understand metabolic changes in the cell biology of *P. brasiliensis*, can mask the identification of the multiple collection of molecules, such as proteases, surface receptors, and adhesins, that are essential for cell shape, invasion, and virulence of the species (Mendes-Giannini et al. 2008).

It has also been shown that adhesion of *P. brasiliensis* to host cells varies among strains and correlates with their virulence. Strains more virulent in animal models, such as PB18, exhibit enhanced adhesion "in vitro" (Hanna et al. 2000). It is well known that along with the morphological change triggered by the temperature variation during dimorphic transition, extensive changes in cell wall also occur. This fact can support further studies on these early events of *P. brasiliensis* internalization which depict an interesting example of pathogen adaptation to colonize the host environment.

With this piece of information, it is clear that there are several points related to the fungus lifestyle outside or inside the host that can still be investigated to unravel particularities of this pathogen species at the molecular level. Some of them have been analyzed having as premise previous observations on the *P. brasiliensis* lifestyle or biochemical properties, but others were not investigated up to now. In the following sections, we present the main lines of investigation that have been exploited in *P. brasiliensis* for global and functional analysis.

## *9.3.1 Transcriptome and Functional Genomics in* P. brasiliensis

Majority of the studies in *P. brasiliensis* have identified and analyzed putative genes involved in virulence and morphogenesis, attempting to provide phase-specific factors in the systemic dimorphic fungi based on evidences from other dimorphic fungi or from known metabolic pathways involved in virulence traits. Some of these genetic factors identified up to now may be indispensable for pathogenicity. Undoubtedly, there are many more that remain to be identified and analyzed. While on the one hand the *P. brasiliensis* research community cannot rely on true virulence factors for cell biology and "in vivo" studies, on the other hand, they balance this scarcity by a number of candidate genes identified (Rappleye and Goldman 2006).

In order to circumvent the methodological limitation for gene function study, *P. brasiliensis* global gene expression analysis using expressed sequence tags (ESTs), library collection, DNA microarrays, subtrative supression hybridization (SSH), "in silico" determination of overexpressed genes, and cDNA representational difference analysis (RDA) has proved to be powerful genetic tools to analyze the fine regulation of complex regulatory networks potentially involved in morphogenesis and virulence. In particular, the possibility of simultaneously capturing the expression profile of a whole fungal genome has become a very attractive approach, and several genes and metabolic pathways could be explored through this methodology in various fungal organisms. The genomes of several dimorphic fungi, including *P. brasiliensis*, are in varying stages of being sequenced and completed, and this information along with functional studies should provide new insights into the identity and function of phase-regulated genes.

Initial transcriptional profiling has been carried out in *P. brasiliensis* and helped to shed light on phase-regulated genes and some specific traits or requirements for the fungus in each morphological phase. Microarray analysis depends on comparing gene expression in two well-defined conditions, making this technique very suitable for detecting differences in the contrasting conditions of the thermodimorphic transition in *P. brasiliensis*. Therefore, relevant studies in *P. brasiliensis* have so far compared gene expression under different nutritional conditions (e.g., complete and minimal medium), at different stages of development (i.e., mycelium-to-yeast transition), and in the presence or absence of stressors (i.e., drugs, high temperature, or phagocytes). These studies have helped to draw conclusions in the areas such as overall intermediary metabolism regulation, virulence determinants, pathogenicity, and signal transduction pathways. Up to now, global gene expression data in *P. brasiliensis* have been extracted either from cDNA array spotted onto glass slides or from macroarray filter experiments. In both cases, probes were obtained using the EST approach. Since 2003, seven microarray studies in *P. brasiliensis* have been published (Table 9.1).

Before the first transcriptional profiling studies indicated in Table 9.1, two independent groups have analyzed different EST collections in *P. brasiliensis* (Felipe et al. 2003; Goldman et al. 2003). The EST approach established by these two groups proved to be a very efficient method to gather large-scale genetic information from *P. brasiliensis*. Goldman et al. (2003) used the strain PB18 and obtained 4,692 genes that were identified and annotated in functional categories. In this study, the main putative virulence and pathogenicity genes were divided into groups of metabolic, cell wall/adhesion, and signal transduction genes. Likewise, Felipe et al. (2003) performed a similar analysis of 6,022 genes from *P. brasiliensis* strain PB01 obtained from two non-normalized cDNA libraries from mycelial and yeast phases of *P. brasiliensis*. The coincident results from both studies highlighted (1) important metabolic pathways and the main eukaryotic signaling pathways controlling dimorphism and virulence, (2) thermal and oxidative stress transcriptional response, and (3) the differential metabolism and energy acquisition between mycelium and yeast cells. One of the most interesting findings explored by Felipe et al. (2003) indicated that mycelial cells present a preferential aerobic metabolism. This was verified considering the group of gene found to be upregulated belonging to the main regulatory points of the Krebs cycle such as isocitrate dehydrogenase and succinyl-CoA synthetase. Conversely, yeast pathogenic cells depict enriched gene expression of transcripts related to anaerobic fermentation and ethanol production represented mainly by alcohol dehydrogenase I and pyruvate dehydrogenase E1 subunit (Felipe et al. 2005). In addition, yeast phase also induces the expression of the glyoxylate cycle main enzymes malate synthase and isocitrate lyase. These enzymes are not present in humans and are thus attractive targets for drug development and chemotherapy.

These first transcriptome profiling analysis highlighted the importance of differential expression genes in the mycelial and yeast phases of *P. brasiliensis*. These differences in cell metabolism between the two morphological conditions were further explored by Nunes et al. (2005) who carried out microarray hybridizations using the EST collection obtained by Goldman et al. (2003) to probe gene expression at several time points of the mycelium-to-yeast switch ranging from 5 to 120 h

**Table 9.1** Microarray studies in *P. brasiliensis* where the differential expression on mycelial and yeast phase was investigated[a]

| Examples of genes and pathways shown as differentially expressed in *P. brasiliensis*[a] | Experimental design | Source | Array design | Reference |
|---|---|---|---|---|
| Genes associated with assimilation of sulfur-containing amino acids<br>Sterol metabolism: C-4 sterol methyl oxidase (*ERG25*) homolog<br>α-1,3-Glucan synthase (*AGS1*)<br>PKC/MAPK pathway-related genes: *RHO1*, *SEP1*, *FLB1*<br>Glutathione-S-transferase (*GST1*)<br>Thiol-specific antioxidant (TSA1 homolog)<br>Methione permease (*MEP1*)<br>APS kinase (*APS1*)<br>Choline sulfatase (*CHS1*) | Mycelium grown in liquid Sabouraud media (26°C) for 20 days<br>Yeast recovered from spleens of infected mice, subcultured in liquid YPD (36°C) for 7 days<br>PB18 strain | cDNA PB18 strain | 1,397 ESTs | Marques et al. (2004)[b] |
| Respiratory genes upregulated in mycelial phase<br>Fermenting genes upregulated in yeast phase<br>PAPS reductase<br>Glyoxylate cycle enzymes: malate dehydrogenase and isocitrate lyase<br>Chitin deacetylase (*CDA*) | Mycelium grown at 22°C for 14 days<br>Yeast grown at 36°C for 7 days in Fava Neto's semi-solid medium<br>PB01 strain | cDNA PB01 strain | 1,152 ESTs | Felipe et al. (2005)[c] |
| Chitin synthases (*CHS1*, *CHS2*)<br>$Ca^{+2}$/calmodulin/calcineurin catalytic subunit (*CNA*)<br>4-Hydroxyl-phenyl pyruvate dioxygenase (*4-HPPD*)<br>*MET1* transcription factor involved in sulfur assimilation[d]<br>Tiol-specific antioxidant (peroxiredoxin *S. cerevisiae TSA1* homolog)<br>cAMP-dependent protein kinase (*PKA*) | Mycelium was induced to mycelium-to-yeast phase transition and time points were collected at 0 (mycelium reference), 5, 10, 24, 48, 72, and 120 h growth (37°C).<br>PB01 strain | cDNA PB18 strain | 4,692 ESTs | Nunes et al. (2005) |
| β-1,3-Glucosidade (*BGL*) (upregulated in mycelium)<br>α-1,3-Glucan synthase (*AGS*)<br>Chitin deacetylase (*CDA*)<br>Cysteine de novo biosynthesis pathway[d]: ATP sulfurylase, APS kinase, PAPS reductase, choline sulfatase | Mycelium and yeast forms were grown for 7–10 days at 22 or 37°C, respectively, in Fava Neto's semisolid medium<br>PB01 strain | cDNA PB01 strain | 1,152 ESTs | Andrade et al. (2006)[c] |
| Glyoxylate cycle enzymes: malate dehydrogenase<br>cAMP-dependent protein kinase catalytic subunit (*PKA*)<br>Cysteine de novo biosynthesis pathway[d]: choline sulfatase (*CHS1*), APS kinase, PAPS reductase, sulfite reductase (*SUR1*) | Mycelium was induced to mycelium-to-yeast phase transition in minimal and complete medium. Several time points were collected | cDNA PB18 strain | 4,692 ESTs | Ferreira et al. (2006) |

| | | | | |
|---|---|---|---|---|
| Organic sulfate assimilation: cysteine dioxygenase (*CDI1*)<br>Sulfur biosynthesis pathway regulating genes: *MET1* and *SCO1* | along the transition (37°C). The mycelium phase (time point 0) was the reference | | | |
| Cu Zn superoxide dismutase (*SOD3*)<br>*HSP60*<br>β-1,3-Glucan synthase (*FKS1*) | Yeast cells cultured "in vitro" were compared to yeast cell internalized by murine macrophages and recovered to probe the early transcriptional events upon internalization | cDNA PB01 strain | 1,152 ESTs | Tavares et al. (2007) |
| Glyoxylate cycle enzymes: malate dehydrogenase and malate synthase<br>Unfolded protein response (UPR) transcription factor: *HAC1* homolog<br>sterol metabolism: *ERG28* homolog<br>*CBP1* (calcineurin-binding protein)<br>Secondary metabolite production: trichothecene C-15 hydroxylase[e] | Mycelial and yeast forms were cultured in McVeigh/Morton medium at 36°C (for yeast) and at room temperature (for mycelium), for 5, 8, or 14 days | genomic DNA PB01 strain | 12,283 elements | Monteiro et al. (2009) |

[a]Genes upregulated in the *P. brasiliensis* yeast form, unless otherwise stated in the table

[b]Selected genes refer to the concomitant experiments of macroarray hybridization and suppression subtraction hybridization (SSH) appearing in the manuscript. See reference for details

[c]Selected genes refer to the concomitant experiments of macroarray hybridization and differential expression analysis "in silico" by electronic subtraction. See references for details

[d]Further details can be found in text

[e]Upregulated in mycelium

in culture. In this report, differentially expressed genes were identified not only at the endpoints of the differentiation but also at different time points of thermodimorphic transition. Among the genes identified in this study, an interesting set of them is related to aminoacid catabolism, particularly aromatic and branched-chain aminoacids. In this context, the homolog of T-cell reactive protein from *C. immitis*, which encodes the enzyme 4-hydrophenyl-pyruvate dehydrogenase (4-HPPD), was highly expressed (about 15-fold increase) at the beginning of the dimorphic transition. The 4-HPPD gene is involved in the second step of aromatic amino acid catabolism, i.e., the conversion of 4-hydroxyphenyl-pyruvate to homogentisate and carbon dioxide in the presence of oxygen and ferrous ion. Additional characterization of this gene using nitizinone (and also its derivative molecules) as the pharmacological probe to inhibit 4-HPPD activity showed that loss of this protein function can inhibit the thermodimorphic switch in a dose-dependent manner (Nunes et al. 2005). Nitisinone has also been used in humans to treat hereditary type I tyrosinemia, an inborn error of metabolism, and although it is capable of inhibiting *P. brasiliensis* growth "in vitro," the potential of this compound as an antifungal molecule or as an auxiliary drug in addition to other antifungals needs to be validated in animal models. The mechanism of action whereby nitizinone kills *P. brasiliensis* cells may be related to the action of 4-HPPD in the production of melanin-like pigments, but it was not determined if nitisinone was directly modulating the fungal growth or was toxic to the cells. Importantly, the inhibitory action of nitisinone was also verified in other fungal pathogens such as *C. albicans* and *A. fumigatus*, but not in *S. cerevisiae* where no 4-HPPD homolog is present, indicating that perhaps the killing effect was not related to toxicity (Nunes et al. 2005).

Both transcriptome analysis reports from Felipe et al. (2005) and Nunes et al. (2005) have shown that central genes belonging to most important signaling cascades known as involved in adaptation, virulence, and pathogencity in other fungi were found to be upregulated in the microarray conditions tested. Among these pathways were genes involved in the *P*rotein *K*inase *C*/*M*itogen-*A*ctivated *P*rotein *K*inase (PKC/MAPK) system, cyclic AMP-Dependent *P*rotein *K*inase *A* (cAMP/PKA system), and calcium calmodulin/calcineurin-related genes.

Calmodulin and calcineurin proteins are present in all eukaryotic cells and were also identified in *P. brasiliensis* (as summarized in Fernandes et al. 2005). In many fungi, $Ca^{2+}$/calmodulin and calcineurin seem to be involved in various aspects of fungal development, environmental sensing, and adaptation in response to external stimuli, hyphal polar growth, and branching (Stie and Fox 2008). Fungal pathogenicity in *C. albicans* is directly linked to $Ca^{2+}$/calmodulin system. Yeast cells subjected to pharmacological inhibition of calmodulin function were unable to emit germ tube, a hallmark of the yeast to the pathogenic filamentous form in this fungus (Sabie and Gadd 1989). Likewise, in *P. brasiliensis*, inhibition of calmodulin pathway by inhibitor compounds such as calmidazolium (R24571), trifluoperazine (TFP), and W7 causes impairment in mycelium-to-yeast transition (de Carvalho et al. 2003). In addition, a transient inhibition of the morphological switch was also observed in the presence of calcium chelators and a stimulatory effect when extracelluar calcium was added to cultures undergoing thermodimorphic transition (de Carvalho et al. 2003; Campos et al. 2008). The role of $Ca^{2+}$-calmodulin serine/threonine-dependent phosphatase calcineurin (*CNA*, catalytic subunit) in *P. brasiliensis* morphogenesis has also been further studied upon pharmacological inhibition. The immunosupressor drugs cyclosporine A (CsA) and tacrolimus (FK506) can bind to the immunophilins cyclophilin A and FK506-binding protein (FKBP), respectively, and ultimately inhibit calcineurin phosphatase activity. Upon treatment with CsA and FK506, the mycelium-to-yeast transition was halted in *P. brasiliensis* (Campos et al. 2008). "In vivo" studies showed that treatment with CsA results in decreased lung fungal load/antigenemia in experimental PCM in BALB/c mice, even though CsA is a immunosupresor agent for the mammalian host (Massuda et al. 2010). Similarly, "in vitro" synergistic fungicidal action has been observed between CsA and azoles in other fungal pathogens such as *Aspergillus fumigatus* and *C. albicans* (Steinbach et al. 2004).

Currently, there is no evidence in *P. brasiliensis* about the function of the calcineurin downstream effectors such as the homolog of the *S. cerevisiae* CRZ1 ("crazy1") transcription factor (human *NFAT* analog). *CRZ1* is one of calcineurin targets and upon calcineurin-dependent dephosphorylation, it is translocated to the nucleus where it induces the transcription of several genes involved in cell survival and calcium homeostasis. No transcriptional study addressing the expression of such genes is available and, therefore, the knowledge of the calcineurin pathway and its connection to other signaling cascades are not known. In *C. albicans*, a large body of data is available reporting the interplay between the $Ca^{+2}$/calcineurin, PKC, and HOG pathways in the expression of chitin synthase genes (Munro et al. 2007). This cross-talk between the most important signaling cascades in eukaryotic cells is thought to enhance the signaling capabilities of signal transduction, and the lateral influences facilitate maintaining the cell wall under stress conitions (for a review, see Fuchs and Mylonakis 2009).

The cyclic AMP (cAMP)-dependent protein kinase (cAMP/PKA) pathway is a signaling cascade whose function has been extensively studied in several fungi (reviewed by Lengeler et al. 2000) and is implicated in controlling morphological changes and virulence of a number of fungi. Both in human and plant pathogens, cAMP/PKA-dependent signaling cascade plays a plethora of several important cellular processes. In *Cryptococcus neoformans*, these include capsule and melanin production, which are two very distinctive virulence factors for *C. neoformans* infection. Mutant strain lacking the Gα subunit of the GPCR–protein complex (*gpa1*) fails to induce capsule and melanin production and is avirulent in animal models of cryptococcosis (Alspaugh et al. 1997). This defect is suppressed by the addition of exogenous cAMP in the presence of a functional adenylate cyclase gene (Alspaugh et al. 1997). Similarly, in the most common human fungal pathogen *C. albicans*, the increase in cAMP concentration induces the transition from the yeast to the filamentous pathogenic form (Sabie and Gadd 1992). A mutant of the catalytic subunit of a protein kinase A (PKA) isoform encoded by *TPK2* blocks morphogenesis, regulates dimorphism, and partially reduces virulence in *C. albicans* (Sonneborn et al. 2000). *P. brasiliensis* presents two catalytic PKA subunits, *TPK1* and *TPK2*, with a highly conserved serine/threonine protein kinase catalytic domain (Fernandes et al. 2005), but the details about the function of these genes is not known. In addition, a puzzling fact about cAMP/PKC pathway in *P. brasiliensis* is that an earlier report described that exogenous cAMP or its analogs inhibit the mycelium-to-yeast switch conversely to what was observed in *C. albicans*, *H. capsulatum*, and *C. neoformans* (Paris and Duran 1985; Gauthier and Klein 2008). These data have recently been carefully confirmed by Chen et al. (2007) showing the same evidence in their results, indicating that exogenous cAMP addition after the thermodimorphic transition was triggered indeed retarded the mycelium-to-yeast morphological transition in *P. brasiliensis*, but not the yeast-to-mycelium switch. In spite of this, the authors conveniently refined the experiments and showed that during the morphological transition, a significant transient peak in *CYR1* transcripts (the adenylate cyclase homolog in *P. brasiliensis*) takes place 24 h after the onset of the morphological transition. This peak in *CYR1* abundance correlates with the peak in mycelia-to-yeast differentiation. In the same monitoring experiments, another further progressive increase in *CYR1* mRNA abundance was found 72–240 h after the beginning of the mycelium-to-yeast transition, which coincides with the stage where the fungus adopted the yeast form definitely. In addition, the increase in *CYR1* transcripts correlated with the increase in cellular cAMP levels. It peaked at about 12 h, decreased, and then progressively increased after 72–340 h upon induction. These results show that increase in both cytosolic cAMP and *CYR1* transcript levels does regulate the morphological transition in *P. brasiliensis*. However, instead of sensing the absolute concentration of cAMP, cells detect and respond to transient changes in cAMP, which is a different feature observed in a dimorphic fungi. In accordance to this hypothesis, exogenous additions of cAMP at 12 h after the onset of the morphological transition, i.e., when cellular cAMP levels are maximal, had less effect in retarding the transition compared to addition at the very beginning of it (Chen et al. 2007). These results are very elucidative and indicate that *P. brasiliensis* also has a functional cAMP/PKA pathway which governs morphological switch at the transcriptional level with a different pattern that was not observed in other fungi.

The intrinsic ability of *P. brasiliensis* to undergo thermodimorphic transition has inspired several other high-throughput studies exploiting this physiological trait, attempting to understand the morphological transition and finding master transcriptional regulator of the process and/or phase-specific genes. In this context, comparison of mycelium and yeast phase cells using macroarray hybridization and SSH by Marques et al. (2004) identified genes potentially involved in the maintenance of the architecture of the pathogenic yeast form of the fungus. Accordingly, these authors observed high levels of induction for the gene encoding α-1,3-glucan synthase (*AGS1*). Interestingly, in *P. brasiliensis*, the most obvious biochemical alteration when it adopts the yeast pathogenic form is the switch from β-1,3- (the sole neutral glucose polymer) to α-1,3-glucan chains in the cell wall. In the yeast form, the β-1,3-glucan content is drastically reduced to a minimum (about 5%) and replaced by α-1,3-glucan (Kanetsuna et al. 1972; San-Blas et al. 1977, 2002). α-1,3-Glucan is recognized as a virulence factor in *P. brasiliensis* and other dimorphic fungal pathogens such as *B. dermatitidis* (Hogan and Klein 1994) and *H. capsulatum* (Rappleye et al. 2004). Consistently, virulent cultures of *P. brasiliensis* isolates grown "in vitro" for long periods have thinner cell walls, low α-1,3-glucan levels, and are consequently less virulent (Andrade et al. 2006). In *P. brasiliensis*, linear glucan polymers constitute 97% of yeast cell wall (Sorais et al. 2010) and thus any interference in cell wall composition through glucan synthases is likely to affect virulence directly (Fernandes et al. 2005). The differential content of α- and β-glucan has also been validated "in vivo" in a study of *P. brasiliensis* transcriptional behavior upon internalization by murine macrophages. Under these conditions, the gene encoding *FKS1* was 3.5 times downregulated (Tavares et al. 2007).

A search for other *AGS* gene homologs in *P. brasiliensis* by Sorais et al. (2010) revealed that *AGS1* is the only gene in the genome of *P. brasiliensis* related to the synthesis of α-1,3-glucan, as is also the case in other pathogenic fungi such as *C. neoformans* and *H. capsulatum*, and different from other fungi such as *Schizosaccharomyces pombe* and *A. niger*, *A. nidulans*, and *A. fumigatus*, where several *AGS* genes are present. Noteworthy is the fact that chitin (the third polysaccharide moiety in *P. brasiliensis*) and β-glucan are recurrent sugars in fungi; however, the same is not true for α-glucan which is of rather infrequent occurrence (San-Blas and Nino-Vega 2008) but strategically exploited by *P. brasiliensis* in the host adaptation and immune response evasion (Gauthier and Klein 2008). It is possible that the absence of *AGS1 in P. brasiliensis* would produce the same phetotypes observed in *H. capsulatum*, where null mutants for α-1,3-glucan synthase eliminates cell wall α-1,3-glucan, retards fungal growth in macrophages following phagocytosis, and attenuates virulence (Gauthier and Klein 2008), but this hypothesis needs to be functionally assessed. Chitin synthase (*CHS*) genes were also found to be transcriptionally upregulated in mycelial saprophytic phase of *P. brasiliensis*, although yeast parasitic phase presents more chitin content (Nino-Vega et al. 2000). This supports observations in other fungi that transcript levels often do not correlate with chitin content and that posttranscriptional regulation of *CHS* gene expression may be important for morphogenesis. The genomes of the three *P. brasiliensis* strains recently sequenced present seven chitin synthase genes, each with one synthase belonging to each of the seven fungal chitin synthase classes. Interesting is the unique presence of a class III chitin synthase in a dimorphic fungus, a class so far reported only in filamentous fungal species (http://www.broadinstitute.org/annotation/genome/paracoccidioides_brasiliensis/MultiHome.html). The cell wall is a specific dynamic structure essential to almost every aspect of the biology and pathogenicity of fungal pathogens. Its structure confers physical protection and shape to fungal cells, and as the most external part of the fungus, it mediates the interaction with the host, including adhesion to host tissues and modulation of the host/pathogen response. Fernandes et al. (2005) presented an "in silico" analysis for the components of known signaling pathways involved in morphogenesis and virulence in other fungal pathogens; some of them were found to be transcriptionally induced in some reports (Table 9.1), including genes belonging the cell wall integrity (CWI) pathway. The CWI pathway utilizes

GTPase-activating proteins and guanyl nucleotide exchange factors to regulate the activation of the MAPK cascade, which ultimately leads to the activation of the transcription factor *RLM1* in *S. cerevisiae*. *RLM1* controls the expression of at least 25 genes which are direct effectors of the CWI, most of which have been implicated in cell wall biogenesis such as chitin and glucan synthases, acting upon MAPK phosphorylation (Jung et al. 2002) (see Fig. 9.1). In several fungi including

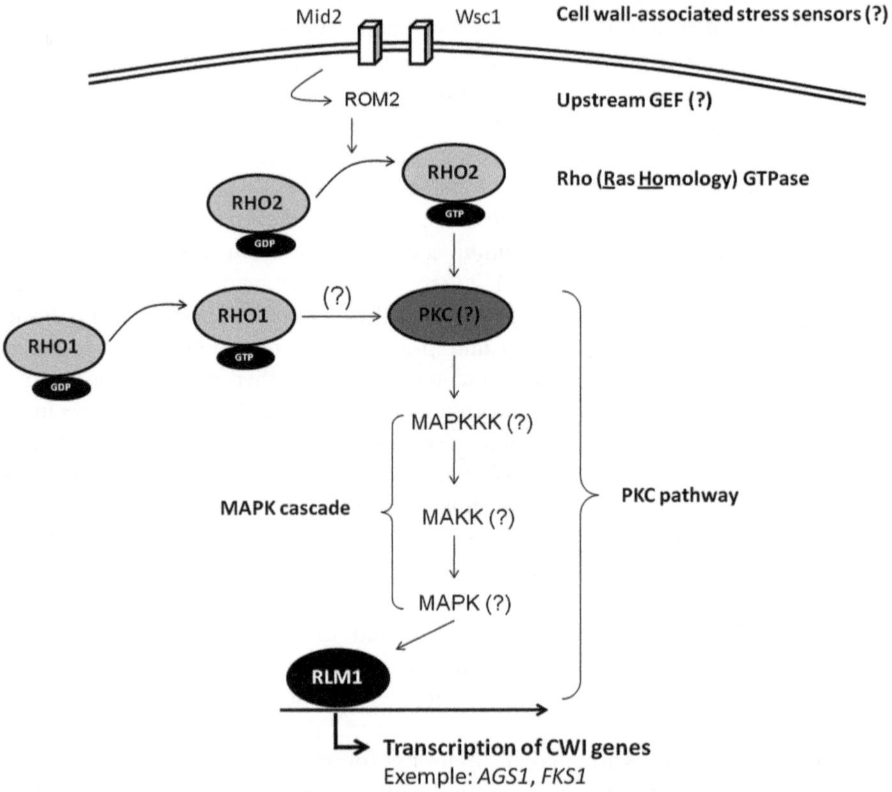


**Fig. 9.1** Proposed Cell Wall Integrity Pathway (CWI) in *P. brasiliensis* based on the model of *S. cerevisiae*. CWI is initiated by cell wall-associated stress sensors Mid2 and Wsc1 (not identified in *P. brasiliensis*). These proteins act as mechanosensors of cell wall stress during growth, morphogenesis, and exposure to high temperatures, chemicals, or other cell wall perturbations, and transmit signals to the downstream signaling pathway. These proteins bind to Rom2 (also not indentified in *P. brasiliensis*), which is a guanyl nucleotide exchange factor (GEF) for *RHO1*. Activated *RHO* binds and activates *PKC*, which in turn regulates the downstream MAPK cascade. In *P. brasiliensis*, it is not known how many PKCs are acting in the pathway. PKC phosphorylates a MAPK kinase kinase (MAPKKK), which transmits the signal to MAPK kinases (MAPKKs). These two kinases finally activate the MAPK. The components and the number of proteins in the MAPK pathway are not known in *P. brasiliensis*. Stimulation of the MAPK leads to phosphorylation of the transcription factors *RLM1* which initiate the expression of cell wall synthesis genes such as chitin syntases (*CHS1*) and glucan synthases (*FKS1* and *AGS1*). The genes represented inside a *circle* in the figure are those having experimental data available about its identity (see text for details). The other genes having a *question mark* were not identified in *P. brasiliensis*

*S. cerevisiae* and other ascomycetes, this is a conserved cascade (Lengeler et al. 2000; Gasch 2007). However, CWI components from some fungi are still being elucidated, and important differences in the pathway have been identified mainly regarding the redundance of the genes in the MAPK cascade. For example, there are four MAPK genes in the genomes of the filamentous fungi *A. nidulans*, *A. fumigatus*, and *Aspergillus oryzae* that are highly conserved across these species and activated by a variety of physiological processes, including growth, differentiation, and high osmolarity responses. However, there are only three genes that code for MAPKKK and MAPKK proteins (May 2008).

In *P. brasiliensis*, there is a total lack of information about the homologs that participate in the CWI, except for the fact that the important components of this pathway *PKC1* and *RHO1* were found to be transcriptionally induced in the yeast phase of *P. brasiliensis* (Marques et al. 2004). Curiously, the *PKC1* role has not been directly targeted in any of the dimorphic fungi, including those in which DNA-mediated transformation and gene deletion can be achieved. A single report in *S. schenckii* confirms the presence of PKC in yeast and mycelial extracts of this fungus and the occurrence of different isoforms of this enzyme in the yeast and mycelial forms (Aquino-Pinero and Rodriguez del Valle 1997).

An experiment-based search for *P. brasiliensis RHO1* homologs recently accomplished by Sorais et al. (2010) identified the Rho (*R*as *Ho*mology)-GTPases subfamily members *RHO1*, *RHO2*, *RAC*, and *CDC42* homologs and shed some light on the transcriptional dependence of these genes with their downstream effectors *FKS1* and *AGS1*. The membrane-bound Rho subfamily is highly conserved in all eukaryotes, while absent in bacteria and archae. Rho GTPases function as on/off molecular switch so that when bound to GTP, the GTPase is activated to affect downstream effectors, whereas when bound to GDP, the GTPase is deactivated. In the budding yeast model, *Rho1* is implicated in the control of morphogenetic events and the activity of *Fks1* (Mazur and Baginsky 1996). However, in other organisms such as the fission yeast, there are two PKC homologs, *PKC1* and *PKC2*. It has been demonstrated that in *S. pombe*, both PKCs interact with *RHO1* and *RHO2* (Arellano et al. 1999a, b; Calonge et al. 2000). However, *RHO1* activates the *FKS1* gene and regulates β-1,3-glucan biosynthesis mainly through *PCK2*, but also through *PCK1*. On the contrary, *RHO2* does not affect *FKS1* but regulates the biosynthesis of α-glucan synthase (via *AGS1*), preferentially through *PCK2*. Interestingly, in *P. brasiliensis*, expression studies showed that mRNAs levels of *RHO2* and *AGS1* kept a direct relationship in the yeast phase, indicating a coherent transcriptional regulation through the PKC-mediated CWI (Fig. 9.1). Whether or not *P. brasiliensis* has one (like in *S. cerevisiae*) or two (like in *S. pombe*) PKC functioning in the CWI needs further investigation through genome analysis. On the contrary, gene expression patterns of both putative subunits of the β-1,3-glucan synthase complex *FKS1* and *RHO1* did not keep a direct relationship with each other when tested against morphological phases and throughout the dimorphic process. Curiously, *RHO1* had a higher expression level during mycelial growth and less when the cells were in the yeast form. *FKS1* presented almost indetectable expression in mycelium and a remarkable higher expression in yeast (Sorais et al. 2010). *P. brasiliensis RHO1* gene was able to complement the *S. cerevisiae Rho1* null mutant fully.

Morphogenetic transition is the essence of *P. brasiliensis* life cycle whereby the fungus reaches its parasitic phase. It is induced upon increasing temperature and it is reasonable to speculate that this change may reflect in the lipid composition of the cell. Lipid signaling in fungal pathogens has been studied and lipid-dependent metabolic pathways have been characterized and related to fungal pathogenesis (for a review, see Rhome and Del Poeta 2009, 2010). Examples are the shingolipid signaling pathways that seem to center around the production of diacylglycerol in the formation of inositol phosphorylceramide. In *C. neoformans*, diacylglycerol activates both melanin production through laccase and transcription of antiphagocytic protein, both of which are involved in fungal virulence. Little is known in *P. brasiliensis* about the involvement of lipid reorganization and metabolism. Transcriptional profiling studies have started to address such issues in *P. brasiliensis* and were successful in indentifying preferentially expressed genes in the yeast pathogenic phase. These genes are related to the ergosterol biosynthesis (*ERG 25*, *ERG28*, *ERG11*, *ERG3*), cholestanol Δ isomerase, lysophospholipase, phospholipase A2, carnitine dehydratase, and fatty acid desaturase (Marques et al. 2004; Bastos et al. 2007; Monteiro et al. 2009; Garcia et al. 2010). The presence of these genes was independently found using different experimental approaches, supporting the idea that membrane lipid reorganization in yeast can be a response to the host higher temperature environment or oxidative stress in comparison to the environmental niche. Changes in membrane fluidity and lipid composition are consequences of temperature increase in biological membranes. Upon transition to yeast phase, it leads to an increase in membrane fluidity which is counteracted by biochemical response for increasing the saturated fatty acid (SFA) content in cell membrane to increase the SFA/unsaturated fatty acid (UFA) ratio. All organisms synthesize only SFA, while UFAs are produced by the action of microsomal desaturases. In *S. cerevisiae*, additions of SFA induced a strong increase in heat shock mRNA accumulation when cells were heat-shocked at 37°C, while treatment with an UFA reduced or eliminated the level of heat shock gene transcription at 37°C. Accordingly, an interesting experiment involving the *H. capsulatum* $\Delta^9$-desaturase (*OLE1*) gene demonstrated the effect of temperature change in this dimorphic fungus (Carratu et al. 1996). *OLE1* is the major enzyme in animal and yeast cells that converts SFA into UFA, catalyzing the insertion of a double bond in palmitoyl- and stearoyl-CoA intermediates. *OLE1* transcription of *H. capsulatum* is upregulated in a temperature-sensitive strain, and a *Ole1* mutant strain of *S. cerevisiae* is complemented with its own *Ole1* coding region under control of either its own promoter or *H. capsulatum Ole1* promoter. The transcriptional regulation of heat shock and desaturase genes are two examples of genes whose expressions are directly controlled by changes in the membrane fluid state, such as physical stresses, pathophysiological conditions (Carratu et al. 1996), and probably fungal dimorphism. In this sense, alteration of the membrane physical state could function as a cellular sensor whereby yeasts would sense the temperature alteration. It is possible that a connection between the membrane sensor and a downstream signal transduction pathway regulates gene expression of effector proteins that ultimately adjust the cell to the new growth condition (Vigh et al. 2005) and possible cell shape. However, this is an open avenue for investigation in dimorphic fungi. Another recent finding regarding this issue shows that overexpression of a $\Delta^9$-desaturase in the highly virulent G217B strain of *H. capsulatum* causes loss of its ability to survive and persist within murine macrophages, impairment of the heat shock response, and lack of virulence in mouse (Porta et al. 2010). This is consistent with the hypothesis that a cell must regulate the SFA/UFA ratio depending on the growth temperature. Currently, few studies in *P. brasiliensis* address the composition and structure of sphingolipids (Levery et al. 1998; Toledo et al. 1999), and the close details about the involvement of these structures in morphogenesis and virulence in this pathogen remain to be determined. One interesting possibility that arises from the observation of the cell membrane plasticity upon thermal induction in dimorphic fungi (besides that related to the fluidity) is that glycosphingolipids can modulate the activity of membrane-bound enzymes, suggesting that they could play a role in dimorphism through their interaction with one or more key enzymes affecting fungal morphology. This is an especially appealing idea since the major cell wall biosynthetic enzymes, chitin synthases and glucan synthases, are membrane bound, and their enzymatic products drastically change during thermodimorphic transition (Cabib et al. 1988; Toledo et al. 1999). Not only that, the cell CWI starts from the activity of stress sensors such as *Wsc1* and *Mid2* in *S. cerevisiae* (Fig. 9.1) which are also membrane proteins. However, neither the functional details about this hypothesis are currently available, nor the *P. brasiliensis* counterparts of *Wsc1* and *Mid2* were identified.

Among the genes involved in lipid metabolism, Monteiro et al. (2009) have also identified the *S. cerevisiae* homolog *HAC1* transcription factor (named as *HAC1*) as upregulated in yeast cells at all times in their experimental design. This transcription factor is involved in the *U*nfolded *P*rotein *R*esponse (UPR) in the endoplasmic reticulum (ER), as well as in phospholipid biosynthesis. The UPR can be caused by stresses such as lipid deprivation, heat, and drug treatment. The UPR has been recently associated with virulence and susceptibility to antifungals in the human opportunistic pathogen *A. fumigatus* (Richie et al. 2009). Upregulation of the gene in *P. brasiliensis* yeast could be a response to growth at higher temperature that responds with an increase in SFA:UFA ratio, or could be related to the marked distinct lipid composition between mycelium and yeast forms (Manocha 1980). It has been shown in *S. cerevisiae*, an organism not routinely exposed to high-fat diets, that UPR is induced under lipotoxic conditions. This suggests that the mechanisms of SFA toxicity are largely conserved throughout eukaryotes. In *S. cerevisiae*, UPR begins with the activation of the ER transmembrane protein *Ire1* in response to high levels of aberrant folded proteins. Ire1 excises an intron from yeast *HAC1* (uninduced) cytoplasmic precursor mRNA. Only the spliced form of *HAC1* mRNA (induced) can be translated once it creates a frame shift in the mRNA molecule. Induced *HAC1* is translocated to the nucleus where it acts as a transcription factor to activate several UPR target genes (Bicknell et al. 2007). Even though there is a transcriptional induction of the *HAC1* gene in *P. brasiliensis*, there is no evidence whether or not thermodimorphic transition affects the amount of the induced and uninduced forms of *HAC1* and if the spliced form of *HAC1* would render contributions to cell shape, morphogenesis, and virulence. Further studies on this issue will be certainly required to address if UPR plays an important role in *P. brasiliensis* and/or other dimorphic fungi.

Coping with the temperature increase during the termodimorphic transition requires from *P. brasiliensis* cells not only the ability to rearrange membrane lipid, but also a sophisticated machinery to manage the oxidative stress in response to temperature. Basically, all sorts of transcriptome analysis so far using several methodologies which aimed establish comparisons between mycelium and yeast cells have demonstrated the upregulation of genes involved in oxidative stress response. This indicates that antioxidant systems in *P. brasiliensis* could be extremely important not only for survival in macrophage phagosomes (Tavares et al. 2007), but also for detoxification of oxygen reactive species (ROS). The work of Tavares et al. (2007) which investigated the early transcriptional response of *P. brasiliensis* to the environment of peritoneal murine macrophages identified that *SOD3*, a putative Cu, Zn superoxide dismutase gene, was involved in the elimination of superoxide anions as the most significantly upregulated gene in the microarray analysis (Table 9.1).

### *9.3.2 Sulfur Metabolism in* **P. brasiliensis** *in the Light of Transcriptional Data*

One of the most interesting features about the growth requirements of some dimorphic fungi, including *P. brasiliensis*, *H. capsulatum*, and *B. dermatitidis*, is related to its selective behavior for sulfur sources. These species are unable to grow using inorganic sulfur in the pathogenic yeast phase and are auxotrophic for organic sulfur such as cysteine, methionine, or cystine. On the contrary, the saprophytic mycelial stage can grow in the presence of either organic or inorganic sulfur (Boguslawski and Stetler 1979; Paris et al. 1985; Medoff et al. 1987; Maresca and Kobayashi 1989). Although it is observed in several species, the organic sulfur growth requirement for the *P. brasiliensis* yeast phase has also been validated in a collection of 79 strains isolated from different environments in South America. All of the tested strains presented the same phenotype (Ferreira et al. 2006). In *H. capsulatum*, the addition of exogenous sulfhydryl reducing agents (dithiothreitol) to the media locks cells in the yeast form independent of temperature, whereas the addition of sulfhydryl oxidizing agents locks cells in the mycelial form independent of temperature (Klein and Tebbets 2007). This sulfur auxotrophy suggests that along with temperature, organic sulfur source is important for driving the morphological switch from mycelium to yeast and a stimulus for maintaining *P. brasiliensis* in the parasitic phase.

The initial observations that yeast cells of *P. brasiliensis* were unable to grow in the presence of inorganic sulfur by Paris et al. (1985) were under intense investigation through the array methodology experiments. Several reports have presented data showing that sulfur metabolism-related proteins such as enzymes involved in extracellular sulfur acquisition and production of cysteine and methione, and membrane permeases were markedly upregulated in the yeast phase of the fungus (Marques et al. 2004; Andrade et al. 2006; Ferreira et al. 2006; Tavares et al. 2007).

The sulfur metabolism in mammals, plants, and fungi is a highly regulated process. Genes encoding the enzymes belonging to the sulfate uptake pathway are subject to transcriptional control and to the action of either positive or negative acting regulatory factors (for reviews, see Marzluf 1997; Thomas and Surdin-Kerjan 1997). The inorganic sulfate assimilatory pathway that is used by these organisms and most likely also by *P. brasiliensis* starts with the sulfate uptake by the cell through several sulfate permeases present in the cell membrane. Once inside the cell, the sulfate is subjected to two phosphorylation reactions yielding the 5′-phosphoadenosine 3′-phosphosulfate molecule (PAPS). This intermediate is then reduced to sulfide (by the enzyme sulfite reductase) which is condensed with *O*-acetyl-serine to generate cysteine (Fig. 9.2). The "de novo" synthesis of cysteine from inorganic sulfate assimilation makes this amino acid available inside the cells to function as a primary source of organic sulfur and an important intermediate for the synthesis of methione and S-adenosylmethionine via the transsulfuration pathway. The transsulfuration pathway consists of reactions that allow the interconversion of homocysteine and cysteine via the intermediary formation of cystathionine (Thomas and Surdin-Kerjan 1997). In several organisms like plants and fungi such as *Neurospora crassa* and *A. nidulans*, there is a lateral branch in the sulfur acquisition metabolic pathway whereby PAPS (which is toxic to cell in high cytosolic concentrations) is converted to choline-*O*-sulfate by the action of PAPS-choline sulfotransferase. Choline-*O*-sulfate is an osmoprotectant and an intracellular additional sulfur source. This metabolite can also be exogenously uptaken by cells through membrane-specific permeases (Marzluf 1997). Choline sulfatase acting on its substrate choline-*O*-sulfate produces an internal pool of inorganic sulfate which can reenter the main pathway depending on the cell needs for cysteine.

As mentioned earlier, the transcriptome analysis reported so far have identified several genes upregulated in the yeast phase encoding for different enzymes acting in the sulfur assimilation pathway, which are gathered here and highlighted in bold letters in Fig. 9.2 (Marques et al. 2004; Andrade et al. 2005; Ferreira et al. 2006; Tavares et al. 2007). Due to the noted similarity, it is predictable that *P. brasiliensis* sulfur assimilation pathway shares the same characteristics, enzymes, and is subjected to regulation as described in well-genetically studied fungi such as *S. cerevisiae* and the filamentous *N. crassa* and *A. nidulans*. Besides the genes involved in inorganic sulfur assimilation, two other genes involved in organic sulfur acquisition – methionine permease and cysteine dioxygenase (the latter not shown in Fig. 9.2) – had increased mRNA accumulation during both mycelium-to-yeast transition and yeast phase growth. Interestingly, genes such as those for choline

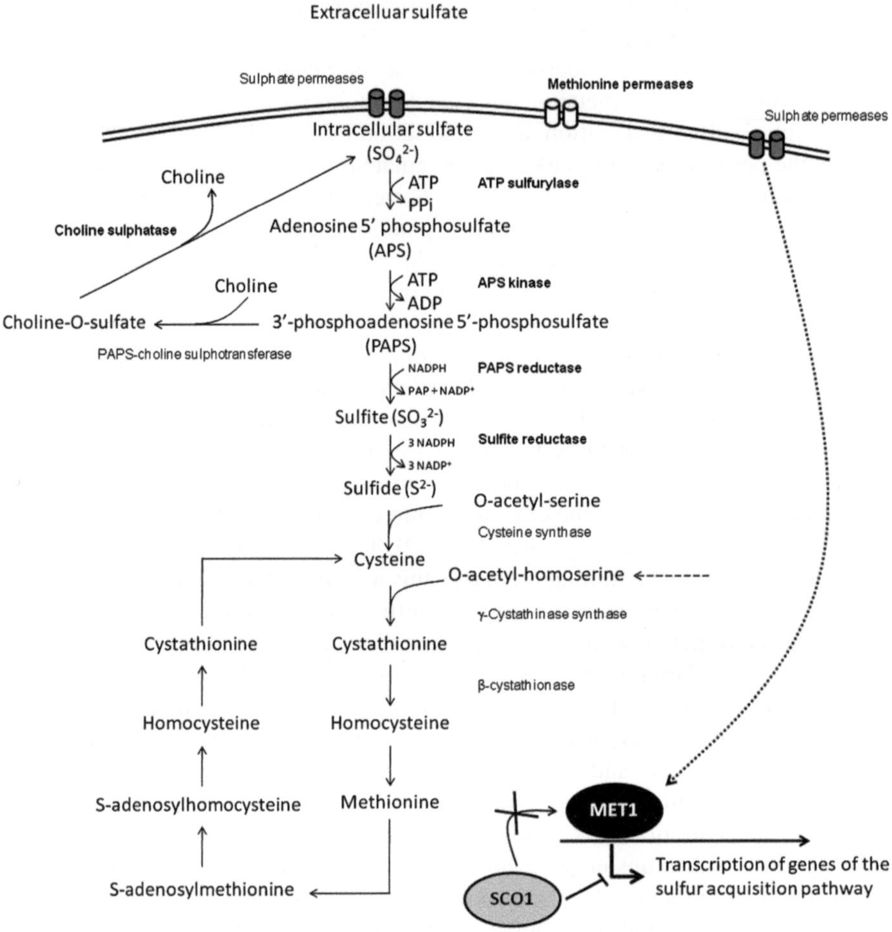


**Fig. 9.2** Proposed sulfur assimilatory pathway following the model of *A. nidulans*, *S. cerevisiae*, and *N. crassa* based on transcriptional data available in *P. brasiliensis*. The enzymes in *bold* indicate those where experimental data show that upregulation in the pathogenic yeast form was obtained from some transcriptional profiling studies (Marques et al. 2004; Andrade et al. 2006; Ferreira et al. 2006). *Dotted line* indicates the role of inorganic sulfur source (1) to induce transcriptional response of *MET1* transcription factor, leading to the mRNA abundance of the enzymes in the assimilatory pathway, or (2) to cause sulfur catabolic repression by stimulating the Skp1 family gene *SCO1* to promote ubiquitin-mediated proteolysis of *MET1* and therefore inhibiting the transcription of the required enzymes for sulfur assimilation. Methionine permease is required for organic sulfur assimilation and is also shown in the figure

sulfatase (*CS1*), APS kinase (*APS1*), and methionine permease reach an induction of about 53-, 8-, and 200-fold, respectively, in *P. brasiliensis* yeast pathogenic phase compared to that in mycelial control (Ferreira et al. 2006).

Several of the genes identified as modulated in the inorganic sulfur assimilation pathway, as described above, are known from other fungal organisms to be subjected to transcriptional control by the action of positive and negative regulatory elements. In *N. crassa*, *S. cerevisiae*, and *A. nidulans*, the main positive *trans*-acting elements for the sulfur assimilatory pathways are *cys-3*, *MET4*, and *metR*, respectively (Natorff et al. 2003). In *P. brasiliensis*, the homolog of *cys-3* and *metR* genes, *MET1*, was also found to be upregulated (35-fold increase) in the yeast phase and was able to complement the *metR* null mutant of *A. nidulans* (Ferreira et al. 2006). These complementation experiments of *MET1* in Δ*metR* mutant show that *MET1* is a member of the bZIP family of DNA-binding proteins. The bZIP domain is bipartite, consisting of a leucine zipper responsible for dimerization and an immediate upstream basic region which is essential for specific binding to DNA. Therefore, under inducing conditions such as abundance of extracellular inorganic sulfur, *MET1* shares the same function observed for *cys-3* and *metR* in sulfur regulation, allowing the expression of an entire set of genes that encode permeases and enzymes involved in the acquisition of sulfur from the environment (Marzluf 1997). In *A. nidulans*, *metR* affects the transcription of genes encoding a sulfur controller, sulfate permease, ATP sulfurylase, homocysteine synthase, and cysteine synthase. Some of these genes were also upregulated in *P. brasiliensis* along with *MET1* (Fig. 9.2).

In contrast, in conditions of plenty of organic sulfur source such as methionine, negative acting elements orchestrate the sulfur metabolite repression that culminates in low levels of the above-described enzymes involved in the sulfur assimilation pathway. Among these negative regulatory elements are the *scon-2*, *MET30*, and *sconC* from *N. crassa*, *S. cerevisiae*, and *A. nidulans*, respectively (Natorff et al. 1998), and *SCO1* from *P. brasiliensis* (Ferreira et al. 2006). These genes are members of the multiprotein ubiquitin ligase SCF complexes. SCF is the best studied class of ubiquitin ligases which are composed of multiple elements including the adaptor protein *S*kp1, the scaffold protein *C*ullin Cdc53/cullin, an *F*-box protein, the RING finger protein Hrt1/Rbx1/Roc1, and the ubiquitin conjugating enzyme Cdc34. In this protein complex, the F box component determines the specificity of the degrading system once it can be replaced depending on the target to be ubiquitynilated and moved to the proteolytic pathway (Marzluf 1997; Thomas and Surdin-Kerjan 1997; Patton et al. 1998). Accordingly, *scon-2*, *MET30*, *sconC*, and SCO1 are homologs of the Skp1 family. The *P. brasiliensis* SCO1 gene was also able to complement the *A. nidulans sconC3* (Ferreira et al. 2006) and thus, its correct assembly in the SCF ubiquitin ligase complex targets the proteolysis of MET1 transcription factor (Fig. 9.2). Although *SCO1* is upregulated at the transcriptional level in the yeast phase in *P. brasiliensis* (about 1.2 fold increase), it is very consistent with the results in the same report by Ferreira et al. (2006), in which the positive regulator of the sulfur acquisition *MET1* was highly upregulated, indicating that the cross-talking of these two regulators are running in opposite direction.

One of the most important observations from these transcriptional data encompassing the sulfur metabolism in *P. brasiliensis* is that even though the fungus cannot grow as yeast or undergo morphological transition in the presence of only inorganic sulfur, the sulfur uptake pathway is fully active as noted by the upregulation of the key enzymes of the pathway highlighted in Fig. 9.2, and for an unknown reason, the fungus uses the two pathways for acquiring sulfur during the growth. Consistently, the *P. brasiliensis MEP1* gene (representing the organic assimilation pathway) was one of the most upregulated genes (200-fold increase) in the condition tested by Ferreira et al. (2006). Interestingly, an antifungal product obtained from *Bacilllus cereus* can inhibit the action of *S. cerevisiae* sulfite reductase (*MET10*) interfering in the transcriptional activation of *MET4* (Pb MET1) and posttranscriptional regulation in *MET10* expression (Aoki et al. 1996). This compound has still not been tested in *P. brasiliensis* or in any other dimorphic fungi which present this sharp regulation between inorganic or organic sulfur source assimilation linked to phase switch. Still regarding the sulfur acquisition pathway, it was demonstrated that *S. cerevisiae* cells grown at 30°C with methionine as sulfur source lose viability upon transfer to 45°C, whereas they survive the transfer in the absence of methionine. This methionine-mediated cell death at high temperature can be explained by the protective effect of intracellular APS. Indeed, APS is elevated after a temperature shift, and cells are unable to synthesize this intermediate [because of repression by methionine or mutation of the *MET3* (ATP sulfuryase) gene] and do not survive the temperature shift. The results demonstrate that APS plays an important role in methionine-mediated cell death at high temperature (Jakubowski and Goldman 1993; Thomas and Surdin-Kerjan 1997). This hypothesis was also not tested in *P. brasiliensis*.

In summary, even before the whole genome sequence, much information has been extracted, validated, or reinterepreted from the transcriptome and DNA array analyses of *P. brasiliensis* and sulfur usage. The details about this important feature of this pathogen still needs to be dissected before a link between its virulence and pathogenicity can be established.

### *9.3.3 Genomic Era of* **P. brasiliensis**

Recently, the genomes of three well-studied *P. brasiliensis* strains, known as Pb01, Pb03, and Pb18, became available by the efforts of the Dimorphic Fungal Genomes Consortium (available at, http://www.broadinstitute.org/annotation/genome/paracoccidioides_brasiliensis/MultiHome.html>).

The strategy of sequencing of the three different *P. brasiliensis* strains is based in previous genomics studies using available molecular tools in the absence of whole genome sequences such as random amplified polymorphic DNA, restriction fragment length polymorphism, partial sequencing of some genes from several isolates, and karyotyping by electrophoresis. Many years of research have provided consistent evidences for the extensive genetic variability in *P. brasiliensis* and the presence of clusters that correlated with geographic locations (Soares et al. 1995; Montoya et al. 1997; Calcagno et al. 1998; Molinari-Madlum et al. 1999; Hahn et al. 2003). Actually, the three *P. brasiliensis* strains are from distinct phylogenetic lineages, exhibit different phenotypes "in vitro," and induce different host immune responses (Carvalho et al. 2005; Matute et al. 2006). For example, PB01 is the most thoroughly studied isolate from a molecular point of view (San-Blas and Nino-Vega 2008). On the contrary, Pb18 has been extensively used in the literature due to its proven virulence in mice when inoculated by intraperitoneal, intratracheal, and intravenous routes (Calich and Kashino 1998). Based on the sequence of eight loci from 65 *P. brasiliensis* isolates, Matute et al. (2006) reported three distinct phylogenetic species within *P. brasiliensis*: S1 (a paraphyletic group formed by 38 isolates of Argentinean, Brazilian, Peruvian, and Venezuelan origins and an isolate from an Antarctic penguin), PS2 (six isolates, five of them of Brazilian origin and one Venezuelan), and PS3 (21 Colombian isolates) (San-Blas and Nino-Vega 2008).

Thus, comparative analysis of the three *P. brasiliensis* strain genomes is being carried out, and data from the genomic analysis encompass the Pb18 strain, representing major S1 group and virulence, Pb3 from phylogenetic species PS2, and Pb01 as a molecular model which alone seems to belong to a phylogenetic group distinct from S1, PS2, and PS3 (Molinari-Madlum et al. 1999; Carrero et al. 2008). Recently, a study reported that the results of the analysis of 13 polymorphic loci from several isolates showed high divergence of the "PB01-like" isolates from the three distinct phylogenetic species identified (S1, PS2, and PS3). Based on the molecular data, exclusive morphological traits, and a possible long period of genetic isolation, the authors suggested a new specie within the *Paracoccidioides* genus containing the "PB01-like" clade, formally named as *Paracoccidioides lutzii* (Teixeira et al. 2009) as tribute to the Brazilian medical mycologist Adolpho Lutz who first described *P. brasiliensis* (Schwarz and Baum 1965).

*P. brasiliensis* genome availability can furnish additional tools for the analysis of the genomic basis of phase-transition and phase-specific virulence determinants. Comparative genome approaches can be applied to the three *P. brasiliensis* strains aiming at the identification of homologous genes to the other dimorphic fungi known for virulence and disease. The analysis of the medically important fungal genomes holds the potential to address clinical issues. In particular, given the complete gene set for a pathogenic fungus, it becomes possible to predict genes necessary for fungal growth that lack human homologs. These may represent targets for antifungal drugs with fewer toxic side effects (Galagan et al. 2005). Genome analysis can provide a detailed list of such genes available for further gene function studies.

Previous genome and transcriptome studies using available molecular tools in the absence of whole genome sequences provided some gene candidates with no human homolog upregulated in either the yeast or mycelial phase [(Felipe et al. 2005), Table III]. Among them are the β-1,3-glucan synthase (*FKS1*) and α-1,3-glucan synthase (*AGS1*) which are differentially upregulated in the mycelial and yeast forms, respectively. Also, in the nonhuman homologs, the main enzymes of the glyoxylate cycle (malate synthase and isocitrate lyase) are more abundant in the yeast phase. Currently, the first reports using the three *P. brasiliensis* sequenced strains are appearing containing data related to genomic analysis. Abadio et al. (2011) have conducted an "in silico" analysis and manual mining. The authors selected initially 57 potential drug targets, experimentally confirmed genes as essential for *C. albicans* or *A. fumigatus*. Orthologs for these 57 potential targets were also identified in eight human fungal pathogens. Of those, ten genes were present in all pathogenic fungi analyzed and absent in the human genome.

These newly released genomic data can favor the design of the whole-genome microarray chips or broader studies in RNA sequencing to support systematic studies at the transcriptional level. For this purpose and in the absence of reliable genetic techniques for *P. brasiliensis* manipulation, pharmacological probes (drugs and enzymatic inhibitors) can be used to study multiple metabolic pathways and help to validate data generated from functional studies derived from planned experimental design in the forthcoming years of research in *P. brasiliensis*. In many other fungal pathogens such as *C. albicans*, the completion of the genome sequence, the availability of whole-genome microarrays, and the development of tools for rapid molecular genetic manipulations of the fungal genomes generated a large body of information about the biology and the molecular mechanisms of virulence. They were also able to reveal the extent of similarities and differences between the pathogenic species in respect to nonpathogen model organisms such as *S. cerevisiae* (Berman and Sudbery 2002). We hope that this panorama can be visualized in the future research.

Although several decades have passed and important observations have been made about the *P. brasiliensis* physiology, several questions are currently partially understood. For example, what are the complete set of genes and pathways regulated during the thermodimorphic transition in the presence of β-stradiol? This experiment has already not been undertaken in *P. brasiliensis*. However, it can be predicted that results arising from this experimental approach could help to elucidate a very important epidemiological data, i.e., the preferred infection of male individuals instead of female in an overall ratio of 13 times higher in men than in women in endemic areas. This unbalanced proportion of incidence can achieve even higher proportion (150:1) in specific areas such as Argentina, Colombia, and Ecuador (Brummer et al. 1993). It has been observed in several reports that female hormones may present a protective unknown mechanism whereby the infection is less represented and/or minimized in women. In this case, the transition from conidia or mycelium to yeast is halted both “in vitro” (Restrepo et al. 1984; Salazar et al. 1988; Clemons et al. 1989) and “in vivo” (Aristizabal et al. 1998). It is known that the mechanisms whereby this inhibition occurs involve the presence of an estradiol-binding protein located in the cytoplasm of *P. brasiliensis* which might regulate the process of mycelium-to-yeast transition and ultimately inhibit or delay the adaptation of the fungus to the host cells (Loose et al. 1983). This estradiol-binding gene was found to be upregulated in the yeast phase of *P. brasiliensis* (Felipe et al. 2005). An understanding of the detailed molecular basis of this hormonal effect could provide novel perspectives on therapeutical targets to prevent the morphological switch.

Another point that genomics and transcriptomics studies may help is on the comprehension of the *P. brasiliensis* mechanisms of host adaptation. A very attractive approach would be to understand which genes and pathways are stimulated (or silenced) during the adaptation of *P. brasiliensis* to different reservoirs. *P. brasiliensis* seems to be a sophisticated fungus with the ability to colonize diverse ecological niches. This organism has been isolated or observed in soil (which appears to be the natural habitat of *P. brasiliensis*), in the intestinal tract of frugivorous bats, in dogs, in dog food contaminated with soil, in the feces of a penguin, during histological examination of squirrel monkey organs, and mainly in nine-banded armadillos *Dasypus novemcinctus* (summarized by Costa et al. 2010). Which mechanisms account for the apparent large capability of *P. brasiliensis* to adapt to various niches? Does *P. brasiliensis* have acquired defined genotypic traits that explain its ubiquitous lifestyle in nature? If so, can these traits be identified within the genome? Can the genetic variability identified in *P. brasiliensis* that allows its classification into three phylogenetic species S1, PS2, and Pb01 be more deeply exploited within the genome? Whole genome comparisons of fungi coming from different environmental niches can help to elucidate similarities and differences that can be linked to the mechanisms of pathogen adaptation.

The genome sequences also can foster a new set of experiments only partially explored in *P. brasiliensis* which are aimed at capturing pathogen gene expression pattern directly from the host niche to understand the highly complex nature of microbial virulence. Although very elucidative, this kind of "ex vivo" transcriptional profiling using fungal samples that had been in intimate contact with the host environment brings together multiple experimental and intellectual challenges, such as difficulties in extracting the pathogen RNA and choice of experimental design (for a review, see Cairns et al. 2010).

## 9.4 Advances in Genetic Manipulation in *P. brasiliensis* to Study Gene Function, Morphogenesis, and Virulence Determinants

The above-mentioned approaches to study the transcriptome of *P. brasiliensis* have recently provided a large amount of data, and much more can be achieved through the whole-genome dissection. However, the study of gene function in *P. brasiliensis* is still limited which represents a bottleneck in virulence analysis. The ability to introduce an engineered mutation in an organism is vital for both forward and reverse genetic analyses, and some questions that arise from the *P. brasiliensis* lifestyle and infective process and disease progression can only be elucidated by means of molecular genetics tools, as observed in *C. albicans* a few years ago (Berman and Sudbery 2002; Magee et al. 2003).

Although several reports have demonstrated that *P. brasiliensis* is suitable for DNA-mediated transformation by both electroporation and *Agrobacterium tumefaciens*-mediated transformation (ATMT), low transformation efficiency and stability were observed. Recently, Almeida et al. (2006) have evaluated factors influencing ATMT efficiency in *P. brasiliensis* such as cocultivation conditions and host cell susceptibility for transformation of the marker gene hygromycin B phosphotransferase. These authors were able to generate GFP-expressing isolates under the control of several promoters from different fungi. In this report, although the transformation efficiency was not equivalent as that observed for other yeasts and filamentous fungi, stable homokaryon progenies could be satisfactorily obtained in a haploid *P. brasiliensis* strain (Almeida et al. 2007). Based on the same transformation strategy, antisense RNA approach was successfully explored by Almeida et al. (2009) to promote the knock down of the *P. brasiliensis* CDC42 homolog, achieving expression decrease rates ranging 49–88% in the selected clones (Almeida et al. 2009). Importantly, the *P. brasiliensis* CDC42 silencing greatly affects the cellular morphology of yeasts, producing cells lacking the typical polymorphism presented by wild-type yeast cells in terms of bud cell size related to the peculiar "steering wheels" phenotype. In addition, the virulence of cells expressing the antisense *P. brasiliensis CDC42* RNA led to a significant increase in survival when compared to the wild-type cells in a mice model (Almeida et al. 2009). Similarly, Boyce et al. (2001) have observed that the *CDC42* homolog in *P. marneffei*, *cflA*, is also required for polarization and determination of cell shape during yeast-like growth, and active CflA is required for the separation of yeast cells. It remains to be determined the role of the complete set of proteins involved in the polar growth in *P. brasiliensis*. Recently, a *P. brasiliensis* 32-kilodalton hydrolase (Had32) was also studied using a knockdown approach, and yeast cells with reduced Had32 expression presented affected cell morphology, impaired capacity of adhesion to human epithelial cell, and decreased virulence in a mouse model of infection (Hernandez et al. 2010). To our knowledge, these are the first studies which tried to address the function of a specific gene in *P. brasiliensis* using a genetic tool rather a biochemical or pharmacological approach.

The value and utility of any model genetic organism rely on many factors, including the basic biological features of the system and the ease with which the organism can be experimentally manipulated (Osmani et al. 2008). Such tool improvements in molecular genetics and cell biology will become necessary to analyze precisely the information emerging from broad-range screenings and the *P. brasilienis* high-quality genome sequences aiming at virulence analysis and fungal pathogenesis traits such as adhesion, in vivo growth, metabolic shift, invasion, and so on.

## References

Abadio AK, Kioshima ES, Teixeira MM et al (2011) Comparative genomics allowed the identification of drug targets against human fungal pathogens. BMC Genomics 12:75

Almeida AJ, Martins M, Carmona JA et al (2006) New insights into the cell cycle profile of *Paracoccidioides brasiliensis*. Fungal Genet Biol 43:401–409

Almeida AJ, Carmona JA, Cunha C et al (2007) Towards a molecular genetic system for the pathogenic fungus *Paracoccidioides brasiliensis*. Fungal Genet Biol 44:1387–1398

Almeida AJ, Cunha C, Carmona JA et al (2009) Cdc42p controls yeast-cell shape and virulence of *Paracoccidioides brasiliensis*. Fungal Genet Biol 46:919–926

Alspaugh JA, Perfect JR, Heitman J (1997) Cryptococcus neoformans mating and virulence are regulated by the G-protein alpha subunit GPA1 and cAMP. Genes Dev 11:3206–3217

Andrade RV, Da Silva SP, Torres FA et al (2005) Overview and perspectives the transcriptome of *Paracoccidioides brasiliensis*. Rev Iberoam Micol 22:203–212

Andrade RV, Paes HC, Nicola AM et al (2006) Cell organisation, sulphur metabolism and ion transport-related genes are differentially expressed in *Paracoccidioides brasiliensis* mycelium and yeast cells. BMC Genomics 7:208

Aoki Y, Yamamoto M, Hosseini-Mazinani SM et al (1996) Antifungal azoxybacilin exhibits activity by inhibiting gene expression of sulfite reductase. Antimicrob Agents Chemother 40:127–132

Aquino-Pinero EE, Rodriguez del Valle N (1997) Different protein kinase C isoforms are present in the yeast and mycelium forms of *Sporothrix schenckii*. Mycopathologia 138:109–115

Arellano M, Coll PM, Perez P (1999a) RHO GTPases in the control of cell morphology, cell polarity, and actin localization in fission yeast. Microsc Res Tech 47:51–60

Arellano M, Valdivieso MH, Calonge TM et al (1999b) *Schizosaccharomyces pombe* protein kinase C homologues, pck1p and pck2p, are targets of rho1p and rho2p and differentially regulate cell integrity. J Cell Sci 112(Pt 20):3569–3578

Aristizabal BH, Clemons KV, Stevens DA, Restrepo A (1998) Morphological transition of *Paracoccidioides brasiliensis* conidia to yeast cells: in vivo inhibition in females. Infect Immun 66:5587–5591

Bagagli E, Theodoro RC, Bosco SM, McEwen JG (2008) *Paracoccidioides brasiliensis*: phylogenetic and ecological aspects. Mycopathologia 165:197–207

Bastos KP, Bailao AM, Borges CL et al (2007) The transcriptome analysis of early morphogenesis in *Paracoccidioides brasiliensis* mycelium reveals novel and induced genes potentially associated to the dimorphic process. BMC Microbiol 7:29

Berman J, Sudbery PE (2002) *Candida albicans*: a molecular revolution built on lessons from budding yeast. Nat Rev Genet 3:918–930

Bialek R, Ibricevic A, Fothergill A, Begerow D (2000) Small subunit ribosomal DNA sequence shows *Paracoccidioides brasiliensis* closely related to *Blastomyces dermatitidis*. J Clin Microbiol 38:3190–3193

Bicknell AA, Babour A, Federovitch CM, Niwa M (2007) A novel role in cytokinesis reveals a housekeeping function for the unfolded protein response. J Cell Biol 177:1017–1027

Boguslawski G, Stetler DA (1979) Aspects of physiology of Histoplasma capsulatum (a review). Mycopathologia 67:17–24

Borges-Walmsley MI, Chen D, Shu X, Walmsley AR (2002) The pathobiology of *Paracoccidioides brasiliensis*. Trends Microbiol 10:80–87

Boyce KJ, Hynes MJ, Andrianopoulos A (2001) The CDC42 homolog of the dimorphic fungus *Penicillium marneffei* is required for correct cell polarization during growth but not development. J Bacteriol 183:3447–3457

Brown AJ, Gow NA (1999) Regulatory networks controlling *Candida albicans* morphogenesis. Trends Microbiol 7:333–338

Brummer E, Castaneda E, Restrepo A (1993) Paracoccidioidomycosis: an update. Clin Microbiol Rev 6:89–117

Cabib E, Bowers B, Sburlati A, Silverman SJ (1988) Fungal cell wall synthesis: the construction of a biological structure. Microbiol Sci 5:370–375

Cairns T, Minuzzi F, Bignell E (2010) The host-infecting fungal transcriptome. FEMS Microbiol Lett 307:1–11

Calcagno AM, Nino-Vega G, San-Blas F, San-Blas G (1998) Geographic discrimination of *Paracoccidioides brasiliensis* strains by randomly amplified polymorphic DNA analysis. J Clin Microbiol 36:1733–1736

Calich VL, Kashino SS (1998) Cytokines produced by susceptible and resistant mice in the course of *Paracoccidioides brasiliensis* infection. Braz J Med Biol Res 31:615–623

Calonge TM, Nakano K, Arellano M et al (2000) *Schizosaccharomyces pombe* rho2p GTPase regulates cell wall alpha-glucan biosynthesis through the protein kinase pck2p. Mol Biol Cell 11:4393–4401

Campos CB, Di Benedette JP, Morais FV, Ovalle R, Nobrega MP (2008) Evidence for the role of calcineurin in morphogenesis and calcium homeostasis during mycelium-to-yeast dimorphism of *Paracoccidioides brasiliensis*. Eukaryot Cell 7:1856–1864

Cao C, Li R, Wan Z et al (2007) The effects of temperature, pH, and salinity on the growth and dimorphism of *Penicillium marneffei*. Med Mycol 45:401–407

Carratu L, Franceschelli S, Pardini CL et al (1996) Membrane lipid perturbation modifies the set point of the temperature of heat shock response in yeast. Proc Natl Acad Sci USA 93:3870–3875

Carrero LL, Nino-Vega G, Teixeira MM et al (2008) New *Paracoccidioides brasiliensis* isolate reveals unexpected genomic variability in this human pathogen. Fungal Genet Biol 45:605–612

Carvalho KC, Ganiko L, Batista WL et al (2005) Virulence of *Paracoccidioides brasiliensis* and gp43 expression in isolates bearing known PbGP43 genotype. Microbes Infect 7:55–65

Chen DL, Janganan TK, Chen GY et al (2007) The cAMP pathway is important for controlling the morphological switch to the pathogenic yeast form of *Paracoccidioides brasiliensis*. Mol Microbiol 65:761–779

Clemons KV, Feldman D, Stevens DA (1989) Influence of oestradiol on protein expression and methionine utilization during morphogenesis of *Paracoccidioides brasiliensis*. J Gen Microbiol 135:1607–1617

Cock AM, Cano LE, Velez D et al (2000) Fibrotic sequelae in pulmonary paracoccidioidomycosis: histopathological aspects in BALB/c mice infected with viable and non-viable *Paracoccidioides brasiliensis* propagules. Rev Inst Med Trop Sao Paulo 42:59–66

Costa PF, Fernandes GF, dos Santos PO, Amaral CC, Camargo ZP (2010) Characteristics of environmental *Paracoccidioides brasiliensis* isolates. Mycopathologia 169:37–46

de Carvalho MJ, Amorim Jesuino RS, Daher BS et al (2003) Functional and genetic characterization of calmodulin from the dimorphic and pathogenic fungus *Paracoccidioides brasiliensis*. Fungal Genet Biol 39:204–210

Egeberg RO, Elconin AE, Egeberg MC (1964) Effect of salinity and temperature on *Coccidioides immitis* and three antagonistic soil saprophytes. J Bacteriol 88:473–476

Felipe MS, Andrade RV, Petrofeza SS et al (2003) Transcriptome characterization of the dimorphic and pathogenic fungus *Paracoccidioides brasiliensis* by EST analysis. Yeast 20:263–271

Felipe MS, Andrade RV, Arraes FB et al (2005) Transcriptional profiles of the human pathogenic fungus *Paracoccidioides brasiliensis* in mycelium and yeast cells. J Biol Chem 280:24706–24714

Fernandes L, Araujo MA, Amaral A et al (2005) Cell signaling pathways in *Paracoccidioides brasiliensis*-inferred from comparisons with other fungi. Genet Mol Res 4:216–231

Ferreira ME, Marques Edos R, Malavazi I et al (2006) Transcriptome analysis and molecular studies on sulfur metabolism in the human pathogenic fungus *Paracoccidioides brasiliensis*. Mol Genet Genomics 276:450–463

Franco M (1987) Host-parasite relationships in paracoccidioidomycosis. J Med Vet Mycol 25:5–18

Franco M, Sano A, Kera K et al (1989) Chlamydospore formation by *Paracoccidioides brasiliensis* mycelial form. Rev Inst Med Trop Sao Paulo 31:151–157

Franco M, Bagagli E, Scapolio S, da Silva LC (2000) A critical analysis of isolation of *Paracoccidioides brasiliensis* from soil. Med Mycol 38:185–191

Fuchs BB, Mylonakis E (2009) Our paths might cross: the role of the fungal cell wall integrity pathway in stress response and cross talk with other stress response pathways. Eukaryot Cell 8:1616–1625

Galagan JE, Henn MR, Ma LJ, Cuomo CA, Birren B (2005) Genomics of the fungal kingdom: insights into eukaryotic biology. Genome Res 15:1620–1631

Garcia AM, Hernandez O, Aristizabal BH et al (2010) Gene expression analysis of *Paracoccidioides brasiliensis* transition from conidium to yeast cell. Med Mycol 48:147–154

Gasch AP (2007) Comparative genomics of the environmental stress response in ascomycete fungi. Yeast 24:961–976

Gauthier G, Klein BS (2008) Insights into Fungal Morphogenesis and Immune Evasion: fungal conidia, when situated in mammalian lungs, may switch from mold to pathogenic yeasts or spore-forming spherules. Microbe Wash DC 3:416–423

Goldman GH, dos Reis ME, Duarte Ribeiro DC et al (2003) Expressed sequence tag analysis of the human pathogen *Paracoccidioides brasiliensis* yeast phase: identification of putative homologues of *Candida albicans* virulence and pathogenicity genes. Eukaryot Cell 2:34–48

Gow NA (1995) Yeast-hyphal dimorphism. In: Gow NA, Gadd GM (eds) The growing fungus. Chapman & Hall, Oxford, pp 403–422

Hahn RC, Macedo AM, Fontes CJ et al (2003) Randomly amplified polymorphic DNA as a valuable tool for epidemiological studies of *Paracoccidioides brasiliensis*. J Clin Microbiol 41:2849–2854

Hanna SA, Monteiro da Silva JL, Giannini MJ (2000) Adherence and intracellular parasitism of *Paracoccidioides brasiliensis* in Vero cells. Microbes Infect 2:877–884

Harris SD (2008) Hyphal morphogenesis in *Aspergillus nidulans*. In: Goldam GH, Osmani SA (eds) The *Aspergilli* genomics, medical aspects, biotechnology, and research methods. CRC, Boca Raton, FL

Hernandez O, Almeida AJ, Gonzalez A et al (2010) A 32-kilodalton hydrolase plays an important role in *Paracoccidioides brasiliensis* adherence to host cells and influences pathogenicity. Infect Immun 78:5280–5286

Hogan LH, Klein BS (1994) Altered expression of surface alpha-1,3-glucan in genetically related strains of *Blastomyces dermatitidis* that differ in virulence. Infect Immun 62:3543–3546

Jakubowski H, Goldman E (1993) Methionine-mediated lethality in yeast cells at elevated temperature. J Bacteriol 175:5469–5476

Jung US, Sobering AK, Romeo MJ, Levin DE (2002) Regulation of the yeast Rlm1 transcription factor by the Mpk1 cell wall integrity MAP kinase. Mol Microbiol 46:781–789

Kanetsuna F, Carbonell LM, Azuma I, Yamamura Y (1972) Biochemical studies on the thermal dimorphism of *Paracoccidioides brasiliensis*. J Bacteriol 110:208–218

Klein BS, Tebbets B (2007) Dimorphism and virulence in fungi. Curr Opin Microbiol 10:314–319

Kurokawa CS, Lopes CR, Sugizaki MF et al (2005) Virulence profile of ten *Paracoccidioides brasiliensis* isolates: association with morphologic and genetic patterns. Rev Inst Med Trop Sao Paulo 47:257–262

Lengeler KB, Davidson RC, D'Souza C et al (2000) Signal transduction cascades regulating fungal development and virulence. Microbiol Mol Biol Rev 64:746–785

Levery SB, Toledo MS, Straus AH, Takahashi HK (1998) Structure elucidation of sphingolipids from the mycopathogen *Paracoccidioides brasiliensis*: an immunodominant beta-galactofuranose residue is carried by a novel glycosylinositol phosphorylceramide antigen. Biochemistry 37:8764–8775

Loose DS, Stover EP, Restrepo A, Stevens DA, Feldman D (1983) Estradiol binds to a receptor-like cytosol binding protein and initiates a biological response in *Paracoccidioides brasiliensis*. Proc Natl Acad Sci USA 80:7659–7663

Magee PT, Gale C, Berman J, Davis D (2003) Molecular genetic and genomic approaches to the study of medically important fungi. Infect Immun 71:2299–2309

Manocha MS (1980) Lipid composition of *Paracoccidioides brasiliensis*: comparison between the yeast and mycelial forms. Sabouraudia 18:281–286

Maresca B, Kobayashi GS (1989) Dimorphism in *Histoplasma capsulatum*: a model for the study of cell differentiation in pathogenic fungi. Microbiol Rev 53:186–209

Marques ER, Ferreira ME, Drummond RD et al (2004) Identification of genes preferentially expressed in the pathogenic yeast phase of *Paracoccidioides brasiliensis*, using suppression subtraction hybridization and differential macroarray analysis. Mol Genet Genomics 271:667–677

Marques ER, Ferreira ME, Drummond RD et al (2006) Identification of genes preferentially expressed in the pathogenic yeast phase of *Paracoccidioides brasiliensis*, using suppression subtraction hybridization and differential macroarray analysis. Mol Genet Genomics 271:667–677

Marzluf GA (1997) Molecular genetics of sulfur assimilation in filamentous fungi and yeast. Annu Rev Microbiol 51:73–96

Massuda TY, Nagashima LA, Leonello PC et al (2010) Cyclosporin A treatment and decreased fungal load/antigenemia in experimental murine paracoccidioidomycosis. Mycopathologia 171(3):161–169

Matute DR, McEwen JG, Puccia R et al (2006) Cryptic speciation and recombination in the fungus *Paracoccidioides brasiliensis* as revealed by gene genealogies. Mol Biol Evol 23:65–73

May GS (2008) Mitogen-activated protein kinase pathways in *Aspergilli*. In: Goldam GH, Osmani SA (eds) The *Aspergilli* genomics, medical aspects, biotechnology, and research methods. CRC, Boca Raton, FL

Mazur P, Baginsky W (1996) In vitro activity of 1,3-beta-D-glucan synthase requires the GTP-binding protein Rho1. J Biol Chem 271:14604–14609

McEwen JG, Bedoya V, Patino MM, Salazar ME, Restrepo A (1987) Experimental murine paracoccidiodomycosis induced by the inhalation of conidia. J Med Vet Mycol 25:165–175

Medoff G, Sacco M, Maresca B et al (1986) Irreversible block of the mycelial-to-yeast phase transition of *Histoplasma capsulatum*. Science 231:476–479

Medoff G, Painter A, Kobayashi GS (1987) Mycelial- to yeast-phase transitions of the dimorphic fungi *Blastomyces dermatitidis* and *Paracoccidioides brasiliensis*. J Bacteriol 169:4055–4060

Mendes-Giannini MJ, Monteiro da Silva JL, De da Fatima Silva J et al (2008) Interactions of *Paracoccidioides brasiliensis* with host cells: recent advances. Mycopathologia 165:237–248

Molinari-Madlum EE, Felipe MS, Soares CM (1999) Virulence of *Paracoccidioides brasiliensis* isolates can be correlated to groups defined by random amplified polymorphic DNA analysis. Med Mycol 37:269–276

Monteiro JP, Clemons KV, Mirels LF et al (2009) Genomic DNA microarray comparison of gene expression patterns in *Paracoccidioides brasiliensis* mycelia and yeasts in vitro. Microbiology 155:2795–2808

Montoya AE, Moreno MN, Restrepo A, McEwen JG (1997) Electrophoretic karyotype of clinical isolates of *Paracoccidioides brasiliensis*. Fungal Genet Biol 21:223–227

Munro CA, Selvaggini S, de Bruijn I et al (2007) The PKC, HOG and $Ca^{2+}$ signalling pathways co-ordinately regulate chitin synthesis in *Candida albicans*. Mol Microbiol 63:1399–1413

Natorff R, Piotrowska M, Paszewski A (1998) The *Aspergillus nidulans* sulphur regulatory gene sconB encodes a protein with WD40 repeats and an F-box. Mol Gen Genet 257:255–263

Natorff R, Sienko M, Brzywczy J, Paszewski A (2003) The *Aspergillus nidulans* metR gene encodes a bZIP protein which activates transcription of sulphur metabolism genes. Mol Microbiol 49:1081–1094

Nelson WJ (2003) Adaptation of core mechanisms to generate cell polarity. Nature 422:766–774

Nemecek JC, Wuthrich M, Klein BS (2006) Global control of dimorphism and virulence in fungi. Science 312:583–588

Nern A, Arkowitz RA (2000) G proteins mediate changes in cell shape by stabilizing the axis of polarity. Mol Cell 5:853–864

Nino-Vega GA, Munro CA, San-Blas G, Gooday GW, Gow NA (2000) Differential expression of chitin synthase genes during temperature-induced dimorphic transitions in *Paracoccidioides brasiliensis*. Med Mycol 38:31–39

Nunes LR, Costa de Oliveira R, Leite DB et al (2005) Transcriptome analysis of *Paracoccidioides brasiliensis* cells undergoing mycelium-to-yeast transition. Eukaryot Cell 4:2115–2128

Osmani SA, Liu HL, Hynes MJ, Oakley BR (2008) Advances in gene manipulations using *Aspergillus nidulans*. In: Goldam GH, Osmani SA (eds) The *Aspergilli* genomics, medical aspects, biotechnology, and research methods. CRC, Boca Raton, FL

Paris S, Duran S (1985) Cyclic adenosine 3′,5′ monophosphate (cAMP) and dimorphism in the pathogenic fungus *Paracoccidioides brasiliensis*. Mycopathologia 92:115–120

Paris S, Duran-Gonzalez S, Mariat F (1985) Nutritional studies on *Paracoccidioides brasiliensis*: the role of organic sulfur in dimorphism. Sabouraudia 23:85–92

Patton EE, Willems AR, Tyers M (1998) Combinatorial control in ubiquitin-dependent proteolysis: don't Skp the F-box hypothesis. Trends Genet 14:236–243

Porta A, Eletto A, Torok Z et al (2010) Changes in membrane fluid state and heat shock response cause attenuation of virulence. J Bacteriol 192:1999–2005

Pringle JR, Bi E, Harkins HA et al (1995) Establishment of cell polarity in yeast. Cold Spring Harb Symp Quant Biol 60:729–744

Rappleye CA, Goldman WE (2006) Defining virulence genes in the dimorphic fungi. Annu Rev Microbiol 60:281–303

Rappleye CA, Engle JT, Goldman WE (2004) RNA interference in *Histoplasma capsulatum* demonstrates a role for alpha-(1,3)-glucan in virulence. Mol Microbiol 53:153–165

Restrepo A, Tobón A (2005) *Paracoccidioides brasiliensis*. In: Mandell GL, Bennett JE, Dollin R (eds) Principles and practice of infectious diseases. Churchill Livingstone, Philadelphia, PA

Restrepo A, Salazar ME, Cano LE et al (1984) Estrogens inhibit mycelium-to-yeast transformation in the fungus *Paracoccidioides brasiliensis*: implications for resistance of females to paracoccidioidomycosis. Infect Immun 46:346–353

Restrepo A, McEwen JG, Castaneda E (2001) The habitat of *Paracoccidioides brasiliensis*: how far from solving the riddle? Med Mycol 39:233–241

Restrepo A, Benard G, de Castro CC, Agudelo CA, Tobon AM (2008) Pulmonary paracoccidioidomycosis. Semin Respir Crit Care Med 29:182–197

Rhome R, Del Poeta M (2009) Lipid signaling in pathogenic fungi. Annu Rev Microbiol 63:119–131

Rhome R, Del Poeta M (2010) Sphingolipid signaling in fungal pathogens. Adv Exp Med Biol 688:232–237

Richie DL, Hartl L, Aimanianda V et al (2009) A role for the unfolded protein response (UPR) in virulence and antifungal susceptibility in *Aspergillus fumigatus*. PLoS Pathog 5:e1000258

Rooney PJ, Klein BS (2002) Linking fungal morphogenesis with virulence. Cell Microbiol 4:127–137

Sabie FT, Gadd GM (1989) Involvement of a $Ca^{2+}$-calmodulin interaction in the yeast-mycelial (Y-M) transition of *Candida albicans*. Mycopathologia 108:47–54

Sabie FT, Gadd GM (1992) Effect of nucleosides and nucleotides and the relationship between cellular adenosine 3′/5′-cyclic monophosphate (cyclic-AMP) and germ tube formation in *Candida albicans*. Mycopathologia 119:147–156

Salazar ME, Restrepo A, Stevens DA (1988) Inhibition by estrogens of conidium-to-yeast conversion in the fungus *Paracoccidioides brasiliensis*. Infect Immun 56:711–713

San-Blas F (1986) Ultrastructure of spore formation in *Paracoccidioides brasiliensis*. J Med Vet Mycol 24:203–210

San-Blas G, Nino-Vega G (2008) *Paracoccidioides brasiliensis*: chemical and molecular tools for research on cell walls, antifungals, diagnosis, taxonomy. Mycopathologia 165:183–195

San-Blas G, San-Blas F (1984) Molecular aspects of fungal dimorphism. Crit Rev Microbiol 11:101–127

San-Blas G, San-Blas F, Serrano LE (1977) Host-parasite relationships in the yeastlike form of *Paracoccidioides brasiliensis* strain IVIC Pb9. Infect Immun 15:343–346

San-Blas G, Nino-Vega G, Iturriaga T (2002) *Paracoccidioides brasiliensis* and paracoccidioidomycosis: molecular approaches to morphogenesis, diagnosis, epidemiology, taxonomy and genetics. Med Mycol 40:225–242

Schwarz J, Baum GL (1965) Pioneers in the discovery of deep fungus diseases. Mycopathol Mycol Appl 25:73–81

Silva SS, Paes HC, Soares CM, Fernandes L, Felipe MS (2008) Insights into the pathobiology of *Paracoccidioides brasiliensis* from transcriptome analysis – advances and perspectives. Mycopathologia 165:249–258

Soares CM, Madlun EE, da Silva SP, Pereira M, Felipe MS (1995) Characterization of *Paracoccidioides brasiliensis* isolates by random amplified polymorphic DNA analysis. J Clin Microbiol 33:505–507

Sonneborn A, Bockmuhl DP, Gerads M et al (2000) Protein kinase A encoded by TPK2 regulates dimorphism of *Candida albicans*. Mol Microbiol 35:386–396

Sorais F, Barreto L, Leal JA et al (2010) Cell wall glucan synthases and GTPases in *Paracoccidioides brasiliensis*. Med Mycol 48:35–47

Steinbach WJ, Schell WA, Blankenship JR et al (2004) In vitro interactions between antifungals and immunosuppressants against *Aspergillus fumigatus*. Antimicrob Agents Chemother 48:1664–1669

Stie J, Fox D (2008) Calcineurin regulation in fungi and beyond. Eukaryot Cell 7:177–186

Svidzinski TI, Miranda Neto MH, Santana RG, Fischman O, Colombo AL (1999) *Paracoccidioides brasilienses* isolates obtained from patients with acute and chronic disease exhibit morphological differences after animal passage. Rev Inst Med Trop Sao Paulo 41:279–283

Tavares AH, Silva SS, Dantas A et al (2007) Early transcriptional response of *Paracoccidioides brasiliensis* upon internalization by murine macrophages. Microbes Infect 9:583–590

Teixeira MM, Theodoro RC, de Carvalho MJ et al (2009) Phylogenetic analysis reveals a high level of speciation in the *Paracoccidioides* genus. Mol Phylogenet Evol 52:273–283

Tercarioli GR, Bagagli E, Reis GM et al (2007) Ecological study of *Paracoccidioides brasiliensis* in soil: growth ability, conidia production and molecular detection. BMC Microbiol 7:92

Thomas D, Surdin-Kerjan Y (1997) Metabolism of sulfur amino acids in *Saccharomyces cerevisiae*. Microbiol Mol Biol Rev 61:503–532

Toledo MS, Levery SB, Straus AH et al (1999) Characterization of sphingolipids from mycopathogens: factors correlating with expression of 2-hydroxy fatty acyl (E)-Delta 3-unsaturation in cerebrosides of *Paracoccidioides brasiliensis* and *Aspergillus fumigatus*. Biochemistry 38:7294–7306

Vigh L, Escriba PV, Sonnleitner A et al (2005) The significance of lipid composition for membrane activity: new concepts and ways of assessing function. Prog Lipid Res 44:303–344

Wedlich-Soldner R, Altschuler S, Wu L, Li R (2003) Spontaneous cell polarization through actomyosin-based delivery of the Cdc42 GTPase. Science 299:1231–1235

Whiteway M, Bachewich C (2007) Morphogenesis in *Candida albicans*. Annu Rev Microbiol 61:529–553

Whiteway M, Oberholzer U (2004) Candida morphogenesis and host-pathogen interactions. Curr Opin Microbiol 7:350–357

# Chapter 10
# Morphogenesis of *Cryptococcus neoformans*

**Elizabeth R. Ballou, J. Andrew Alspaugh, and Connie B. Nichols**

**Abstract** *Cryptococcus neoformans* was first recognized as a human pathogen over 100 years ago when it was independently isolated from a patient with a tibial infection and from environmental sources (peach juice). This basidiomycete has subsequently been isolated from most regions of the world, causing a significant number of lethal infections each year, especially in AIDS patients. Originally described as a yeast-like fungus causing human and animal infections, *C. neoformans* is now known to undergo morphological transitions that are important for its survival and dissemination. Some of the signaling pathways that control yeast and hyphal morphogenesis in this organism are also central regulators of its pathogenesis.

## 10.1 Introduction

### *10.1.1 The Pathogen*

Over the 80 years following its initial clinical identification, cryptococcosis was a rare clinical entity, primarily observed in patients with hematological malignancies. However, during this time, investigators were able to describe the main clinical manifestations of cryptococcal disease. Like many environmental human fungal pathogens, *C. neoformans* enters the host primarily through the lungs (Baker and Haugen 1955). *C. neoformans* infections were reported in many organs, but the main manifestation of the disease was recognized as a meningoencephalitis, or an

E.R. Ballou • J.A. Alspaugh (✉) • C.B. Nichols
Departments of Medicine and Molecular Genetics/Microbiology, Duke University School of Medicine, DUMC 3355, 1543 Duke Hospital, South Durham, NC 27710, USA
e-mail: andrew.alspaugh@duke.edu

J. Pérez-Martín and A. Di Pietro (eds.), *Morphogenesis and Pathogenicity in Fungi*, Topics in Current Genetics 22, DOI 10.1007/978-3-642-22916-9_10, 

infection of the brain and meninges (Hanseman 1905; Levin 1937). Prior to effective antifungal therapy, this infection was uniformly fatal.

During the earliest days of the HIV pandemic, clinicians recognized a striking association between AIDS and cryptococcosis (Chuck and Sande 1989), similar to other previously obscure conditions such as *Pneumocystis* pneumonia and Kaposi's sarcoma. The majority of cases of *C. neoformans* disease now occur in highly immunosuppressed AIDS patients. In fact, recent epidemiological studies in sub-Saharan Africa have demonstrated that *C. neoformans* is estimated to cause more than one million infections each year, resulting in greater than 600,000 deaths, almost exclusively in patients with AIDS. Mortality due to *C. neoformans* now exceeds that due to tuberculosis in this patient population (Park et al. 2009).

### *10.1.2 Morphogenesis In Vivo*

In the host and in the environment, *C. neoformans* exists primarily as a haploid, budding yeast-like fungus. However, under appropriate conditions, these yeast cells can differentiate to form mating structures, including hyphae, clamp connections, terminal basidia, and basidiospores (Kwon-Chung 1975, 1976). Unlike other basidiomycetes such as *Ustilago maydis*, in which mating is an obligate step in the infectious cycle, mating and filamentation are not required for *C. neoformans* to infect or proliferate in the human host. However, the basidiospores produced by the mating process may in fact be the most common infectious agent in the environment. These spores are 1,000-fold more pathogenic than equally delivered yeast cells in animal models of infection (Giles et al. 2009; Sukroongreung et al. 1998). Therefore, gross morphological transitions are more important in the environmental survival of *C. neoformans* and in the generation of infectious spores, rather than in the actual pathophysiology of human disease.

Comprehensive histopathological examinations of cryptococcal infections described the morphological features of this fungus in the setting of the human host. In a systematic review of cases of central nervous system cryptococcosis in 1937, Levin described the histological appearance of round, budding yeasts of varying sizes surrounded by chronic inflammatory cells. The yeast cells tended to form cysts in the infected brain tissue, representing grouped fungal forms surrounded by their polysaccharide capsule (Levin 1937). Very rare cases of cryptococcosis demonstrated elongated yeasts or the early appearance of germ tube-like structures; however, these were presumed to be variant strains not representative of the greater *C. neoformans* population (Lurie and Shadomy 1971; Todd and Herrmann 1936). True hyphae or pseudohyphae were not observed in cryptococcal infections. Together, these extensive studies revealed that the *C. neoformans* yeast cell is the predominant or exclusive morphological form in the setting of human infection (Fig. 10.1).

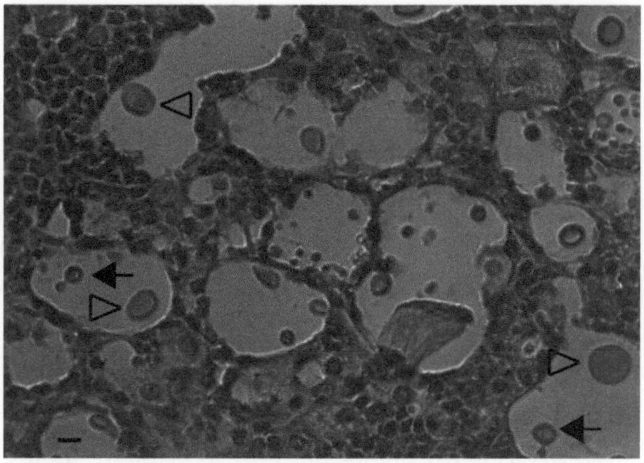

**Fig. 10.1** *C. neoformans* yeast cell morphology in vivo. Mice were infected by inhalation with *C. neoformans*. Histopathological examination of the infected lungs (stained with hematoxylin/eosin) demonstrates murine inflammatory cell infiltration surrounding encapsulated yeast cells, including the typical budding yeast form (*arrows*) and fungal giant/titan cells (*open arrowheads*). *Scale bar* 10 μm

### *10.1.3 Fungal Cell Gigantism*

Early histopathological descriptions of cryptococcosis noted more size variation among *C. neoformans* cells in vivo than encountered when the cells were grown in culture (Fig. 10.1). Although most yeast cells in central nervous system infections were uniformly small (2–8 μm), larger round forms ranging from 40 to 50 μm were also noted, especially in the lungs (Levin 1937). In 1958, Fazekas and Schwartz experimentally inoculated mice with *C. neoformans* and observed the migration of these fungal cells throughout the body. When present within granulomas or intracellularly in macrophages, the yeast cells were universally small. However, the extracellular forms were more varied in size, with giant, round fungal forms noted in the photomicrographs (Fazekas and Schwarz 1958). Feldmesser later noted that such changes in cell size were also associated with altered cell wall thickness and melanin production (Feldmesser et al. 2001).

The significance of giant fungal forms was not known until very recently. Many observers assumed that these large cells were merely injured or dying. However, two groups recently and independently suggested that these fungal "giant" or "titan" cells represent a distinct morphotype of *C. neoformans* that is important in the pathophysiology of infection, especially in the lung (Okagaki et al. 2010; Zaragoza et al. 2010). Not only are these cells able to replicate, but they are also too large to be ingested by pulmonary macrophages. Additionally, they are more resistant than smaller yeast cells to antifungal drugs and cell stresses such as nitric oxide and hydrogen peroxide, perhaps allowing a hardy state for fungal dormancy (Okagaki et al. 2010; Zaragoza et al. 2010). Future studies describing the host signals that induce *C. neoformans* gigantism will be very important to understand the role of this morphological change in the context of infection.

## 10.2 Yeast Cell Morphogenesis

Although *C. neoformans* has several morphological forms, it is clear that the budding yeast cell is the predominant morphotype observed in all stages of human infection. Therefore, recent investigations have attempted to define the molecular machinery required for *C. neoformans* yeast cell polarity and morphogenesis. Many of these studies use paradigms developed in the well-studied model ascomycetes *Saccharomyces cerevisiae* and *Schizosaccharomyces pombe* to better understand morphogenesis in pathogenic yeasts. We therefore describe the similarities and points of departure between these distantly related fungi, focusing on the specifics of basidiomycetous yeast cell growth, and especially how it relates to the pathogenesis of human infections.

*C. neoformans* enters the host as haploid spores. Once inhaled, these spores rapidly germinate into budding yeasts. This model is supported by animal models of infection in which purified spores are highly infectious in inhalation models of cryptococcosis. In vitro, *C. neoformans* spores germinate within 48–72 h of microdissection, even in the absence of nutrients. This germination time is consistent with a slight delay in the virulence of an inoculum of spores when compared to yeasts in the murine model of infection (Velagapudi et al. 2009).

In terms of gross morphology, *C. neoformans* and *S. cerevisiae* share a number of common features. In both species, haploid mother cells produce daughters by budding (Moore 2000). Initial polarized growth at the bud selection site transitions to isotropic expansion of the bud. Septal structures composed of chitin and septin filaments are deposited at the bud neck and facilitate cytokinesis following nuclear division. However, despite this broad pattern of similarities, the details of yeast morphogenesis are strikingly different between these divergent species. While *S. cerevisiae* buds are placed adjacent to the previous bud site in an axial pattern, *C. neoformans* buds emerge consecutively from a single bud site, requiring repeated digestion of the cell wall at this position (Adams 2004; Moore 2000). Chitin deposition is increased in *C. neoformans* compared to *S. cerevisiae*, and the deacetylated form of chitin, chitosan, is required for cell integrity during vegetative growth (Baker et al. 2007). Conversely, septins, which are essential proteins in *S. cerevisiae*, are dispensable for yeast cell growth in *C. neoformans* (Kozubowski and Heitman 2010). Moreover, the nuclear dynamics of the two are dissimilar, with the basidiomycete nucleus moving into the bud prior to mitotic division. Finally, pathways regulating growth under various environmental conditions have been extensively rewired between the two species.

### *10.2.1 Bud Site Selection*

Whereas haploid *S. cerevisiae* yeast place new buds adjacent to the old bud scar, basidiomycetes repeatedly bud from a single site. This results in a single bud scar and the accumulation of cell wall material at this position (Moore 2000). While proteins controlling various aspects of bud site selection and maintenance, such as Bud3, Bud4, Axl1, and Axl2, have been well characterized in *S. cerevisiae*, functional orthologs have not yet been reported in *C. neoformans* (Chant and Pringle 1991; Chant 1999; Fujita et al. 2004; Roemer et al. 1998; Sanders and Herskowitz 1996). A survey of the *C. neoformans* genome identified homologs for *AXL1* and *AXL2*, and a low homology match for *BUD4*, but failed to identify *BUD3*. This loss of bud site selection genes has also been observed in the basidiomycete *U. maydis* (Banuett et al. 2008). It is likely that these protein changes contribute in part to the different patterns of bud emergence between ascomycetes and basidiomycetes. Moreover, these changes have implications for differences in the mechanisms of polarized growth between *C. neoformans* and *S. cerevisiae*.

### *10.2.2 Cell Cycle Events*

In the absence of stress conditions, *C. neoformans* cells in log phase growth proceed through G1, S, G2, and M phases in approximately 2 h (Takeo et al. 1995), and budding is initiated following completion of DNA synthesis (Yamaguchi et al. 2007). Unlike *S. cerevisiae*, budding *C. neoformans* cells form an early polarized protrusion known as a sterigma before commencing isotropic bud growth. Once the bud has reached approximately half the size of the mother, mitosis begins (Kopecka et al. 2001; Yamaguchi et al. 2007).

The spindle pole body (SPB) duplicates early in G1, and it can be observed in interphase in both G1 and G2 cells as two dumb-bell shaped structures on the outer nuclear envelope, where it associates with cytoplasmic microtubules (Yamaguchi et al. 2007, 2009). Similar to *S. pombe*, during interphase *C. neoformans* maintains a complex network of cytoplasmic microtubules that is polarized to the site of new growth during sterigma formation and early bud growth. This network disassembles during mitosis, concomitant with the formation of the mitotic spindle across the bud neck, and reassembles following cytokinesis (Kopecka et al. 2001).

The mitotic nucleus is translocated into the bud before nuclear division, with one nucleus sent back into the mother over the course of mitosis, an event typical of basidiomycetous yeast (Mochizuki et al. 1987). As the nucleus enters prophase and begins movement into the bud, the globular elements of the SPB separate and move along the outer nuclear envelope. They invade the nucleus during prometaphase, and associate with nuclear microtubules. At this point the nucleus has moved from the bud compartment back into the bud neck. By metaphase, spindle pole bodies occupy opposite sides of the nucleus along the neck axis. Nuclear division occurs primarily in the daughter cell, with half of the nucleus sent back into the mother cell upon completion of anaphase (Yamaguchi et al. 2009).

As in *Saccharomyces cerevisiae*, the cryptococcal yeast buds via polarization of its actin cytoskeleton (Kopecka et al. 2001). Before budding, actin patches are distributed across the cell surface. As budding commences, patches localize to the

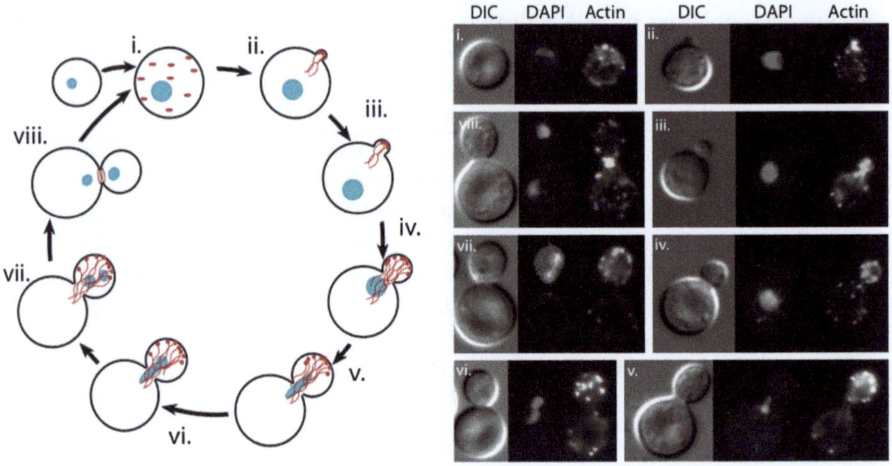


**Fig. 10.2** Budding cycle of the *C. neoformans* yeast cell. Photomicrographs of *C. neoformans* cells in logarithmic phase of growth depict sequential stages of the cell budding and nuclear division cycles. The cells were stained with rhodamine-conjugated phalloidin to demonstrate filamentous actin, as well as DAPI to stain the nucleus. Images correspond to the adjacent schematic. Prior to budding, (i) actin is distributed across the cell. Following S phase, actin is polarized to the nascent bud (ii), facilitating apical (iii) and isotropic (iv) expansion of the daughter cell. The nucleus is translocated into the bud (v) during prophase and is positioned for division across the neck (vi). Division of the nucleus occurs primarily in the daughter (vii). Following nuclear division, actin filaments coalesce into a contractile ring at the neck (viii), facilitating cytokinesis

site of the emerging bud, and actin cables appear pointing to the growing sterigma. Actin cables accumulate in the mother cell, and can be seen encircling the nucleus prior to its translocation into the daughter cell. Following mitosis, actin cables form a cytokinetic ring at the site of septum formation (Fig. 10.2).

## *10.2.3 Cell Cycle Control Under Stress Conditions*

*C. neoformans* has adapted basic aspects of the cell cycle and morphogenesis pathways to survive host stress conditions. For instance, under glucose or oxygen limiting conditions or when exposed to changes in temperature, *C. neoformans* yeast cells that have completed DNA synthesis may delay budding by arresting in the G2 phase of the cell cycle (Ohkusu et al. 2001a, b; Takeo et al. 1995). Interestingly, this implies that the signal for the start of budding is separate from that of DNA synthesis, and Takeo et al. suggest that this signal be termed the "B-factor" or "Start-$2_{Bud}$" in their model of the *C. neoformans* cell cycle (Virtudazo et al. 2010). Additionally, newly separated daughter cells experience a delay in doubling time relative to mother cells, implying that there is a minimum cell volume restriction for DNA synthesis and budding, although this has not yet been directly examined (Yamaguchi et al. 2007).

Given the unusual G2 delay in *C. neoformans* budding, Takeo, Virtudazo and colleagues have undertaken a characterization of the molecular mechanisms regulating DNA synthesis and budding (Virtudazo et al. 2010). They determined that the *C. neoformans* genome (strain B4500) encodes a single Cdc28/Cdc2 homolog, termed Cyclin-Dependent Kinase 1 (Cdk1), which is able to complement loss of Cdc28 function in *S. cerevisiae* (Takeo et al. 2004).

The *C. neoformans* genome also encodes a single G1-type cyclin, Cln1, as well as two putative B-type cyclins and several Pho-type cyclins whose function has not yet been characterized. Although not essential, loss of Cln1 appears to result in a delay in DNA synthesis and budding, as mutants accumulate in the G1 phase and increase to a volume more typical of wild type G2 phase cells before initiating DNA synthesis (Virtudazo et al. 2010). However, budding was also delayed relative to the volume of wild type cells, indicating that bud emergence is not dictated by cell size alone in this organism.

### *10.2.4 Polarized Growth in the* C. neoformans *Yeast Cell*

Cell polarity, or subcellular asymmetry, allows the *C. neoformans* yeast cell to direct the site of new bud emergence and to support the growth of the developing daughter cell. Initial investigations into the mechanisms of polarized growth in *C. neoformans* determined that the Ras1 protein is a major regulator of this process. The Ras1 GTPase is required for high temperature growth of the yeast phase, where it plays a role in the polarization of actin to the bud site. When yeast cells encounter stress conditions such as elevated temperature, the actin cytoskeleton becomes transiently depolarized. If the stress is nonlethal, the actin rapidly repolarizes, allowing the cell to resume growth (Ho and Bretscher 2001; Nichols et al. 2007). *C. neoformans* strains lacking *RAS1* are unable to repolarize actin after temperature stress, resulting in a failure to bud at 37°C, and instead arresting as large, apolar cells. The *ras1Δ* mutants are also unable to undergo the polarized growth involved in the hyphal morphogenesis of mating.

Unlike *S. cerevisiae* but similar to *S. pombe*, *C. neoformans* Ras proteins are not major regulators of cAMP pathways. Mutations in cAMP signaling proteins are associated with distinct deficits in capsule and melanin, phenotypes not observed in *ras1Δ* strains. Also, defective Ras1 signaling does not affect cAMP levels in the cell. Indeed, a recent report comparing transcriptional changes in cAMP and Ras1 pathway mutants identified few overlapping transcriptional targets of these two pathways (Maeng et al. 2010). In contrast, *C. neoformans* Ras proteins appear to activate two distinct signaling pathways. The first is the mitogen-activated protein (MAP) kinase/pheromone response pathway that mediates the mating response (mating pathway). Second, Ras1 controls morphogenesis and thermotolerance through its regulation of polarity establishment proteins such as Cdc24, Cdc42, Rac, and septins (morphogenesis pathway).

Ras1 activity depends on its localization to specific subcellular membranes, a process mediated by posttranslational modifications of C-terminal cysteine residues (Nichols et al. 2009). Point mutations in Ras1 residues targeted for modification are sufficient to direct protein activity toward either the mating or the morphogenesis pathways by altering the localization of Ras1 to either endomembranes or the plasma membrane, respectively (Nichols et al. 2009). This localization-dependent separation of function provides functional specificity, allowing the Ras1 protein to potentially activate these two pathways independently.

### *10.2.5 Ras1 Signaling and Morphogenesis/Thermotolerance of the Yeast Cell*

*C. neoformans* Ras1 activity is likely transduced to the actin cytoskeleton via Cdc24, a highly conserved regulator of the Rho-GTPase Cdc42. First, Ras1 interacts with Cd24 in yeast two-hybrid assays, suggesting physical interactions between these proteins. Second, loss of *CDC24* results in *ras1Δ*-like mutant phenotypes, such as a failure to proliferate at elevated temperatures, with yeasts arresting as large, apolar cells (Nichols et al. 2007). Similar to *ras1Δ* mutants, *cdc24Δ* cells also delay repolarization of the actin cytoskeleton upon exposure to temperature stress (Nichols et al. 2007). In *S. pombe*, the interaction of Ras1 with Cdc24 has been shown to regulate polar growth and cytokinesis via the actin cytoskeleton as well as spindle formation via the microtubule network (Chang et al. 1994; Li and Chang 2003). *C. neoformans* cells lacking *CDC24* have increased sensitivity to the actin destabilizing agent Latrunculin B, suggesting that the interaction of these signaling proteins in *C. neoformans* also controls actin localization. However, unlike the corresponding mutants in *S. pombe*, the *C. neoformans ras1Δ* and *cdc24Δ* strains are no more sensitive than wild type to the microtubule destabilizers benomyl or nocodazole (Ballou et al. 2010).

Studies in *C. neoformans* Cdc24 suggest that conserved Cdc24-interacting proteins might also regulate cell polarity and yeast cell morphogenesis. Cdc24 is a Guanine-nucleotide Exchange Factor (GEF) that is highly conserved throughout eukaryotes, facilitating the activation of the Cdc42 family of proteins. The *C. neoformans* genome encodes two functional paralogs, Cdc42 and Cdc420, which appear to constitute major and minor versions of the protein (Ballou et al. 2010). The major paralog *CDC42* is more highly transcribed and is further transcriptionally induced at 37°C. Consistent with a model in which the effects of *cdc24Δ* are attributed to a decrease in the availability of active, GTP-bound Cdc42, loss of *CDC42* also results in increased sensitivity to Latrunculin B and temperature-sensitive growth, similar to *cdc24Δ* mutants (Ballou et al. 2010).

Despite initial similarities to homologous proteins in other fungal species, *C. neoformans* Cdc42 seems to play a very distinct role in yeast cell morphogenesis compared to model ascomycetes. In contrast to *S. cerevisiae CDC42*, which is an essential gene, neither *C. neoformans CDC42* nor *CDC420* is required for growth in the absence of elevated temperatures or other cell stress. Moreover, intact actin structures and budding are evident in the *cdc42Δ* single mutant and the *cdc42Δ cdc420Δ* double mutant under permissive growth conditions. This suggests that other proteins can direct the events required for actin polarization and yeast cell budding. In contrast, the *C. neoformans* Cdc42 proteins play a primary role in the recruitment and organization of septin proteins at the bud neck (Ballou et al. 2010). These results suggest that *C. neoformans* Cdc42 proteins play a direct role in mediating cytokinesis, and perhaps a secondary or redundant role in yeast cell polarity.

### *10.2.6 Cytokinesis and Septin Proteins*

In *S. cerevisiae*, the septum acts as a scaffold for the recruitment of proteins to the bud neck and to separate the mother and bud compartments (reviewed in Weirich et al. 2008). The proteins which make up the septum in *C. neoformans*, Cdc3, Cdc10, Cdc11, and Cdc12, as well as an apparent cryptic septin, Cns5, undergo a series of functional reorganizations over the course of budding and cytokinesis (Ballou et al. 2010; Kozubowski and Heitman 2010). Septins first form a ring at the site of future bud growth. As isotropic growth commences, this ring forms a collar around the neck that then separates into an hourglass structure spanning the neck during cytokinesis (Fig. 10.3). Unlike *S. cerevisiae*, the septins are not essential for budding in the absence of external stress. These proteins, and the Rho GTPases which direct their organization, are required for growth under human host physiological conditions, such as incubation at temperatures above 37°C, or in the presence of cell wall stressors (SDS and caffeine) (Ballou et al. 2010; Kozubowski and Heitman 2010). In the absence of *CDC3*, *CDC11*, or *CDC12*, *C. neoformans* yeast cells are slightly larger and more elongated than wild type cells, consistent with defects in septin assembly (Gladfelter et al. 2002; Kozubowski and Heitman 2010). These defects become more pronounced at 30°C, and result in a complete failure to proliferate at 37°C. Loss of *CDC3*, *CDC11*, or *CDC12* results in defects in proliferation and bud neck formation at 37°C (Kozubowski and Heitman 2010).

Cdc42, and to a lesser extent Cdc420, are required for the recruitment and organization of the septin proteins at the bud neck, and their loss results in defects similar to loss of the septins themselves. Loss of both *CDC42* and *CDC420* results in the accumulation of cytokinetic defects. At 30°C this results in severe cell morphology defects, although these cells remain viable in the absence of temperature stress. At 37°C the *cdc42Δ cdc420Δ* strain is not viable.

It is possible that other GTPases compensate for the loss of Cdc42 paralogs at 30°C. In addition to the two *CDC42* genes, the *C. neoformans* genome also encodes

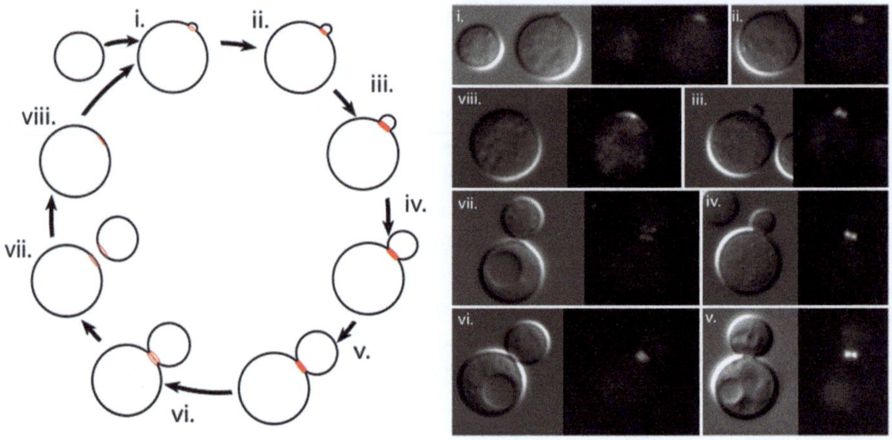


**Fig. 10.3** Septins direct cytokinesis of *C. neoformans* yeast cells. In order to visualize *C. neoformans* septins, a strain was designed to express a Cdc10–Gfp fusion protein (Kozubowski and Heitman 2010). Septin localization can be followed throughout the replication cycle of the yeast cell, from early bud emergence through cytokinesis. Images correspond to the adjacent schematic. Septins localize in a ring (i) to the bud site and persist as a collar during formation of the sterigma (ii) and apical (iii) and isotropic (iv and v) expansion of the bud. Following nuclear division, septins are reorganized into the classic hourglass structure across the neck (vi). Two separate rings become visible during cytokinesis (vii). Following separation of the mother and daughter cells, the septin ring may persist (viii) or may disassemble until the next cycle of budding begins

two Rac paralogs. Rac proteins are very similar to Cdc42. In *C. albicans* and in other fungi, Rac proteins are required for polarized growth and are synthetically lethal with the loss of *CDC42* (Bassilana and Arkowitz 2006). Also, the specific induction of Cdc42 upon exposure to temperature stress, coupled with the ability of *C. neoformans* to delay bud emergence following DNA synthesis under stress conditions, raise the intriguing possibility that this organism initiates a condition-specific budding program upon exposure to host physiological conditions.

Determining the downstream effectors of Cdc42 function is an area of ongoing interest. Homologs of the p21-activated protein kinases (PAKs), which contain conserved Cdc42/Rac Interaction Binding (CRIB) domains, have been demonstrated to play roles in polarized growth during hyphal and yeast phase growth. Loss of Ste20α, the homolog of the *S. cerevisiae* Cdc42 effector Cla4 (Cvrckova et al. 1995; Longtine et al. 2000), results in the accumulation of hyperpolarized cells at 39°C, indicative of a failure to switch from polarized to isotropic growth, while Pak1 is required for the polarization of actin during fusion (Nichols et al. 2004; Wang et al. 2002). Although the CRIB domains of these proteins have been shown to physically interact with Cdc42 by yeast-two hybrid studies (Wang et al. 2002), these defects in the polarization of actin, coupled with a lack of septin localization defects in the absence of Ste20α, provide hints that *C. neoformans* may have rewired the control of polarized growth compared to *S. cerevisiae*. The Rac paralogs are strong candidates for additional effectors of cell polarity given their known roles in polarized growth in other basidiomycetes (Leveleki et al. 2004).

## 10.3 Hyphal Transitions and Posthyphal Forms

*C. neoformans* undergoes a dramatic change in morphology during the transition from a unicellular yeast cell to a hyphal form. Many fungi exist exclusively as hyphae, and some switch between yeast and filamentous forms via specific cues such as temperature or pH. The yeast-to-hyphae transition in *C. neoformans* is sex driven, and inducing signals include nutrient deprivation and pheromone. Although the filamentous form of *C. neoformans* is not found in vivo and is not required for disease, many of the genes involved in the establishment of this form are important for the pathogenicity of this organism.

### *10.3.1 The First Step: Pheromone Response and Cell Fusion*

As in many fungi, sexual identity in *C. neoformans* is determined by the genetic information located on *MAT* (mating type) loci. *C. neoformans* has a bipolar mating system consisting of a single *MAT* locus and two *MAT* idiomorphs, *MAT*α (alpha) and ***MATa*** (Kwon-Chung 1975). Mating most commonly occurs between unlike *MAT* idiomorphs (α-**a** mating) but has also been shown recently to occur between like *MAT*α idiomorphs (α-α mating) (Lin et al. 2005). Interest in the connection between the *C. neoformans MAT* locus and virulence began with two observations: most clinical isolates are *MAT*α (Kwon-Chung and Bennett 1978) and *MAT*α strains are significantly more virulent than congenic ***MATa*** strains in animal models of cryptococcosis (Kwon-Chung et al. 1992). Also interesting was the finding that the *C. neoformans MAT* locus was very large (approximately 120 kb), incorporating genes that had not been previously associated with fungal mating loci, including MAP kinase cascade genes, in addition to genes normally associated with *MAT* loci such as homeodomain transcription factors, pheromone receptors, and pheromone genes (Fraser et al. 2004).

In response to nutrient deprivation and pheromone, cryptococcal cells undergo several morphogenic changes that end with cell–cell fusion, including cell cycle arrest, cessation of budding, establishment of a new growth axis, conjugation tube formation, and cell wall remodeling (Kwon-Chung 1976). Several signaling cascades have been found to be important for cell fusion in *C. neoformans*, including the MAP kinase pheromone-induced signaling cascade that induces the production of pheromone along with other pheromone responsive genes (Fig. 10.4).

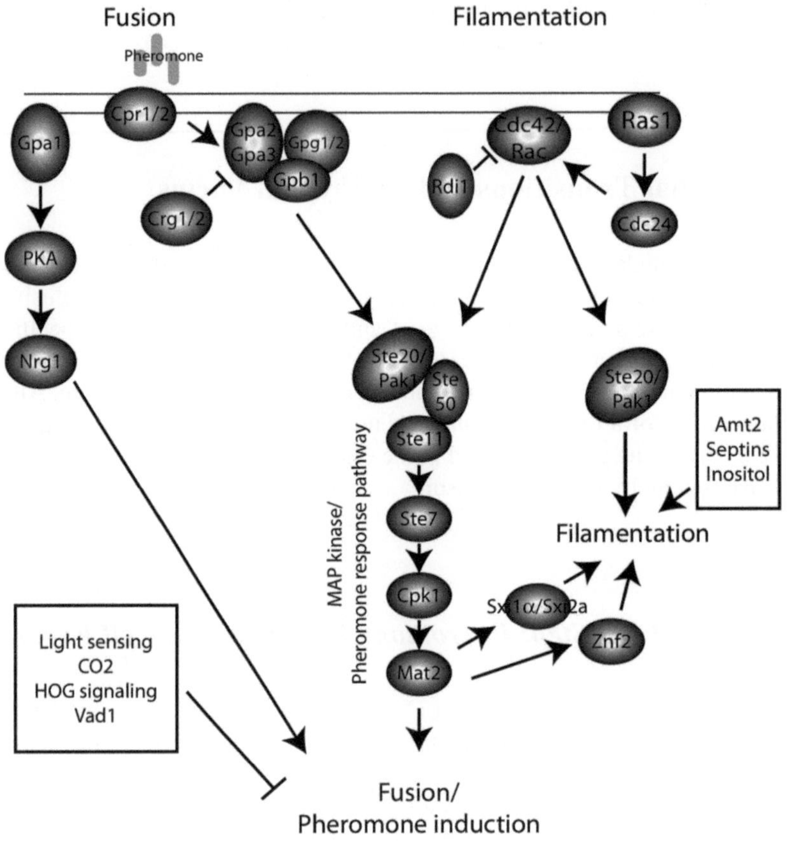


**Fig. 10.4** A model of the regulatory pathways predicated to control the morphological transitions of mating and haploid fruiting in *C. neoformans*

#### 10.3.1.1 Pheromones and Pheromone Receptors

Both *C. neoformans MAT* idiomorphs contain one pheromone receptor gene (*STE3a*/*α*), four pheromone genes in the *MATα* idiomorph (*MFα1*, *MFα2*, *MFα3*, *MFα4*), and three pheromone genes in the *MAT**a*** idiomorph (*MF**a**1*, *MF**a**2*, *MF**a**3*) (reviewed in Kozubowski et al. 2009) (Fig. 10.4). The *STE3α*/*STE3**a*** alleles encode a seven transmembrane domain G-protein-coupled receptor (GPCR) that is activated by binding of pheromone from the opposite mating type (Li et al. 2007). Both *MFα* and *MF**a*** pheromone genes encode hydrophobic pheromones that are posttranslationally modified and secreted from the cell by the ATP-binding cassette transporter protein Ste6 (Hsueh and Shen 2005). *MFα1* was the first gene characterized in the *C. neoformans* MAT locus (Moore and Edman 1993). Perturbation of the pheromone, pheromone export, and pheromone receptor genes attenuates cell fusion, but mating is not completely abolished in these strains (Davidson et al. 2000; Moore and Edman 1993). Ste3**a** and MFα1 have also been implicated in virulence, however their roles appear to be anecdotal rather than causative. For example, ***ste3a*** mutant strains are attenuated for virulence in mouse models of cryptococcosis, however this mutant also has an additional defect in capsule production, a trait essential for cryptococcal pathogenicity (Chang et al. 2003). Also, *MFα* pheromone gene expression is induced in vivo, although the physiological consequence of this induction is unknown (del Poeta et al. 1999).

#### 10.3.1.2 Pheromone-Induced G-Protein and MAP Kinase Signaling

GPCRs such as Ste3α/**a** stimulate GDP–GTP exchange on the alpha unit of heterotrimeric guanine nucleotide-binding proteins (G-proteins). In *C. neoformans*, there are two Gα subunits, Gpa2 and Gpa3; one Gβ subunit, Gpb1; and two Gγ subunits, Gpg1 and Gpg2, that comprise the G-protein complexes stimulated by Ste3α/**a** and pheromone (reviewed Xue et al. 2008a) (Fig. 10.4). Mutational analyses of these multiple subunits have uncovered both overlapping and distinct roles in cell fusion in response to pheromone. Cell fusion is abolished in *gpb1* (Gβ) or *gpg2* (Gγ) mutant strains and is decreased in *gpg1* (Gγ) mutant strains, indicating that the Gβ–Gγ complexes consisting of these gene products promote cell fusion in response to pheromone (Hsueh et al. 2007; Li et al. 2007). In contrast, there are multiple levels of specificity conferred by the Gα subunits. Various combinations of *gpa2* and *gpa3* mutant crosses are able to fuse during mating with only slight decreases in efficiency. Fusion is only abolished when *GPA2* and *GPA3* are deleted in both mating partners (double mutant bilateral cross). In addition, both Gpa2 and Gpa3 interact physically with the pheromone receptor Ste3α/**a**, indicating that Gpa2 and Gpa3 share overlapping roles in promoting cell fusion. Finally, analysis of wild type and dominant active alleles of *GPA2* and *GPA3* demonstrated that Gpa2 has an additional positive role in pheromone signaling while Gpa3 has an inhibitory role (Hsueh et al. 2007; Li et al. 2007).

The regulator of G-protein signaling (RGS) family of proteins mediates an additional level of G-protein regulation by acting as GTPase-activating proteins (GAP) (Xue et al. 2008b). Three RGS homologs have been identified in *C. neoformans*; two of those, Crg1 and Crg2, negatively regulate pheromone sensing by inactivating Gα subunits Gpa2 and Gpa3 (Wang et al. 2004; Xue et al. 2008b).

The effector pathway of pheromone-induced signaling is the activation of a linear MAP kinase signaling cascade. The MAP kinase signaling modules are highly conserved and consist of a core of three kinases activated sequentially by phosphorylation. The final kinase in the module, the MAP kinase, is usually transported into the nucleus after activation where it in turn activates a transcription factor effector. In *C. neoformans*, the MAP kinase module that is induced by and regulates pheromone consists of Ste11, a MAP kinase kinase kinase (MAP3K); Ste7, a MAP kinase kinase (MAP2K); and Cpk1, a MAP kinase (MAPK) (Davidson et al. 2003). The gene encoding Ste11 is MAT-specific with α and **a** alleles. Transcription effector targets of the pheromone MAP kinase include the recently described HMG transcription factor Mat2 (Lin et al. 2010). Mutation of genes encoding any of the MAP kinase components, including *MAT2*, abolishes pheromone production and inhibits cell fusion during mating.

In many MAP kinase signaling cascades, a PAK family kinase activates the MAP3K. Two PAK kinases have been identified in *C. neoformans*, Pak1 and the MAT-specific alleles Ste20α and Ste20**a** (Nichols et al. 2004; Wang et al. 2002). *ste20α/**a*** mutant strains do not exhibit any cell fusion defects during mating, while fusion is dramatically reduced in *pak1* mutant strains, suggesting that Pak1 is the activator of Ste11 (Nichols et al. 2004). However, pheromone production is normal in *pak1* mutant strains (as well as in *ste20α/**a*** mutant strains) indicating that Pak1 and Ste20α/**a** share overlapping roles activating Ste11 and the pheromone MAP kinase cascade. So why do *pak1* mutants fail to fuse in response to the proper cues? In response to pheromone, cells repolarize their actin cytoskeleton in order to grow toward pheromone secreted from cells of the opposite mating type. However, *pak1* mutant cells are not able to repolarize in response to pheromone (Nichols et al. 2004). Interestingly, both PAK kinases (as well as Cdc24 and Ras1) are required for virulence in *C. neoformans*, suggesting a correlation between cell polarity and pathogenicity (Alspaugh et al. 2000; Ballou et al. 2010; Nichols et al. 2004; Wang et al. 2002).

Another conserved component of MAP kinase signaling that has been recently elucidated in *C. neoformans* is a Ste50 ortholog that functions as an adapter protein bridging MAP3K and PAK kinases. In *C. neoformans*, Ste50 is required for cell fusion and interacts with Ste11 and the Ste20 PAK kinase in two-hybrid assays (Fu et al. 2011; Jung et al. 2011). However, as previously described, neither *ste20* nor *pak1* mutant strains have true MAP kinase signaling defects. Interestingly, the two-hybrid analysis presented by Fu et al. used the N-terminal portion of Ste20 that contained the CRIB domain, a highly conserved domain that is shared by both Ste20 and Pak1 (Fu et al. 2011). Although not yet determined, it is likely that Pak1 also interacts with Ste50, and both PAK kinases activate Ste11.

#### 10.3.1.3 Nutritional Signaling and Cell Fusion: G-Proteins and cAMP-PKA

Fungi utilize G-protein signaling to sense nutrients such as glucose and amino acids. In *C. neoformans*, nutritional signals also mediate pheromone production and cell fusion. The first G-protein identified in *C. neoformans* was the nutritional-specific Gα subunit Gpa1. The *gpa1* mutant strain was required for pathogenesis and had defects in the virulence traits melanin and capsule, but it also exhibited a cell fusion defect (Alspaugh et al. 1997). Subsequently, a nutritional heterotrimeric G-protein signaling cascade was elucidated that consists of a GPCR Gpr4, Gpa1 (Gα), Gib2 (Gβlike), Gpg1 (Gγ), Gpg2 (Gγ), and the RGS inhibitor Crg2 (reviewed in Pukkila-Worley and Alspaugh 2004). Several of these components also function in the pheromone G-protein signaling cascade (see above). Like *gpa1*, cell fusion is decreased in *gpr4* mutant strains, and the Gpa1 and Gpr4 proteins physically interact (Xue et al. 2006). However, Gpr4 is probably not the only receptor signaling to Gpa1. While Gpa1 responds to glucose, Grp4 is not glucose-responsive but instead responds to the amino acid methionine (Xue et al. 2006).

The effector of Gpa1 signaling in *C. neoformans* is adenylyl cyclase encoded by *CAC1*. *cac1* mutant strains exhibit defects in cell fusion in addition to defects in melanin and capsule (Alspaugh et al. 2002). The mechanistic function of adenylyl cyclase is highly conserved, converting ATP to cAMP which in turn activates Protein Kinase A (PKA). In *C. neoformans* PKA consists of the regulatory (inhibitory) and catalytic subunits Pkr1, Pka1, and Pka2 (Pukkila-Worley and Alspaugh 2004). cAMP inhibits the regulatory subunit Pkr1 causing the release and activation of the catalytic subunits Pka1 and Pka2. Consistent with this signaling model, pheromone production is increased in *pkr1* mutants and decreased in *pka1* and *pka2* mutant strains (D'Souza et al. 2001). In addition to Gpa1, the adenylyl cyclase-associated protein Aca1 also positively regulates Cac1 in addition to Gpa1 demonstrating that multiple inputs regulate cAMP-PKA signaling (Bahn et al. 2004).

Although the signaling cascade components that mediate pheromone production in response to nutrient limitation have been characterized, the downstream targets have yet to be elucidated. Microarray experiments have identified several different downstream targets of Gpa1-cAMP-PKA signaling, but only the cAMP-responsive Nrg1 transcription factor has been identified as having a role in mating (Cramer et al. 2006). However, while mating is decreased in *nrg1* mutant strains, pheromone production is normal indicating that other targets exist that control pheromone in response to nutritional signaling.

#### 10.3.1.4 Ras1 Regulation of Pheromone and Cell Fusion

In addition to heterotrimeric G-protein signaling, the monomeric G-protein Ras1 is essential for cell fusion in response to pheromone (Waugh et al. 2003). As described previously, Ras1 mediates morphology during vegetative growth through a signaling pathway consisting of Cdc24, Cdc42/Cdc420, Rac1/Rac2, and Ste20α/**a** (Nichols et al. 2007; Vallim et al. 2005). Pheromone production and cell fusion is abolished in *ras1* mutant strains (Waugh et al. 2003). While this mechanism is not well understood, it appears to be distinct from the *C. neoformans* Ras protein signal transduction pathway regulating morphogenesis since *cdc42*, *cdc420*, and *ste20α/**a*** mutant strains do not exhibit any fusion defects during mating (Ballou et al. 2010; Nichols et al. 2004). However, *cdc24* mutant crosses are not able to produce filaments, indicating that Cdc24 may be required for cell–cell fusion (Nichols, unpublished data). Also, *rac2* mutant bilateral crosses exhibit a fusion defect (Ballou and Alspaugh, unpublished data). Rac2 is the second Rac homolog identified in *C. neoformans* and is a Rho-type GTPase. It is possible that Ras1 signals through Cdc24 and Rac2 to mediate pheromone response and fusion via the PAK kinases. Interestingly, a *ras1* mutant engineered to be restricted to internal cell membranes is able to undergo cell–cell fusion and proceed through mating, demonstrating that Ras1 does not have to interact with plasma membrane-bound proteins to mediate cell fusion in *C. neoformans* (Nichols et al. 2009).

#### 10.3.1.5 Negative Regulators of Pheromone and Cell Fusion

In addition to the RGS proteins that negatively regulate Gα subunits, other negative regulators of pheromone and cell fusion have been characterized in *C. neoformans*. For example, the molecular mechanism of Vad1, a negative regulator of mating, has been recently elucidated at the molecular level (Park et al. 2010). Vad1 belongs to a family of DEAD box proteins that typically regulate the degradation of mRNA transcripts (Panepinto et al. 2005). Pheromone is expressed in *vad1* mutant cells during nonmating conditions and overexpression of *VAD1* inhibits cell–cell fusion during mating. Park et al. found that Vad1 functions by promoting the constitutive degradation of pheromone transcripts during vegetative growth. These data suggest that pheromone production is constitutive and only when Vad1 activity is repressed (during mating) do the pheromone levels accumulate (Park et al. 2010).

The stress-responsive Hog1 MAP kinase pathway negatively regulates mating. Cell fusion and pheromone expression are enhanced in strains mutated for genes encoding Hog1 (MAPK), Pbs1 (MAP3K), and Ssk2 (MAP3K). Pheromone expression is also fivefold higher in *hog1* cells under nonmating conditions indicating that pheromone expression is constitutive yet regulated (Bahn et al. 2005b). This is similar to *vad1* mutants in which pheromone expression was also high during nonmating conditions. However, the Hog1 pathway does not appear to signal to Vad1 since MFα was degraded in a Vad1-dependent manner in a *hog1* mutant strain. Instead Vad1 and the Hog1 signaling cascade appear to regulate pheromone in parallel pathways (Park et al. 2010).

Although it has long been known that *C. neoformans* mating mixtures were sensitive to light, it was recently demonstrated that pheromone expression and cell fusion is specifically inhibited by white light. This inhibition is mediated by white collar genes *BWC1* and *BWC2* which encode light-responsive photoreceptors. Accordingly, pheromone expression and cell fusion were not inhibited by light in either the *bwc1* or *bwc2* mutant strains (Idnurm and Heitman 2005).

Another physiological inhibitor of pheromone expression and cell fusion is carbon dioxide. High (4–10%) levels of carbon dioxide inhibit cell fusion and MFα production. This inhibition is relieved in *can2* mutant strains. *CAN2* encodes a carbonic anhydrase and coverts carbon dioxide to bicarbonate (Bahn et al. 2005a; Mogensen et al. 2006). These results suggest that bicarbonate signals inhibit cell fusion. Interestingly, bicarbonate was also shown to stimulate adenylyl cyclase and PKA signaling, which in turn stimulates pheromone expression and cell fusion (Klengel et al. 2005). More experimentation will be necessary to resolve these paradoxical results.

## *10.3.2 The Second Step: Making a Dikaryotic Filament*

In basidiomycete fungi such as *C. neoformans*, nuclei do not fuse immediately after **a**–α cell fusion. Instead, an initial heterokaryon is formed that contains both parental nuclei. The fused cell transitions into a dikaryotic hyphal form. Within this filament, the identity of each nucleus is maintained and propagated throughout each hyphal cell using structures unique to basidiomycetes: clamp cells and pegs cells (Kwon-Chung 1976). A clamp cell is a hook-like projection that forms on the apical cell or hyphal tip. As the clamp cell is formed, both **a** and α nuclei undergo mitosis synchronously but with different spindle lengths so that one nucleus produced from the shorter spindle pair enters the clamp cell while its sister nucleus migrates toward the apical tip. Meanwhile, a nucleus from the longer spindle pair migrates into the apical cell while its sister nucleus remains in the subapical cell. The clamp cell with its nuclear cargo polarizes back toward the subapical cell where the peg cell has formed. The peg cell is a small bulge that forms from the subapical cell. The clamp cell and peg cell fuse, releasing the nucleus into the subapical cell that already contains one of the nuclei from the other spindle pair (Kwon-Chung 1976). Septa form between the apical and subapical cell and also at the point of fusion between clamp cell and peg cell. This complex morphological process ensures that each new hyphal cell contains one nucleus from each original parent (Kwon-Chung and Popkin 1976). The cycle of hyphal cell formation continues until the final stages of the sexual cycle: basidium formation, nuclear fusion, meiosis, and spore production (Kwon-Chung 1976).

While the mechanics of these events (filament formation, clamp cell formation and fusion, etc.) are well described in other basidiomycetes, the regulatory mechanisms and signaling pathways have yet to be well characterized. In other basidiomycete species, clamp cell fusion is regulated by pheromone signaling while clamp cell formation and nuclear progression are regulated by homeodomain transcription factors (Casselton and Olesnicky 1998).

#### 10.3.2.1 Establishment of the Dikaryon

Recent work has demonstrated that C. *neoformans* sexual identity is dependent on the *MAT*-specific homeodomain proteins Sxi1α and Sxi2**a**. Transcriptionally induced by Mat2 and the Cpk1 MAP kinase signaling cascade, Sxi1α and Sxi2a form a heterodimer after cell fusion and are required for the establishment of the dikaryotic filament. *sxi1α* and *sxi2**a*** mutant strains are able to fuse during mating but are not able to form postfusion filaments. Instead, the cells immediately diploidize (Hull et al. 2002). Interestingly, the Sxi1α/Sxi2**a** proteins repress pheromone levels in postfusion cells (Hull et al. 2002).

In other basidiomycetes, the transcription factor Clp1 is a target of homeodomain transcription factors and is required for clamp formation. The Clp1 homolog has been identified in *C. neoformans* and is a target of Sxi1α/Sxi2**a**

(Ekena et al. 2008). Similar to the *C. neoformans* Sxi1α/Sxi2**a** proteins, Clp1 does not appear to be involved in clamp cell formation. Instead Clp1 is required for postfusion filamentation. In bilateral crosses, *clp1* mutant fusion products arrest as dumbbell-shaped cells that contained two nuclei. The exact role of Clp1 is not known, but it is hypothesized that it functions to reengage the cell cycle after cell–cell fusion (Ekena et al. 2008).

Sxi1α/Sxi2**a** also induces the postfusion transcription of Cpr2, a constitutive pheromone receptor. Cpr2 functions with and competes with the pheromone-induced receptor Ste3α/**a** to regulate G-protein signaling in *C. neoformans* during mating (Hsueh et al. 2009). Transcription of these two pheromone receptors is temporally distinct: *STE3α* transcription peaks 24 h after mating, while *CPR2* peaks at 72 h after mating. Fusion is decreased in *cpr2* mutant crosses, and the surviving fusion products are smaller than wild type fusions, suggesting that Cpr2 is required for fusion and/or for the survival of postfusion products (Hsueh et al. 2009).

Znf2 is a zinc-finger transcription factor that is also required for postfusion filamentation. Similar to *sxi1α* and *sxi2**a***, pheromone expression and cell fusion are enhanced in *znf2* mutants. However, while the Sxi1α/Sxi2**a** complex is required for establishing cell identify, Znf2 appears to function exclusively as a hyphal regulator. *ZNF2*, but not *SXI1α* or *SXI2**a***, is required for virulence in mice (Hull et al. 2004; Lin et al. 2010).

The $Ca^{+}$-calcineurin signaling pathway is also required for postfusion filamentation. In *C. neoformans*, calcineurin is comprised of a catalytic (Cna1) and a regulatory (Cnb1) subunit (Fox et al. 2003; Odom et al. 1997). Mutation of either gene inhibits the ability of cells to undergo cell–cell fusion during mating (Cruz et al. 2001). In addition, a calcineurin-binding protein, Cbp1, functions with calcineurin to promote filamentation (Görlach et al. 2000). Interestingly, a few basidia were observed in *cbp1* mutant crosses suggesting that, while filamentation is absent in these strains, the final stages of sexual development do occur.

#### 10.3.2.2 Filament and Clamp Cell Morphogenesis

Once the filamentous dikaryotic form has been established, it must be maintained. Postfusion *C. neoformans* filamentation mutant strains can be classified by morphology. Mutants completely unable to filament make up one class and include *sxi1α*, *sxi2**a***, *clp1*, and *znf2* (Ekena et al. 2008; Hull et al. 2002; Lin et al. 2010). A second class consists of mutants with reduced filamentation. These include many of the cell fusion mutants discussed in the previous section. Although reduced in numbers, some heterokaryons are generated in these mutant crosses and produce isolated mating foci. Often the filaments are morphologically normal within these foci and produce viable spores. The third class includes mutant strains with filament morphology defects, clamp cell fusion and morphology defects, and combinations of both. Not surprisingly, many of the genes characterized in filament morphology have roles in cell polarity and cytokinesis and several of these also have implications in pathogenesis. Interestingly, many of the genes involved in filament morphology are dose-dependent and the defects are revealed only when both mating partners are mutated (bilateral crosses).

As previously mentioned, Cdc42 and Rac1 are Rho-family GTPases. Both *cdc42* and *rac1* mutant strains exhibit clamp cell morphology and fusion defects in bilateral crosses (Ballou et al. 2010; Vallim et al. 2005). *cdc42* mutant strain crosses also have spore production and morphology defects (Ballou et al. 2010). The filaments produced in *rac1* mutant crosses are thicker than wild type but produce normal basidia and spores, indicating that the clamp cell defect does not inhibit the progression of nuclei. Based on phenotype, the role of Rac1 is restricted to mating, while Cdc42 is also required for high temperature growth and cytokinesis of the yeast cell (Ballou et al. 2010; Vallim et al. 2005).

*ste20α/**a*** mutant strains exhibit filament morphology defects, but clamp cell formation and fusion appear normal. These strains also have defects in high temperature growth and cytokinesis (Nichols et al. 2004; Wang et al. 2002). In many model systems, Ste20 functions as an effector of Cdc42. However, the filament morphology defects exhibited by *ste20α/**a*** mutant strains are quite distinct from either *cdc42* or *rac1* mutant strains. Ste20α**a** appears to be required to maintain hyphal tip polarity, and the filaments produced in *ste20α/**a*** mutant crosses undergo consecutive tip splitting. As with *rac1*, normal basidia and spores are produced in *ste20* crosses even though nuclei are often mis-sorted and lost during tip-splitting (Nichols et al. 2004).

The ability to produce normal basidia and spores in spite of impaired nuclear progression defects is also evident in *bim1* mutant strains. Bim1 is a microtubule-binding protein that is required for filament structural integrity and nuclear distribution. Filaments produced in *bim1* crosses are deformed and appear collapsed. In addition, the nuclei are misplaced within the filament. The formation of basidia and basidiospores is accelerated in *bim1* crosses, however the structures appear normal and produce viable progeny (Staudt et al. 2010).

Septins are another family of proteins involved in *C. neoformans* high temperature growth and cytokinesis that are also required for the morphology of the filament. Similar to the *bim1* mutant strain, septin mutant crosses displayed defects in nuclear distribution (Kozubowski and Heitman 2010). In other organisms septins associate with microtubules and actin, and *C. neoformans* septins also colocalize with microtubules (Kozubowski and Heitman 2010). These data suggest that Bim1 may function with septins and microtubules to maintain the distribution of nuclei in filaments. In addition, the filaments produced by septin mutant strains also exhibited defects in clamp cell fusion, basidia morphology, and basidiospore morphology.

Certain nutrients enhance or decrease filamentation in *C. neoformans*. In the laboratory, plant-derived materials are used to induce *C. neoformans* mating and include the commercially available V8 juice. Recently, it was found that the defined Musashige and Skoog (MS) mating medium also stimulated *C. neoformans* mating and that the inducing agent was myo-inositol (Xue et al. 2007). Myo-inositol is also a component of V8 juice. The *C. neoformans* genome contains ten or more myo-inositol transporters, and two transporters, Itr1 and Itr1A, are important for postfusion filamentation and spore production suggesting that they are specific to myo-inositol-mediated stimulation of mating (Xue et al. 2010b).

Nitrogen deprivation also stimulates mating and is dependent on the high-affinity ammonium transporter Amt2. Mating filamentation is decreased in *amt2* mutant crosses but appears normal on medium containing myo-inositol (Rutherford et al. 2008). However, at this time it is not known if the Amt2 protein is required for the pre- or postfusion stage of mating. Since filamentation is restored on myo-inositol medium, it is suggested that the mating defect of the *amt2* mutant occurs postfusion.

### *10.3.3 Monokaryotic Fruiting*

In *C. neoformans*, haploid cells can filament in response to nitrogen deprivation in a process first described as haploid fruiting. The filaments produced during haploid fruiting are monokaryotic with unfused clamp cells (Tscharke et al. 2003; Wickes et al. 1996). Recent studies have determined that while haploid filamentation does occur, the nuclei are often diploid and are the result of endoduplication or self-mating. Same sex mating usually occurs between α and α strains, but **a** and **a** matings can also occur (Lin et al. 2005). Regardless of ploidy, clamp cells are not able to fuse with the subapical cell, and peg cells are not observed. Many of the gene and gene products necessary for the formation of dikaryotic filaments postfusion are also required for monokaryotic filamentation. For example, the transcription factor Znf2 is required for both α-**a** and monokaryotic filamentation (Lin et al. 2010). In contrast, sex-identity transcription factors Sxi1α and Sxi2**a** are not required for monokaryotic fruiting but are required for α-**a** filamentation (Hull et al. 2002). The Ste12 transcription factor is required for monokaryotic fruiting but is not required for mating (Wickes et al. 1997; Yue et al. 1999).

### *10.3.4 More Post-hyphal Structures: Blastospores, Basidia, and Basidiospores*

The end result is the same for both dikaryotic filaments and monokaryotic filaments: basidia formation and basidiospore production (Kwon-Chung 1976) (Fig. 10.5). The basidium is the spore-forming cell that forms at the end of the filament. In most basidiomycetes, including *C. neoformans*, the basidium is club-shaped. Within the basidium, the nuclei undergo fusion, followed by a single meiosis and multiple rounds of mitosis. Nuclei are packaged into spore units and are ejected from the basidia at four spatially distinct locations. The basidiospores remain loosely attached to each other after ejection, forming four long chains of basidiospores. The four chains can remain distinct from each other, or can fuse at

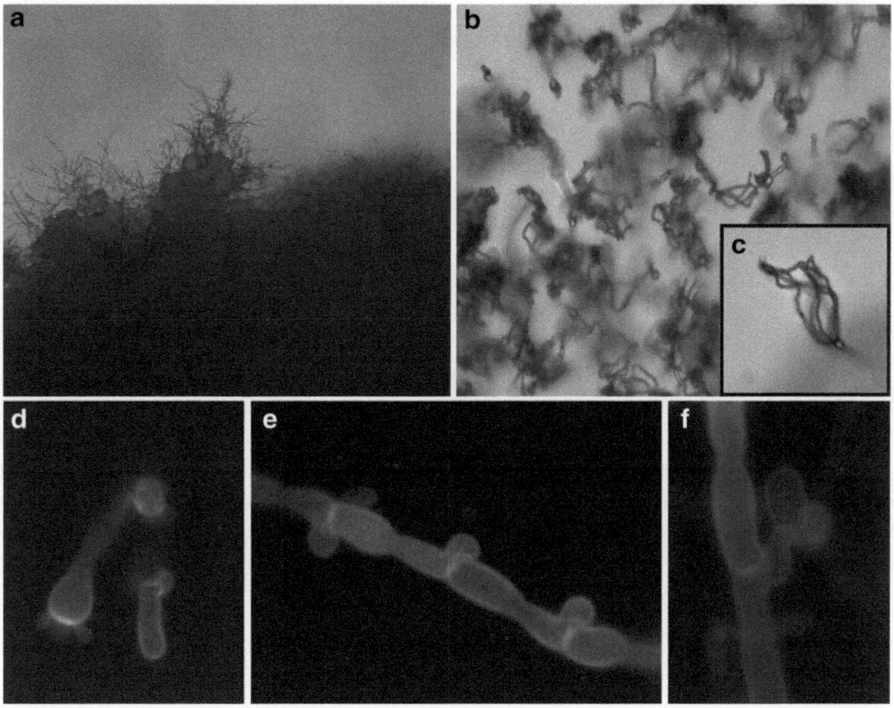


**Fig. 10.5** *C. neoformans* mating structures. Congenic mating partners were coincubated on MS mating medium for 1 week. (**a**) Filamentous differentiation at the edge of the culture patch represents the perfect form of this fungus, *Filobasidiella neoformans* (×50). (**b**) At higher magnification, mating hyphae are evident, terminating in flask-shaped basidia with chains of basidiospores (×200). (**c**) A single basidium with four long chains of basidiospores (×400). Mating hyphae were stained with calcofluor white and visualized by epifluorescence microscopy (×1,000) to demonstrate a basidium with early meiotic spores (**d**), septate hyphae with clamp cells (**e**), and rare branches of the hyphae with mitotic blastospores (**f**)

the apex and form a spore ball (Kwon-Chung 1976). Several proteins with roles in morphology have also been implicated in basidiospore production and/or morphology. These include Cdc42 (Ballou et al. 2010), Gdi1 (a GTPase dissociation factor and negative regulator of Cdc42) (Price et al. 2008), and septins (e.g., Cdc3, Cdc10) which are thought to function downstream of Cdc42 (Ballou et al. 2010; Kozubowski and Heitman 2010).

Blastospores and haustoria are two other morphogenic forms observed in *C. neoformans* filaments (Kwon-Chung 1976). Both forms originate from clamp cells. Blastospores refer to yeast cells that bud from clamp cells (Fig. 10.5) while haustoria are long thin filaments that form from the clamp cells. In other fungi, including members of the genus Filobasidiella, haustoria are indicative of parasitism and function to penetrate the host cell (Golubev and Golubev 2003). However, the role of haustoria in *C. neoformans* has not been defined. Interestingly, *cpr2* mutant filaments produce haustoria and budding cells in place of clamp cells, indicating that the constitutive pheromone receptor also regulates clamp cell morphology (Xue et al. 2010a).

## 10.4 Summary

*C. neoformans* uses conserved, interacting signaling pathways to control its morphogenesis in the yeast and hyphal forms. Yeast cell budding displays many similarities to distantly related ascomycete and basidiomycete species. However, *C. neoformans* has adapted these processes to direct its unique survival in the host, making it a formidable human pathogen.

A model of the regulatory signaling proteins involved in *C. neoformans* hyphal transitions is presented in Fig. 10.4. These complex protein interactions suggest that multiple signals (nutrients, pheromone, cell stress) play important roles in inducing and maintaining the hyphal state. Although hyphal formation is not required for invasion of the host or for virulence, this morphological transition is likely very important for survival and dispersal in the environment.

## References

Adams DJ (2004) Fungal cell wall chitinases and glucanases. Microbiology 150:2029–2035

Alspaugh JA, Perfect JR, Heitman J (1997) *Cryptococcus neoformans* mating and virulence are regulated by the G-protein alpha subunit GPA1 and cAMP. Genes Dev 11:3206–3217

Alspaugh JA, Cavallo LM, Perfect JR, Heitman J (2000) RAS1 regulates filamentation, mating and growth at high temperature of *Cryptococcus neoformans*. Mol Microbiol 36:352–365

Alspaugh JA, Pukkila-Worley R, Harashima T, Cavallo LM, Funnell D, Cox GM, Perfect JR, Kronstad JW, Heitman J (2002) Adenylyl cyclase functions downstream of the G-alpha protein Gpa1 and controls mating and pathogenicity of *Cryptococcus neoformans*. Eukaryot Cell 1:75–84

Bahn YS, Hicks JK, Giles SS, Cox GM, Heitman J (2004) Adenylyl cyclase-associated protein Aca1 regulates virulence and differentiation of *Cryptococcus neoformans* via the cyclic AMP-protein kinase A cascade. Eukaryot Cell 3:1476–1491

Bahn YS, Cox GM, Perfect JR, Heitman J (2005a) Carbonic anhydrase and $CO_2$ sensing during *Cryptococcus neoformans* growth, differentiation, and virulence. Curr Biol 15:2013–2020

Bahn YS, Kojima K, Cox GM, Heitman J (2005b) Specialization of the HOG pathway and its impact on differentiation and virulence of *Cryptococcus neoformans*. Mol Biol Cell 16:2285–2300

Baker RD, Haugen RK (1955) Tissue changes and tissue diagnosis in cryptococcosis: a study of twenty-six cases. Am J Clin Pathol 25:14–24

Baker LG, Specht CA, Donlin MJ, Lodge JK (2007) Chitosan, the deacetylated form of chitin, is necessary for cell wall integrity in *Cryptococcus neoformans*. Eukaryot Cell 6:855–867

Ballou ER, Nichols CB, Miglia KJ, Kozubowski L, Alspaugh JA (2010) Two CDC42 paralogues modulate *Cryptococcus neoformans* thermotolerance and morphogenesis under host physiological conditions. Mol Microbiol 75:763–780

Banuett F, Quintanilla RH Jr, Reynaga-Pena CG (2008) The machinery for cell polarity, cell morphogenesis, and the cytoskeleton in the Basidiomycete fungus *Ustilago maydis* – a survey of the genome sequence. Fungal Genet Biol 45:S3–S14
Bassilana M, Arkowitz RA (2006) Rac1 and Cdc42 have different roles in *Candida albicans* development. Eukaryot Cell 5:321–329
Casselton LA, Olesnicky NS (1998) Molecular genetics of mating recognition in basidiomycete fungi. Microbiol Mol Biol Rev 62:55–70
Chang EC, Barr M, Wang Y, Jung V, Xu HP, Wigler MH (1994) Cooperative interaction of S. pombe proteins required for mating and morphogenesis. Cell 79:131–141
Chang YC, Miller GF, Kwon-Chung KJ (2003) Importance of a developmentally regulated pheromone receptor of *Cryptococcus neoformans* for virulence. Infect Immun 71:4953–4960
Chant J (1999) Cell polarity in yeast. Annu Rev Cell Dev Biol 15:365–391
Chant J, Pringle JR (1991) Budding and cell polarity in *Saccharomyces cerevisiae*. Curr Opin Genet Dev 1:342–350
Chuck SL, Sande MA (1989) Infections with *Cryptococcus neoformans* in the acquired immunodeficiency syndrome. N Engl J Med 321:794–799
Cramer KL, Gerrald QD, Nichols CB, Price MS, Alspaugh JA (2006) The transcription factor Nrg1 mediates capsule, stress response, and pathogenesis in *Cryptococcus neoformans*. Eukaryot Cell 5:1147–1156
Cruz MC, Fox DS, Heitman J (2001) Calcineurin is required for hyphal elongation during mating and haploid fruiting in *Cryptococcus neoformans*. EMBO J 20:1020–1032
Cvrckova F, Virgilio CD, Manser E, Pringle JR, Nasmyth K (1995) Ste20-like protein kinases are required for normal localization of cell growth and for cytokinesis in budding yeast. Genes Dev 9:1817–1830
D'Souza CA, Alspaugh JA, Yue C, Harashima T, Cox GM, Perfect JR, Heitman J (2001) Cyclic AMP-dependent protein kinase controls virulence of the fungal pathogen *Cryptococcus neoformans*. Mol Cell Biol 21:3179–3191
Davidson RC, Moore TD, Odom AR, Heitman J (2000) Characterization of the MFalpha pheromone of the human fungal pathogen *Cryptococcus neoformans*. Mol Microbiol 38:1017–1026
Davidson RC, Nichols CB, Cox GM, Perfect JR, Heitman J (2003) A MAP kinase cascade composed of cell type specific and non-specific elements controls mating and differentiation of the fungal pathogen *Cryptococcus neoformans*. Mol Microbiol 49:469–485
del Poeta M, Toffaletti DL, Rude TH, Sparks SD, Heitman J, Perfect JR (1999) *Cryptococcus neoformans* differential gene expression detected in vitro and in vivo with green fluorescent protein. Infect Immun 67:1812–1820
Ekena JL, Stanton BC, Schiebe-Owens JA, Hull CM (2008) Sexual development in *Cryptococcus neoformans* requires CLP1, a target of the homeodomain transcription factors Sxi1alpha and Sxi2a. Eukaryot Cell 7:49–57
Fazekas G, Schwarz J (1958) Histology of experimental murine *Cryptococcosis*. Am J Pathol 34:517–529
Feldmesser M, Kress Y, Casadevall A (2001) Dynamic changes in the morphology of *Cryptococcus neoformans* during murine pulmonary infection. Microbiology 147:2355–2365
Fox DS, Cox GM, Heitman J (2003) Phospholipid-binding protein Cts1 controls septation and functions coordinately with calcineurin in *Cryptococcus neoformans*. Eukaryot Cell 2:1025–1035
Fraser JA, Diezmann S, Subaran RL, Allen A, Lengeler KB, Dietrich FS, Heitman J (2004) Convergent evolution of chromosomal sex-determining regions in the animal and fungal kingdoms. PLoS Biol 2:e384
Fu J, Mares C, Lizcano A, Liu Y, Wickes BL (2011) Insertional mutagenesis combined with an inducible filamentation phenotype reveals a conserved STE50 homologue in *Cryptococcus neoformans* that is required for monokaryotic fruiting and sexual reproduction. Mol Microbiol 79:990–1007

Fujita A, Lord M, Hiroko T, Hiroko F, Chen T, Oka C, Misumi Y, Chant J (2004) Rax1, a protein required for the establishment of the bipolar budding pattern in yeast. Gene 327:161–169

Giles SS, Dagenais TR, Botts MR, Keller NP, Hull CM (2009) Elucidating the pathogenesis of spores from the human fungal pathogen *Cryptococcus neoformans*. Infect Immun 77:3491–3500

Gladfelter AS, Bose I, Zyla TR, Bardes ES, Lew DJ (2002) Septin ring assembly involves cycles of GTP loading and hydrolysis by Cdc42p. J Cell Biol 156:315–326

Golubev VI, Golubev NV (2003) A new basidiomycetous yeast species, *Cryptococcus mycelialis*, related to Holtermannia Saccardo et Traverso. Mikrobiologiia 72:822–827

Görlach J, Fox DS, Cutler NS, Cox GM, Perfect JR, Heitman J (2000) Identification and characterization of a highly conserved calcineurin binding protein, CBP1/calcipressin, in *Cryptococcus neoformans*. EMBO J 19:3618–3629

Hanseman DV (1905) Uber eine bisher nicht beobachtete Gehriner Krankung durch Hefen. Verh Dtsch Ges Pathol 9:21–24

Ho J, Bretscher A (2001) Ras regulates the polarity of the yeast actin cytoskeleton through the stress response pathway. Mol Biol Cell 12:1541–1555

Hsueh YP, Shen WC (2005) A homolog of Ste6, the a-factor transporter in *Saccharomyces cerevisiae*, is required for mating but not for monokaryotic fruiting in *Cryptococcus neoformans*. Eukaryot Cell 4:147–155

Hsueh YP, Xue C, Heitman J (2007) G protein signaling governing cell fate decisions involves opposing Galpha subunits in *Cryptococcus neoformans*. Mol Biol Cell 18:3237–3249

Hsueh YP, Xue C, Heitman J (2009) A constitutively active GPCR governs morphogenic transitions in *Cryptococcus neoformans*. EMBO J 28(9):1220–1233

Hull CM, Davidson RC, Heitman J (2002) Cell identity and sexual development in *Cryptococcus neoformans* are controlled by the mating-type-specific homeodomain protein Sxi1alpha. Genes Dev 16:3046–3060

Hull CM, Cox GM, Heitman J (2004) The alpha-specific cell identity factor Sxi1alpha is not required for virulence of *Cryptococcus neoformans*. Infect Immun 72:3643–3645

Idnurm A, Heitman J (2005) Light controls growth and development via a conserved pathway in the fungal kingdom. PLoS Biol 3:e95

Jung KW, Kim SY, Okagaki LH, Nielsen K, Bahn YS (2011) Ste50 adaptor protein governs sexual differentiation of *Cryptococcus neoformans* via the pheromone-response MAPK signaling pathway. Fungal Genet Biol 48:154–165

Klengel T, Liang WJ, Chaloupka J, Ruoff C, Schroppel K, Naglik JR, Eckert SE, Mogensen EG, Haynes K, Tuite MF, Levin LR, Buck J, Muhlschlegel FA (2005) Fungal adenylyl cyclase integrates $CO_2$ sensing with cAMP signaling and virulence. Curr Biol 15:2021–2026

Kopecka M, Gabriel M, Takeo K, Yamaguchi M, Svoboda A, Ohkusu M, Hata K, Yoshida S (2001) Microtubules and actin cytoskeleton in *Cryptococcus neoformans* compared with ascomycetous budding and fission yeasts. Eur J Cell Biol 80:303–311

Kozubowski L, Heitman J (2010) Septins enforce morphogenetic events during sexual reproduction and contribute to virulence of *Cryptococcus neoformans*. Mol Microbiol 75:658–675

Kozubowski L, Lee SC, Heitman J (2009) Signalling pathways in the pathogenesis of *Cryptococcus*. Cell Microbiol 11:370–380

Kwon-Chung KJ (1975) A new genus, *Filobasidiella*, the perfect state of *Cryptococcus neoformans*. Mycologia 67:1197–1200

Kwon-Chung KJ (1976) Morphogenesis of *Filobasidiella neoformans*, the sexual state of *Cryptococcus neoformans*. Mycologia 68:821–833

Kwon-Chung KJ, Bennett JE (1978) Distribution of a and a mating types of *Cryptococcus neoformans* among natural and clinical isolates. Am J Epidemiol 108:337–340

Kwon-Chung KJ, Popkin TJ (1976) Ultrastructure of septal complex in *Filobasidiella neoformans* (*Cryptococcus neoformans*). J Bacteriol 126:524–528

Kwon-Chung KJ, Edman JC, Wickes BL (1992) Genetic association of mating types and virulence in *Cryptococcus neoformans*. Infect Immun 60:602–605

Leveleki L, Mahlert M, Sandrock B, Bolker M (2004) The PAK family kinase Cla4 is required for budding and morphogenesis in *Ustilago maydis*. Mol Microbiol 54:396–406

Levin EA (1937) Torula infection of the central nervous system. Arch Intern Med 59:667–684

Li Y, Chang EC (2003) *Schizosaccharomyces pombe* Ras1 effector, Scd1, interacts with Klp5 and Klp6 kinesins to mediate cytokinesis. Genetics 165:477–488

Li L, Shen G, Zhang ZG, Wang YL, Thompson JK, Wang P (2007) Canonical heterotrimeric G proteins regulating mating and virulence of *Cryptococcus neoformans*. Mol Biol Cell 18:4201–4209

Lin X, Hull CM, Heitman J (2005) Sexual reproduction between partners of the same mating type in *Cryptococcus neoformans*. Nature 434:1017–1021

Lin X, Jackson JC, Feretzaki M, Xue C, Heitman J (2010) Transcription factors Mat2 and Znf2 operate cellular circuits orchestrating opposite- and same-sex mating in *Cryptococcus neoformans*. PLoS Genet 6:e1000953

Longtine MS, Theesfeld CL, McMillan JN, Weaver E, Pringle JR, Lew DJ (2000) Septin-dependent assembly of a cell cycle-regulatory module in *Saccharomyces cerevisiae*. Mol Cell Biol 20:4049–4061

Lurie HI, Shadomy HJ (1971) Morphological variations of a hypha-forming strain of *Cryptococcus neoformans* (Coward strain) in tissues of mice. Sabouraudia 9:10–14

Maeng S, Ko YJ, Kim GB, Jung KW, Floyd A, Heitman J, Bahn YS (2010) Comparative transcriptome analysis reveals novel roles of the Ras and cyclic AMP signaling pathways in environmental stress response and antifungal drug sensitivity in *Cryptococcus neoformans*. Eukaryot Cell 9:360–378

Mochizuki T, Tanaka S, Watanabe S (1987) Ultrastructure of the mitotic apparatus in *Cryptococcus neoformans*. J Med Vet Mycol 25:223–233

Mogensen EG, Janbon G, Chaloupka J, Steegborn C, Fu MS, Moyrand F, Klengel T, Pearson DS, Geeves MA, Buck J, Levin LR, Muhlschlegel FA (2006) *Cryptococcus neoformans* senses $CO_2$ through the carbonic anhydrase Can2 and the adenylyl cyclase Cac1. Eukaryot Cell 5:103–111

Moore RT (2000) Cytology and ultrastructure of yeasts and yeastlike fungi. In: Kurtzman CP, Fell JW (eds) The yeasts: a taxonomic study. Elsevier Science, Amsterdam, pp 33–44

Moore TD, Edman JC (1993) The alpha-mating type locus of *Cryptococcus neoformans* contains a peptide pheromone gene. Mol Cell Biol 13:1962–1970

Nichols CB, Fraser JA, Heitman J (2004) PAK Kinases Ste20 and Pak1 govern cell polarity at different stages of mating in *Cryptococcus neoformans*. Mol Biol Cell 15:4476–4489

Nichols CB, Perfect Z, Alspaugh JA (2007) A Ras1-Cdc24 signal transduction pathway mediates thermotolerance in the fungal pathogen *Cryptococcus neoformans*. Mol Microbiol 63:1118–1130

Nichols CB, Ferreyra J, Ballou ER, Alspaugh JA (2009) Subcellular localization directs signaling specificity of the *Cryptococcus neoformans* Ras1 protein. Eukaryot Cell 8:181–189

Odom A, Muir S, Lim E, Toffaletti DL, Perfect J, Heitman J (1997) Calcineurin is required for virulence of *Cryptococcus neoformans*. EMBO J 16:2576–2589

Ohkusu M, Hata K, Takeo K (2001a) Bud emergence is gradually delayed from S to G2 with progression of growth phase in *Cryptococcus neoformans*. FEMS Microbiol Lett 194:251–255

Ohkusu M, Raclavsky V, Takeo K (2001b) Deficit in oxygen causes G(2) budding and unbudded G (2) arrest in *Cryptococcus neoformans*. FEMS Microbiol Lett 204:29–32

Okagaki LH, Strain AK, Nielsen JN, Charlier C, Baltes NJ, Chretien F, Heitman J, Dromer F, Nielsen K (2010) Cryptococcal cell morphology affects host cell interactions and pathogenicity. PLoS Pathog 6:e1000953

Panepinto J, Liu L, Ramos J, Zhu X, Valyi-Nagy T, Eksi S, Fu J, Jaffe HA, Wickes B, Williamson PR (2005) The DEAD-box RNA helicase Vad1 regulates multiple virulence-associated genes in *Cryptococcus neoformans*. J Clin Invest 115:632–641

Park BJ, Wannemuehler KA, Marston BJ, Govender N, Pappas PG, Chiller TM (2009) Estimation of the current global burden of cryptococcal meningitis among persons living with HIV/AIDS. AIDS 23:525–530

Park YD, Panepinto J, Shin S, Larsen P, Giles S, Williamson PR (2010) Mating pheromone in *Cryptococcus neoformans* is regulated by a transcriptional/degradative "futile" cycle. J Biol Chem 285:34746–34756

Price MS, Nichols CB, Alspaugh JA (2008) The *Cryptococcus neoformans* Rho-GDP dissociation inhibitor mediates intracellular survival and virulence. Infect Immun 76:5729–5737

Pukkila-Worley R, Alspaugh JA (2004) Cyclic AMP signaling in *Cryptococcus neoformans*. FEMS Yeast Res 4:361–367

Roemer T, Vallier L, Sheu Y, Snyder M (1998) The Spa2-related protein, Sph1p, is important for polarized growth in yeast. J Cell Sci 111:479–494

Rutherford JC, Lin X, Nielsen K, Heitman J (2008) Amt2 permease is required to induce ammonium-responsive invasive growth and mating in *Cryptococcus neoformans*. Eukaryot Cell 7:237–246

Sanders SL, Herskowitz I (1996) The BUD4 protein of yeast, required for axial budding, is localized to the mother/BUD neck in a cell cycle-dependent manner. J Cell Biol 134:413–427

Staudt MW, Kruzel EK, Shimizu K, Hull CM (2010) Characterizing the role of the microtubule binding protein Bim1 in *Cryptococcus neoformans*. Fungal Genet Biol 47:310–317

Sukroongreung S, Kitiniyom K, Nilakul C, Tantimavanich S (1998) Pathogenicity of basidiospores of *Filobasidiella neoformans* var. *neoformans*. Med Mycol 36:419–424

Takeo K, Tanaka R, Miyaji M, Nishimura K (1995) Unbudded G2 as well as G1 arrest in the stationary phase of the basidiomycetous yeast *Cryptococcus neoformans*. FEMS Microbiol Lett 129:231–235

Takeo K, Ogura Y, Virtudazo E, Raclavsky V, Kawamoto S (2004) Isolation of a CDC28 homologue from *Cryptococcus neoformans* that is able to complement cdc28 temperature-sensitive mutants of *Saccharomyces cerevisiae*. FEMS Yeast Res 4:737–744

Todd RL, Herrmann WW (1936) The life cycle of the organism causing yeast meningitis. J Bacteriol 32:89–101

Tscharke RL, Lazera M, Chang YC, Wickes BL, Kwon-Chung KJ (2003) Haploid fruiting in *Cryptococcus neoformans* is not mating type alpha-specific. Fungal Genet Biol 39:230–237

Vallim MA, Nichols CB, Fernandes L, Cramer KL, Alspaugh JA (2005) A Rac homolog functions downstream of Ras1 to control hyphal differentiation and high-temperature growth in the pathogenic fungus *Cryptococcus neoformans*. Eukaryot Cell 4:1066–1078

Velagapudi R, Hsueh YP, Geunes-Boyer S, Wright JR, Heitman J (2009) Spores as infectious propagules of *Cryptococcus neoformans*. Infect Immun 77:4345–4355

Virtudazo EV, Kawamoto S, Ohkusu M, Aoki S, Sipiczki M, Takeo K (2010) The single Cdk1-G1 cyclin of *Cryptococcus neoformans* is not essential for cell cycle progression, but plays important roles in the proper commitment to DNA synthesis and bud emergence in this yeast. FEMS Yeast Res 10:605–618

Wang P, Nichols CB, Lengeler KB, Cardenas ME, Cox GM, Perfect JR, Heitman J (2002) Mating-type-specific and nonspecific PAK kinases play shared and divergent roles in *Cryptococcus neoformans*. Eukaryot Cell 1:257–272

Wang P, Cutler J, King J, Palmer D (2004) Mutation of the regulator of G protein signaling Crg1 increases virulence in *Cryptococcus neoformans*. Eukaryot Cell 3:1028–1035

Waugh MS, Vallim MA, Heitman J, Alspaugh JA (2003) Ras1 controls pheromone expression and response during mating in *Cryptococcus neoformans*. Fungal Genet Biol 38:110–121

Weirich CS, Erzberger JP, Barral Y (2008) The septin family of GTPases: architecture and dynamics. Nat Rev Mol Cell Biol 9:478–489

Wickes BL, Mayorga ME, Edman U, Edman JC (1996) Dimorphism and haploid fruiting in *Cryptococcus neoformans*: association with the alpha-mating type. Proc Natl Acad Sci USA 93:7327–7331

Wickes BL, Edman U, Edman JC (1997) The *Cryptococcus neoformans STE12alpha* gene: a putative *Saccharomyces cerevisiae STE12* homologue that is mating type specific. Mol Microbiol 26:951–960

Xue C, Bahn YS, Cox GM, Heitman J (2006) G protein-coupled receptor Gpr4 senses amino acids and activates the cAMP-PKA pathway in *Cryptococcus neoformans*. Mol Biol Cell 17:667–679

Xue C, Tada Y, Dong X, Heitman J (2007) The human fungal pathogen Cryptococcus can complete its sexual cycle during a pathogenic association with plants. Cell Host Microbe 1:263–273

Xue C, Hsueh YP, Chen L, Heitman J (2008a) The RGS protein Crg2 regulates both pheromone and cAMP signalling in *Cryptococcus neoformans*. Mol Microbiol 70:379–395

Xue C, Hsueh YP, Heitman J (2008b) Magnificent seven: roles of G protein-coupled receptors in extracellular sensing in fungi. FEMS Microbiol Rev 32:1010–1032

Xue C, Liu T, Chen L, Li W, Liu I, Kronstad JW, Seyfang A, Heitman J (2010a) Role of an expanded inositol transporter repertoire in *Cryptococcus neoformans* sexual reproduction and virulence. MBio 1(1):e00084-10

Xue C, Wang Y, Hsueh YP (2010b) Assessment of constitutive activity of a G protein-coupled receptor, CPR2, in *Cryptococcus neoformans* by heterologous and homologous methods. Methods Enzymol 484:397–412

Yamaguchi M, Ohkusu M, Biswas SK, Kawamoto S (2007) Cytological study of cell cycle of the pathogenic yeast *Cryptococcus neoformans*. Nihon Ishinkin Gakkai Zasshi 48:147–152

Yamaguchi M, Biswas SK, Ohkusu M, Takeo K (2009) Dynamics of the spindle pole body of the pathogenic yeast *Cryptococcus neoformans* examined by freeze-substitution electron microscopy. FEMS Microbiol Lett 296:257–265

Yue C, Cavallo LM, Alspaugh JA, Wang P, Cox GM, Perfect JR, Heitman J (1999) The *STE12alpha* homolog is required for haploid filamentation but largely dispensable for mating and virulence in *Cryptococcus neoformans*. Genetics 153:1601–1615

Zaragoza O, Garcia-Rodas R, Nosanchuk JD, Cuenca-Estrella M, Rodriguez-Tudela JL, Casadevall A (2010) Fungal cell gigantism during mammalian infection. PLoS Pathog 6:e1000945

# Chapter 11
# Morphogenesis and Infection in *Botrytis cinerea*

**Julia Schumacher and Paul Tudzynski**

**Abstract** *Botrytis cinerea*, the gray mold fungus, is a ubiquitous pathogen of high economic importance. Hence, its development and infection cycle have been well characterized over the years. Modern approaches using molecular methods and "omics" data now have opened new and fascinating perspectives on the molecular mechanisms involved in morphogenesis and development, and their relationship to the highly efficient pathogenic strategies used by this pathogen. This chapter focuses on recent data obtained by analyzing signaling cascades which influence morphogenesis and virulence, highlighting the plethora of open questions that still remain. The light-dependent regulation of development is discussed as a particular example of a highly interesting area of research in which the broad classical analyses are not yet substantiated by molecular investigations.

## Abbreviations

| | |
|---|---|
| ROS | Reactive oxygen species |
| CWDE | Cell wall degrading enzyme |
| MAPK | Mitogen-activated protein kinase |

## 11.1 *Botrytis cinerea*: The Gray Mold Fungus

*Botrytis cinerea* Persoon: Fries [teleomorph *Botryotinia fuckeliana* (de Bary) Whetzel] causes the rotting of plant material accompanied by the formation of gray conidiophores and conidia (gray mold disease). *B. cinerea* is an ascomycetous

J. Schumacher (✉) • P. Tudzynski
Institut für Biologie und Biotechnologie der Pflanzen, Westfälische Wilhelms-Universität Münster, Hindenburgplatz 55, Münster 48143, Germany
e-mail: jschumac@uni-muenster.de

J. Pérez-Martín and A. Di Pietro (eds.), *Morphogenesis and Pathogenicity in Fungi*, Topics in Current Genetics 22, DOI 10.1007/978-3-642-22916-9_11, 

fungus belonging to the class Leotiomycetes and the family Sclerotiniaceae. A close relative is the soil-borne fungus *Sclerotinia sclerotiorum* [(Lib.) de Bary] that causes white mold disease in more than 400 different plant species (reviewed in Bolton et al. 2006). Both pathogens are typical necrotrophs, which induce host cell death, leading to damage of plant tissue and finally to rotting of the whole plant or of the harvested organ. More detailed molecular and comparative analyses have become feasible as the genomes of both pathogens were sequenced in recent years (Amselem et al. 2011).

*B. cinerea* is known to infect more than 200 plant species worldwide, including a range of agronomically important crops. Primary hosts are dicotyledonous plants, such as grape vine, strawberry, tomato, cucumber, and ornamental flowers. Monocots are generally considered as poor hosts, but some are susceptible under particular conditions. Almost all above-ground parts of the plant, including the stems, leaves, and flowers, unripe and ripe fruits can be infected. Even though *B. cinerea* is able to cause disease on healthy plants, the pathogen is most destructive on ripening or harvested vegetables, fruits, ornamental flowers, and senescent leaves in which host defense reactions are much reduced, or negligible compared to actively growing tissues.

The life cycle of *B. cinerea* is not that complex: mycelia can form macroconidia for propagation and dispersal, sclerotia for survival and as a basis for sexual development, microconidia as male gametes, and apothecia producing ascospores. The pathogenic relationship with the host further includes the development of infection structures for penetration and the active secretion of factors for killing cells and for obtaining nutrition from the dead tissue. In addition, *B. cinerea* counteracts host defense reactions, for instance by enzymatic degradation and/or secretion of antifungal compounds. Primary sources of infections by *B. cinerea* are the conidia that are ubiquitously distributed in the air dispersed by wind, rainfall, or insects. After landing on the plant surface, the conidia germinate under favorable conditions (free surface water and high relative humidity support germination, but are not essential) and form short germ tubes which directly penetrate the plant surface (Williamson et al. 1995). After penetration, the epidermal and underlying cells are killed and *B. cinerea* establishes a primary restricted infection. At this time point, necrosis and plant defense responses can occur; and there are two possibilities for ongoing pathogenesis. Occasionally, e.g., in infected flowers, the mycelium remains quiescent or latent in the infected tissue and a pathogenic relationship is not established until the fruit ripens (Holz et al. 2004). Different changes occurring during the ripening process may affect the transition from a quiescent to an active pathogenic relationship. Hence, fungitoxic compounds in unripe fruits decline in concentration, sugars increase and the chemical composition of cell walls change. In the pathogenic interaction, the fungus overcomes the plant defense barriers and starts a massive outgrowth (spreading lesion, secondary infection) that results finally in total maceration of plant tissue (soft rot) and formation of reproductive structures for over-wintering (sclerotia), or for inoculum production for further infections (macroconidia). Depending on the tissue attacked, and the environment, one infection cycle starting from the arrival of the conidium on the host, through to conidiation may be completed within 3–5 days. Therefore, such rapid growth and propagation rates mark this fungus as an aggressive pathogen that causes serious damage in the field after its first appearance in spring. *B. cinerea* field populations are known for high genetic variation regarding their aggressiveness on different plant species, their arsenals of weapons, and their preferred method of reproduction (Grindle 1979). Hence, the outcome of experiments using different isolates should be carefully discussed. The *Vitis* isolate B05.10 (Büttner et al. 1994) has now become the general recipient strain for molecular studies allowing for comparative studies of virulence and differentiation. The life cycle of this isolate is described in Fig. 11.1.

Detailed descriptions of *Botrytis* and its biology have been given by Coley-Smith et al. (1980) and Elad et al. (2004) and have been updated by reviews by van Kan (2006), Choquer et al. (2007), Williamson et al. (2007), and Tudzynski and Kokkelink (2009). The emphasis of this review is placed upon the regulation of the morphological changes *B. cinerea* is undergoing during infectious and saprophytic growth. By combining observations made by researchers in the last century and data arising from molecular analyses of key signaling components (Fig. 11.2), as well as from the *Botrytis*/*Sclerotinia* genome project, we would like to draw attention to some interesting issues for future research.

### *11.1.1 Conidial Germination and Differentiation of Germ Tubes In Vitro*

Macroconidia are the predominant reproductive structures of *B. cinerea* and are formed by conidiophores at the end of the infection cycle. The conidia may remain in an ungerminated state for long periods, held in check by lack of moisture and nutrients. As they have very little stored food resources, the initiation of the germination process is a life-and-death issue for the fungus: the conidia are only allowed to germinate when there is the option for propagation, i.e., when the conidia reach a living host plant or another suitable source of food. The first visible sign of germination is the swelling of the conidium followed by the emergence of the germ tube. Differences between isolates were observed with regard to the induction of this process; some *B. cinerea* isolates germinate readily in distilled water, while others require sugars and amino acids (Blakeman 1975).

Döhlemann et al. (2006a) demonstrated that germination of *B. cinerea* B05.10 conidia can be induced in vitro in two ways. When conidia are incubated on hydrophilic surfaces in presence of nutrients (fructose is more effective that glucose or sucrose), developing germ tubes are relatively thick and fast growing. A second inducing signal is a hydrophobic surface: when conidia are incubated on polypropylene foil without nutrients, the conidia form small nose-like germ tubes, usually one germ tube per conidium. Conidial germination induced by nutrients is regulated by the MAP kinase BMP1, the G$\alpha$ subunit BCG3 and the adenylate cyclase BAC

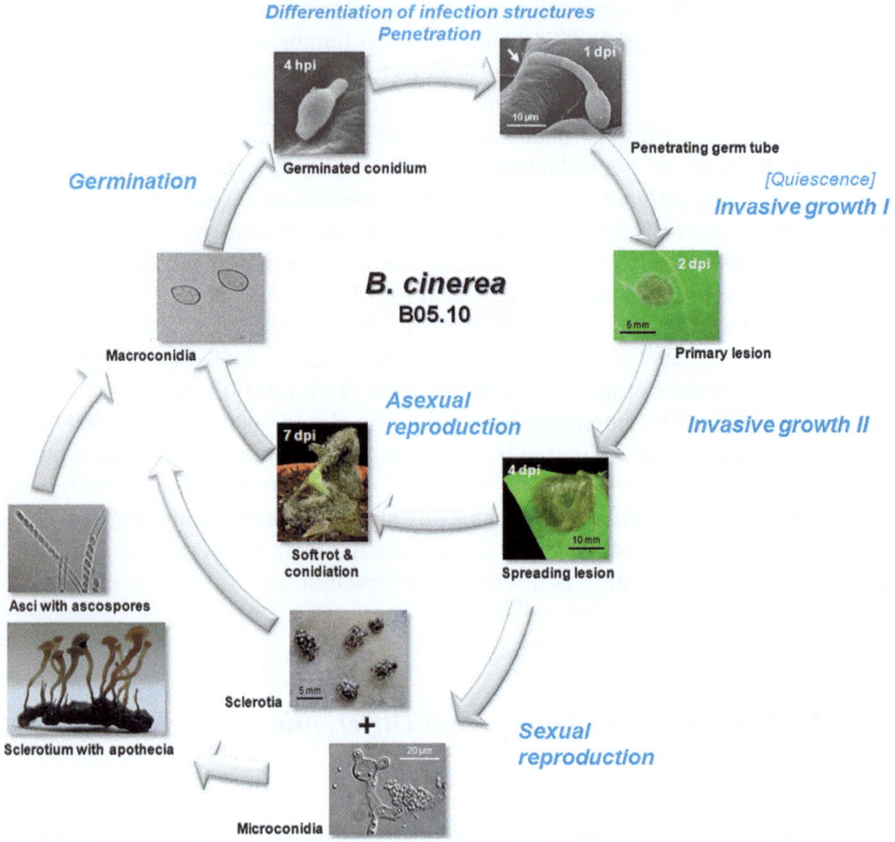


**Fig. 11.1** Life cycle of *B. cinerea* B05.10 under laboratory conditions. The germination of conidia on primary leaves of the French bean *Phaseolus vulgaris* begins within hours. The short germ tubes quickly penetrate the epidermis and invade the plant tissue, causing collapse and disintegration of the cells. Two days postinoculation, first macroscopically visible symptoms appear: small necrotic spots (primary lesions) that spread quickly, reaching diameters of 25 mm after 4 days. After 1 week the whole leaf is infected: the plant tissue collapses and becomes watery (soft rot) accompanied by formation of *gray* conidiophores and conidia. Conidiation is strongly dependent on light; in darkness conidiation is abolished and after 3–4 weeks sclerotia appear. Sclerotia are *dark* resting structures which may germinate by forming conidiophores or mycelia under appropriate conditions. Furthermore, they can act as female (sclerotial) parents after fertilization with suspensions of microconidia from a male (spermatial) parent carrying the opposite mating type. After several weeks of incubation under diurnal illumination, apothecia can be found containing asci with eight ascospores. B05.10 and derivative strains possess the *MAT1-1* locus and can be crossed for instance with strain SAS405, carrying the *MAT1-2* locus (Faretra et al. 1988)

(Döhlemann et al. 2006a), but not by the cAMP-dependent protein kinase (PKA) (Schumacher et al. 2008b). Germination in the presence of nutrients is accompanied by the rapid degradation of trehalose, an important carbon storage compound and stress protectant. Accordingly, mutants either defective in trehalose synthesis or in

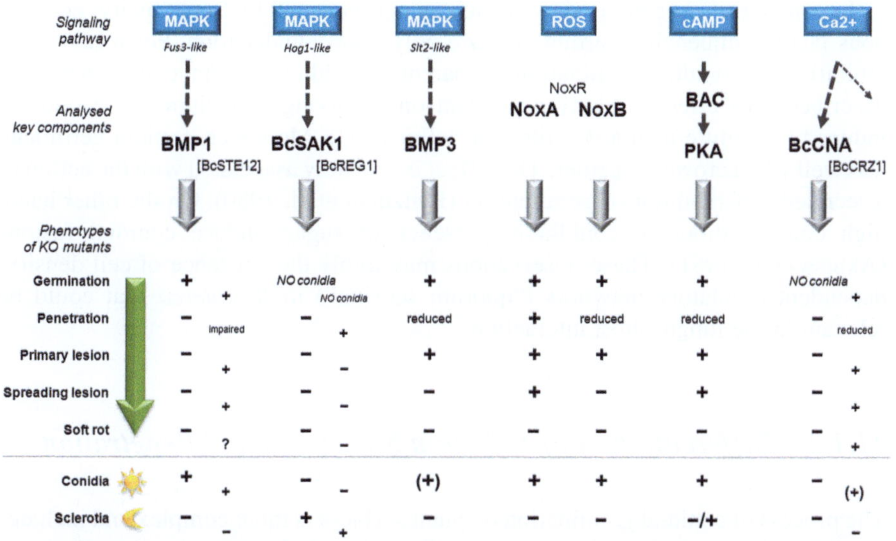


**Fig. 11.2** The major signal transduction pathways in *B. cinerea*. The phenotypes, i.e., virulence and light-dependent development in vitro of the respective replacement mutants are listed. BMP1 – Fus3-like MAP kinase (Zheng et al. 2000; Döhlemann et al. 2006a); BcSTE12 – Ste12-like transcription factor (Schamber et al. 2010); BcSAK1 – Hog1-like MAP kinase (Segmüller et al. 2007); BcREG1 – putative transcriptional regulator (Michielse et al. 2011), BMP3 – Slt2-like MAP kinase (Rui and Hahn 2007); BcNOXA, BcNOXB and BcNOXR – catalytic subunits and regulatory subunit of the NADPH oxidase complex, respectively (Segmüller et al. 2008); BAC – adenylate cyclase (Klimpel et al. 2002); PKA – catalytic subunit 1 of the cAMP dependent protein kinase (Schumacher et al. 2008b); BcCNA – catalytic subunit of the $Ca^{2+}$-regulated calcineurin phosphatase (Viaud et al. 2003); BcCRZ1 – calcineurin-responsive transcription factor (Schumacher et al. 2008a)

trehalose degradation exhibit impaired germination rates, but they are fully virulent (Döhlemann et al. 2006b).

Induction of germination by hydrophobic surfaces is controlled by the MAP kinase BMP1 in a cAMP-independent manner, as *Δbmp1* conidia fail to germinate (Döhlemann et al. 2006a). Mutants in which the Rho-GTPase BcCDC42 or the β subunit of heterotrimeric G proteins (BcGB1) are deleted are not impaired in germination per se but conidia form elongated germ tubes with an abnormal shape (Kokkelink et al. in press; J. Schumacher and B. Tudzynski, unpublished). The role of the $Ca^{2+}$-dependent signal transduction in regulation of germination is not yet clear. Neither the inhibition of the calcineurin phosphatase (Viaud et al. 2003) nor the deletion of the downstream transcription factor BcCRZ1 resulted in impaired germination rates (Schumacher et al. 2008a). However, conidia of deletion mutants of the phospholipase C (BcPLC1), which is supposed to be an upstream regulator of $Ca^{2+}$ signaling, failed to germinate under all tested conditions (Schumacher et al. 2008c).

In contrast to exogenous signals such as nutrients and hydrophobicity, endogenous factors influencing germination are only poorly understood. For instance, an inhibition of conidial germination by parental conidiophores/mycelium has been described; this effect is likely dependent on the living mycelium and cannot be induced by culture media (Carlile and Sellin 1963). Moreover, conidia germinate less well when crowed together. This effect is probably associated with the action of a secreted self-inhibitor of germination (Kritzman et al. 1980). On the other hand, high concentrations of conidia in presence of sugars induce conidial fusions (Akutsu et al. 1981). These observations may imply the existence of cell density-dependent regulatory networks ("quorum sensing") in *B. cinerea* that could be relevant in the fungus–host interaction.

### 11.1.2 Differentiation of Infection Structures and Penetration

The process of conidial germination on plant surfaces is more complex and includes the differentiation of infection structures. Even though the plant signal that induces germination is unknown, germination on onion and bean leaf epidermis resembles the in vitro germination program induced by hydrophobic surfaces rather than that induced by nutrients, i.e., the conidia merely form short germ tubes which immediately penetrate the epidermal cells. The shape of infection structures forming the infection hyphae varies from simple apical swellings of germ tubes (appressoria) to multibranched structures called infection cushions (Backhouse and Willetts 1987).

Several signaling pathways are implicated in the early stages of plant infection by germinating conidia. Several mutant phenotypes have been observed: mutants that do not form any infection structures or nonfunctional ones, and hence are unable to cause any necrotic lesions, mutants that form exclusively primary lesions and mutants that are (only) retarded in the infection process. The MAP kinase module BcSTE11–BcSTE7–BMP1 is essential for penetration; Δ*bmp1* conidia germinate on plant surfaces, produce long germ tubes, but are unable to penetrate and to cause disease (Zheng et al. 2000; Döhlemann et al. 2006a; Schamber et al. 2010). Another example is the MAP kinase BcSAK1; mycelia of the nonsporulating deletion mutants are unable to penetrate the intact plant surface (Segmüller et al. 2007). The effects of other mutations (Δ*bcg3*, Δ*bac*, Δ*pka*, Δ*bmp3*) are less severe compared to those caused by mutations of the two MAP kinases, as only the number of successful penetration events is reduced resulting in delayed primary lesion formation. Nonpenetrating germlings are marked by elongated straight-growing germ tubes (Döhlemann et al. 2006a; Schumacher et al. 2008b; Rui and Hahn, 2007). On the other hand, deletion of *bcnoxB* encoding one of two catalytic subunits of the NADPH oxidase (NOX) complex, or of *bcnoxR* encoding the regulatory subunit, resulted in another interesting phenotype: the conidia germinate, form germ tubes and appressoria, and try to penetrate. However, penetration is unsuccessful, new outgrowths appear, followed by further attempts to penetrate. Thus, the mutants are specifically blocked in the penetration process, because they recognize the surface and form infection structures, but fail finally to enter the plant tissue (Segmüller et al. 2008; Heller and Tudzynski 2011). Similar phenotypes have been described for mutants of the tetraspanin BcPLS1 (Gourgues et al. 2004) and the transcription factor BcSTE12 that is supposed to be target of the BMP1 MAP kinase (Schamber et al. 2010).

An interesting observation was made by Akutsu and co-workers (1981) who reported on conidial fusions induced by sugars and high conidial densities that resulted in netted structures on cucumber leaves. While "primary" appressoria formed by separated conidia failed to penetrate, the appressoria formed by the netted structures were able to penetrate resulting in lesion development. Conidial fusions have also been observed in B05.10, and the underlying regulatory networks are currently under investigation (M.G. Roca, U. Siegmund, P. Tudzynski, A. Fleißner, unpublished data).

There is some evidence that conidia and mycelia of the same strains may have different capabilities with regard to penetration. Hence, mutants of the BcCRZ1 transcription factor and the Slt2-homologous MAP kinase BMP3 are able to infect the plant when conidial suspensions are used, but mycelia of these strains are unable to induce visible disease symptoms on intact surfaces (Rui and Hahn 2007; Schumacher et al. 2008a). This observation might be explained by two different kinds of infection structures, appressoria or infection cushions that are formed by germ tubes and hyphae, respectively.

Taken together, germination and penetration are strongly regulated processes and involve multiple signaling pathways. Still the mechanism by which *B. cinerea* penetrates the host is not well understood. The availability of several signaling mutants that are disturbed at distinct stages of the penetration process represents a treasure box for a detailed analysis of the underlying molecular mechanisms.

### *11.1.3 Invasive Growth*

After penetration of the plant surface, *B. cinerea* kills the epidermal and underlying cells and establishes a primary infection that is characterized by collapsed brownish tissue and by defined margins; at this stage of the infection process fungal tissue is probably restricted to this region. Then, *B. cinerea* apparently overcomes the plant defense barriers and starts a massive outgrowth; hyphae are growing invasively and decompose the plant tissue very rapidly (secondary infection). Plants react to pathogen attack with a hypersensitive response (HR) that is characterized by cell death. The HR might be effective against biotrophic fungi by depriving them of nutrition and a livelihood, but is less effective against necrotrophic fungi which subsist on dead tissue. In turn, the HR may be utilized by the pathogen for killing the host. In fact, *B. cinerea* may even require the hypersensitive response of the host plant to achieve full pathogenicity (Govrin and Levine 2000).

Cell death-inducing compounds can be proteins or low molecular weight compounds, including toxins, oxalic acid and ROS which are secreted by the fungus. Remarkably, the expression of these factors might be induced during the fungus–host interaction but is not confined to it as these compounds may also be produced during saprophytic growth. *B. cinerea* produces two families of nonhost-selective phytotoxic metabolites: a family of sesquiterpenoids (botryanes; reviewed in Collado et al. 2007), and a family of polyketide derivatives formerly known as botcinolides (now referred to botcinins; Tani et al. 2005, 2006). Reino et al. (2004) described a correlation between in vitro production of botryanes and botcinins and the degree of virulence in 11 *B. cinerea* isolates. While all isolates produced botryanes, botcinins were only detected in the more aggressive isolates. Accordingly, the prevention of botrydial production by deletion of biosynthetic genes results in reduced virulence in T4, a strain that does not produce botcinins, but does not affect virulence in botcinin-producing strains (Siewers et al. 2005; Pinedo et al. 2008). However, the lack of both toxins led to severely reduced virulence (Dalmais et al. 2011), indicating that the toxins are important for the pathogenic interaction.

ROS play an ambivalent role in the pathogenic relationship as they are produced by both interacting partners. $H_2O_2$ generation has been observed in and around the penetrated host cell wall (Tenberge et al. 2002). Fungal ROS production may involve the NOX complex and superoxide dismutases (SOD) that convert (fungal or host) $O_2^-$ to $H_2O_2$. Components of the NOX complex as well as a Cu–Zn–superoxide dismutase are crucial for the fungus–plant interaction (Rolke et al. 2004; Segmüller et al. 2008); however, loss of the NOX complexes has no significant impact on intracellular and extracellular ROS production, so there must be an alternative ROS source, as postulated also for other fungal pathogens (Heller and Tudzynski 2011). Apart from that, detoxification of plant ROS by fungal enzymes does not seem to be essential for a successful infection, since loss of the major oxidative stress response system (triggered by the transcription factor BAP1) has no effect on virulence. Obviously, the fungus does not suffer from oxidative stress in planta (Temme and Tudzynski 2009).

Hydrolytic enzymes may be involved in different stages of plant infection; in penetration by lysing the cuticle or epidermal cell walls, in manifestation of infection by degrading components of the plant defense, or in nutrition by effective use of the dead tissue. The fact that several genes, including those encoding endopolygalacturonases, pectin methylesterases and aspartic proteases, play only a minor, or no, role in virulence (Tudzynski and Kokkelink 2009), correlates with the finding that multiple genes for these enzyme activities exist in the *B. cinerea* genome. However, CWDEs can also show necrotizing activity that is either independent (endo-β-1,4-xylanase XYN11A; Brito et al. 2006; Noda et al. 2010) or dependent (endopolygalacturonase BcPG2; Kars et al. 2005) on the enzymatic activity. Both proteins are essential for lesion spreading and represent therefore bona-fide virulence factors.

A couple of signaling mutants are affected in virulence. Many of them fail to penetrate (see above), or are retarded during the whole infection process but are finally able to complete the life cycle. Only a few mutants exist which stop at defined stages of plant colonization. A central signaling component in regulation of invasive growth is the Gα subunit BCG1: the mutant penetrates and causes primary lesions, but is unable to undergo further growth and development. *Bcg1* mutants lost several functions that may contribute to invasive growth, namely the expression of proteases and xylanases and the biosynthesis of toxic secondary metabolites (Schulze Gronover et al. 2001, 2004). By contrast, mutants of the cAMP/PKA cascade (Δ*bac*, Δ*bcpka1*) are more aggressive than Δ*bcg1*: they are blocked within (not before) the spreading of lesions accompanied by strong chlorosis of plant tissue. Both Δ*bac* and Δ*bcpka1* continue to produce the toxins but showed reduced expression levels of xylanase-encoding genes (Schumacher et al. 2008b); a detailed transcriptomics approach could help to identify downstream genes which are important for the specific phenotypes of these mutants.

### *11.1.4 Differentiation of Reproductive Structures*

*B. cinerea* can reproduce asexually by forming macroconidia for dispersal or sclerotia for survival in adverse conditions. In general, reproduction starts when the fungus has colonized the whole plant organ or agar plate.

The multinucleate macroconidia are formed at denticles that arise from the spherical ampulla located at the tip of a conidiophore. The development of conidia starting from initial conidiophores through to the presence of mature conidia may be completed within 8 h (Suzuki et al. 1977), and is dependent on light. A detailed investigation of the effect of light on one single spore isolate was reported by Tan and Epton (1973). They reported that an alternation of black light (near-UV) and white light is most effective in promoting conidiation, while almost no conidiation can be observed after irradiation with blue, green, yellow, and red light, or in the dark. Blue light can partially reverse the near-UV effect (and vice versa), indicating that the conidiation response is photoreversible (Tan 1974, 1975; Tan and Epton 1974). Suzuki et al. (1977) observed morphological changes in the various stages of conidiogenesis when conidiophores were subjected to blue light irradiation: mature conidiophores, ampullae at the tips of conidiophores, denticles and conidial initials dedifferentiated into sterile hyphae. In addition, a red/far-red photoreversible system has been described that comes into play after the action of the near-UV/blue photoreversible system (Tan and Epton 1975).

Sclerotia are characteristically dark pigmented structures of limited growth. Sclerotial development consists of three stages, namely initiation (formation of dichotomous branches), development (growth to the full size), and maturation (surface delimitation, internal changes, and pigmentation of the peripheral hyphae) (Willetts 1972). They can germinate in three ways, by forming mycelia, conidiophores or apothecia (=sexual stage "*Botryotinia*") (Coley-Smith and Cooke 1971). The formation of sclerotia may be influenced by different factors, for instance by temperature, pH, nutrition, and growth against mechanical barriers (Honda and Mizumura 1991; Townsend 1957). Light requirements for the formation of sclerotia are very different from those for production of conidia. Sclerotia are readily formed when *B. cinerea* is incubated in continuous darkness. Tan and Epton (1973) showed that small dosages of white light (60 min) are sufficient to inhibit the production of sclerotia, whereas sclerotia are promoted by yellow, red, and far-red light, but blue light suppresses their formation (Tan and Epton 1973; Suzuki and Oda 1979). Sclerotia are also involved in the sexual reproduction cycle of *B. cinerea* because they function as the female parent. The male parent represents the uninucleate microconidia called spermatia: the latter have only a sexual function and are likely only able to germinate in presence of a female parent carrying the opposite mating type. Microconidia may be formed throughout the life cycle by phialides developed on germ tubes, on more mature hyphae and sclerotia. There is little published information about the physiology of apothecia, although they are apparently only formed in the light. In *S. sclerotiorum*, normal apothecial development is strictly dependent on near-UV light or daylight; other qualities of light result in misshaped apothecia (Thaning and Nilsson 2000). Several apothecia may arise simultaneously from a single sclerotium, and conidiophores may also be present at the same time. The upper part of the apothecium consists of a hymenial layer composed of asci with eight uninucleate ascospores and paraphyses.

Little is known about genes that are directly involved in conidiogenesis, sclerotial, and apothecial development. However, several signaling mutants are affected in either conidiation, sclerotia formation, or both (Fig. 11.2). Mutants impaired in the stress-activated MAPK module (e.g., the MAP kinase BcSAK1) are completely blocked in conidiation while growth rate, biomass production and sclerotia formation are almost comparable to those of the wild type (Segmüller et al. 2007). More detailed microscopic analyses of mutants deleted for the transcriptional regulator BcREG1, that is supposed to be a downstream target of this MAP kinase, revealed a block in a defined stage of conidiogenesis: conidiophores with the ampulla and denticles, but no mature conidia were formed (Michielse et al. 2011). The ability to form sclerotia is impaired in all other signaling mutants displayed in Fig. 11.2. However, it is not conclusive as to whether the signaling components are directly involved in induction of sclerotial development or whether the inability to produce sclerotia is a consequence of impaired cell wall integrity and/or melanin production (e.g., Δ*bmp3*, Δ*bccrz1*). The fact that the components of the NOX complex are essential for sclerotial formation suggests that ROS are involved in this distinct differentiation process. Mutants that are unable to form sclerotia are female-sterile, but this does not imply male sterility. For instance, the NOX mutations had no effect on male fertility: the formation of apothecia containing ascospores, ascospore germination, and meiotic segregation in crosses with NOX mutants as male parents were normal (Segmüller et al. 2008). A "fluffy" phenotype associated with prevention of sclerotia formation, almost no conidiation, and significantly increased aerial hyphae formation is caused by modulating heterotrimeric G protein signaling, e.g., by deletion of the Gβ subunit or by expression of a constitutively active Gα1 (J. Schumacher and B. Tudzynski, unpublished data) or by deletion of the bZIP transcription factor BcATF1 (N. Temme, P. Tudzynski, unpublished data). Elucidation of the complex activities leading to conidiation, sclerotial, and apothecial formation is still a long way off. Comparative analyses with the close relative *S.*

*sclerotiorum* based on comparative genomics approaches may be helpful as the fungus lacks a conidiation cycle and reproduces exclusively by undergoing the sexual cycle.

## 11.2 Shedding Light on Light Regulation in *B. cinerea*

Evidently, *B. cinerea* responds to light by induction of asexual reproduction. However, other steps in the life cycle of *B. cinerea* also might be regulated by light. Jaffe and Etzold (1962) found that the germ tube emerged on the illuminated side of the conidium and suggested that the direction of germination is controlled by photoreceptors sensitive to blue light. Islam et al. (1998) reported on the effect of light on germ tube growth and differentiation. Thus, near-UV, blue light and far-red light slightly inhibited germ tube growth and induced negative phototropism, whereas red light promoted germ tube elongation and induced a positive phototropic response. Furthermore, the authors provided evidence for the importance of the phototropic response in early stages of plant infection: negative phototropism induced by near-UV and blue light promoted the formation of infection hyphae on both onion and broad bean epidermis, while positive phototropism induced by red light suppressed it, resulting in a high proportion of germ tubes without infection hyphae (Islam et al. 1998).

Taken together, at least four qualities of light are sensed by the fungus, namely near-UV, blue, far-red and red light. Basing on these observations, Tan (1975) postulated a two-photoreceptor model that suggests a close interaction of a near-UV/blue reversible and a far-red/red reversible photoreceptor. Studies in *Neurospora crassa* and *Aspergillus nidulans* revealed several proteins that are involved in the sensing of blue light. A fast response is mediated via the white collar complex (WCC) which senses light leading to the transient binding to promoters of light-induced genes. A second blue light receptor (VIVID) is involved in photo-adaption upon prolonged light exposure (is not present in *A. nidulans*), and a third group consists of cryptochromes that are photolyase-like receptors mediating light responses different from a DNA repair function. Red light is sensed by plant-like phytochromes that represent light-modulated histidine kinases. Fungal genomes also include potential green-light sensors: opsins are membrane-bound proteins with seven transmembrane domains that may bind retinal as a chromophore (for more details see Bayram et al. 2010; Chen et al. 2010; Idnurm et al. 2010; Rodriguez-Romero et al. 2010). In accordance with the response of *B. cinerea* to blue and red light, we found homologues of the recorded photoreceptors in the genome of *B. cinerea* (Table 11.1) and *S. sclerotiorum* (data not shown). In common with *N. crassa*, both *B. cinerea* and *S. sclerotiorum* possess the white collar transcription factors, a VIVID homologue and two cryptochrome-like proteins. Strikingly, the numbers of phytochrome-encoding genes vary among species: *B. cinerea* and *S. sclerotiorum* comprise three while *N. crassa* possesses two and *A. nidulans* only one phytochrome. Two opsins have been found in the

**Table 11.1** Analysis of the genome of *B. cinerea* for the presence of different photoreceptors and related proteins

| Name (description) | *B. cinerea* B05.10[a] | *B. cinerea* T4[b] | *A. nidulans* | *N. crassa* |
|---|---|---|---|---|
| Potential blue light sensors | | | | |
| BcWCL1 (White collar-like 1) | BC1G_13505 | BofuT4_P091830/40[c] | LreA | WC-1 |
| BcWCL2 (White collar-like 2) | BC1G_01840 | BofuT4_P135970 | LreB | WC-2 |
| BcVVD1 (Vivid-like) | BC1G_04348 | BofuT4_P126250 | – | VVD |
| BcCRY1 (Cryptochrome-like) | BC1G_13162 | BofuT4_P103490 | CryA | NCU08626 |
| BcCRY2 (Cryptochrome-like) | BC1G_08145 | BofuT4_P159580/90[c] | – | NCU00582 |
| Potential green light sensors | | | | |
| BOP1 (Opsin-like 1) | BC1G_02456 | BofuT4_P110210 | NopA | NOP-1 |
| BOP2 (Opsin-like 2) | BC1G_13906 | BofuT4_P163470 | – | ORP-1 |
| Potential red light sensors | | | | |
| BcPHY1 (Phytochrome-like 1) | BC1G_13369 | BofuT4_P078780 | FphA | PHY-1 |
| BcPHY2 (Phytochrome-like 2) | BC1G_08283 | BofuT4_P014010 | – | PHY-2 |
| BcPHY3 (Phytochrome-like 3) | BC1G_01106 | BofuT4_P030530 | – | – |
| Proteins implicated in photoresponses (components of the Velvet protein complex) | | | | |
| BcVEL1 (Velvet-like 1) | BC1G_02976/7[c] | BofuT4_P003460 | VeA | VE-1 |
| BcVEL2 (Velvet-like 2) | BC1G_11858 | BofuT4_P161180 | VelB | NCU02775 |
| BcVEL3 (Velvet-like 3) | BC1G_06127 | BofuT4_P017230 | VosA | NCU05964 |
| BcVEL4 (Velvet-like 4) | BC1G_11619 | BofuT4_P157800 | VelC | NCU07553 |
| BcLAEA | BC1G_15168/9[c] | BofuT4_P148930 | LaeA | NCU01148 |
| Proteins implicated in the circadian clock | | | | |
| BcFRQ1 (Frequency-like) | BC1G_13940 | BofuT4_P067640 | – | FRQ |

Proteins in *B. cinerea* were identified by performing BlastP analyses using sequences from *Aspergillus nidulans* and *Neurospora crassa* as queries

[a]Annotated proteins in the B05.10 genome database (http://www.broadinstitute.org/annotation/genome/botrytis_cinerea/Home.html)

[b]Annotated proteins in the T4 genome database (http://urgi.versailles.inra.fr/index.php/urgi/Species/Botrytis)

[c]Proteins were incorrectly annotated due to discontinuous nucleotide sequences

*B. cinerea* genome: *bop1* has been shown to be dispensable for light-development and virulence (J. Heller and P. Tudzynski, unpublished); *bop2* is located in the carotenoid gene cluster, and has not been analyzed yet.

Within the scope of the *B. cinerea* genome project, the genomes of two different wild-type isolates were sequenced. B05.10 was isolated from *Vitis* (Büttner et al. 1994) and strain T4 was isolated from tomato (Levis et al. 1997). Strikingly, both isolates exhibited different phenotypes when incubated in the dark: while B05.10 undergoes sclerotial development, strain T4 produces abundant conidia. Thus, conidiation in T4 does not rely on photoinduction. A release of sporulation from photocontrol was also previously reported for *A. nidulans veA1* mutants by Käfer (1965), and this finding has led later to the identification of Velvet (VeA) as a bridging protein in a protein complex which coordinates light signals with fungal development and secondary metabolism. The responsiveness to light is achieved by the while collar homologues LreA/LreB and the single phytochrome FphA, while the output relies on further VeA-interacting proteins, i.e., LaeA, VelB, and VosA (for review see Calvo 2008; Bayram et al. 2010). However, homologues of the four velvet domain-containing proteins have been found in both B05.10 and T4. Further analyses are in progress and will provide evidence if the function of the velvet complex is conserved in *A. nidulans* and *B. cinerea* and if the "blind" phenotype of T4 is due to altered light-sensing properties or to altered downstream signaling events (J. Schumacher, M. Viaud, B. Tudzynski, unpublished). A correlation between the ability to sense light and pathogenesis is conceivable for several reasons. For instance, the "blind" wild-type isolate T4 is less aggressive on bean plants when compared to the wild-type isolate B05.10. Furthermore, we observed that several of the less virulent T-DNA insertion mutants display additional defects in light-dependent development in vitro (J. Schumacher, P. Tudzynski, unpublished). Moreover, reduced virulence of a *B. cinerea* isolate on broad bean has been observed after incubation in red light in comparison to near-UV and blue light (Islam et al. 1998). However, the latter experiment emphasizes the problem of studying different qualities of light on virulence as both the fungus and the plant may react to the same signal. Hence, studies on the *B. cinerea*–broad bean interaction showed that red light induces plant resistance toward *B. cinerea* (Islam et al. 1998; Rahmann et al. 2002), and therefore, the question remains open as to whether or not red light really affects virulence of *B. cinerea*. The use of color-blind mutants generated by targeted inactivation of the different photoreceptors will allow for specification of the impact of light in the fungus–host interaction without inferring the host's metabolism. Another interesting aspect of light is its influence on the circadian clock, an endogenous time keeping device that provides temporal control of physiology in accordance with predicted daily changes in the environment. Jarvis (1962) observed a circadian periodicity in the dispersal of conidia of *B. cinerea* in a raspberry plantation that is dependent on environmental conditions; dispersal takes place around noon, a period that is marked by daylight, increasing temperatures and decreasing relative humidity. Accordingly, conidial germination (and penetration?) can be estimated during night when the relative humidity rises again or in the early morning when dew forms. Thus, running a circadian clock might be very useful for *B. cinerea* to be well prepared for the daily changes and resulting differentiation processes. *B. cinerea* possesses homologues of the key components of the circadian clock, such as the WC complex, and the VVD and FRQ proteins, but a clear output of a circadian rhythm in continuous darkness is not detectable. We expect that the future studies on the mechanisms of light responsiveness of *B. cinerea* by both reverse and forward genetics will gain new insights in the fungus–host interaction and in the survival strategies of this robust plant pathogen.

**Acknowledgments** We are grateful to Brian Williamson for critical reading of the manuscript. Experimental work performed in our lab was funded by the Deutsche Forschungsgemeinschaft (DFG).

# References


Akutsu K, Ko K, Misato T (1981) Role of conidial fusion in infection by *Botrytis cinerea* on cucumber leaves. Ann Phytopath Soc Japan 47:15–23

Amselem J, Cuomo CA, van Kan JA, Viaud M, et al. (2011) Genomic analysis of the necrotrophic fungal pathogens Sclerotinia sclerotiorum and Botrytis cinerea. PLoS Genet 7:e1002230

Backhouse D, Willetts HJ (1987) Development and structure of infection cushions of *Botrytis cinerea*. Trans Br Mycol Soc 89:89–95

Bayram O, Braus GH, Fischer R, Rodriguez-Romero J (2010) Spotlight on *Aspergillus nidulans* photosensory systems. Fungal Genet Biol 47:900–908

Blakeman JP (1975) Germination of *Botrytis cinerea* conidia in vitro in relation to nutrient conditions on leaf surfaces. Trans Br Mycol Soc 65:239–247

Bolton MD, Thomma BPHJ, Nelson BD (2006) Pathogen profile - *Sclerotinia sclerotiorum* (Lib.) de Bary: biology and molecular traits of a cosmopolitan pathogen. Mol Plant Pathol 7:1–16

Brito N, Espino JJ, González C (2006) The endo-beta-1,4-xylanase xyn11A is required for virulence in *Botrytis cinerea*. Mol Plant Microbe Interact 19:25–32

Büttner P, Koch F, Voigt K, Quidde T, Risch S, Blaich R, Brückner B, Tudzynski P (1994) Variations in ploidy among isolates of *Botrytis cinerea*: implications for genetic and molecular analyses. Curr Genet 25:445–450

Calvo AM (2008) The VeA regulatory system and its role in morphological and chemical development in fungi. Fungal Genet Biol 45:1053–1061

Carlile MJ, Sellin MA (1963) An endogenous inhibition of spore germination in fungi. Trans Br Mycol Soc 46:15–18

Chen CH, Dunlap JC, Loros JJ (2010) *Neurospora* illuminates fungal photoreception. Fungal Genet Biol 47:922–999

Choquer M, Fournier E, Kunz C, Levis C, Pradier JM, Simon A, Viaud M (2007) *Botrytis cinerea* virulence factors: new insights into a necrotrophic and polyphageous pathogen. FEMS Microbiol Lett 277:1–10

Coley-Smith JR, Cooke RC (1971) Survival and germination of fungal sclerotia. Annu Rev Phytopathol 9:65–92

Coley-Smith JR, Verhoeff K, Jarvis WR (1980) The biology of *Botrytis cinerea*. Academic, London

Collado IG, Sánchez AJ, Hanson JR (2007) Fungal terpene metabolites: biosynthetic relationships and the control of the phytopathogenic fungus *Botrytis cinerea*. Nat Prod Rep 24:674–686

Dalmais B, Schumacher J, Moraga J, Le Pêcheur P, Tudzynski B, Collado IG, Viaud M (2011) The *Botrytis cinerea* phytotoxin botcinic acid requires two polyketide synthases for production and has a redundant role in virulence with botrydial. Mol Plant Pathol 12:564–579

Döhlemann G, Berndt P, Hahn M (2006a) Different signalling pathways involving a Galpha protein, cAMP and a MAP kinase control germination of *Botrytis cinerea* conidia. Mol Microbiol 59:821–836

Döhlemann G, Berndt P, Hahn M (2006b) Trehalose metabolism is important for heat stress tolerance and spore germination of *Botrytis cinerea*. Microbiology 152:2625–2634

Elad Y, Williamson B, Tudzynski P, Delen N (2004) *Botrytis*: biology, pathology and control. Kluwer, Dordrecht

Faretra F, Antonacci E, Pollastro S (1988) Sexual behaviour and mating system of *Botryotinia fuckeliana*, teleomorph of *Botrytis cinerea*. J Gen Microbiol 134:2543–2550

Gourgues M, Brunet-Simon A, Lebrun MH, Levis C (2004) The tetraspanin BcPls1 is required for appressorium-mediated penetration of *Botrytis cinerea* into host plant leaves. Mol Microbiol 51:619–629

Govrin EM, Levine A (2000) The hypersensitive response facilitates plant infection by the necrotrophic pathogen *Botrytis cinerea*. Curr Biol 10:751–757

Grindle (1979) Phenotypic differences between natural and induced variants of *Botrytis cinerea*. J Gen Microbiol 111:109–120

Heller J, Tudzynski P (2011) ROS in phytopathogenic fungi: signaling, development and disease. Annu Rev Phytopathol 49:369–390

Holz G, Coertze S, Williamson B (2004) The ecology of Botrytis on plant surfaces. In: Elad Y, Williamson B, Tudzynski P, and Delen N (eds) Botrytis, biology, pathology and control. Kluwer, Dordrecht pp. 9–27.

Honda Y, Mizumura Y (1991) Light and temperature dependent conidium and sclerotium in *Botrytis* spp. Bull Fac Agric Shimane Univ 25:27–35

Idnurm A, Verma S, Corrochano LM (2010) A glimpse into the basis of vision in the fungal kingdom Mycota. Fungal Genet Biol 47:881–892

Islam SZ, Honda Y, Sonhaji M (1998) Phototropism of conidial germ tubes of *Botrytis cinerea* and its implication in plant infection processes. Plant Dis 82:850–856

Jaffe L, Etzold H (1962) Orientation and locus of tropic photoreceptor molecules in spores of *Botrytis* and *Osmunda*. J Cell Biol 13:13–31

Jarvis WR (1962) The dispersal of spores of *Botrytis cinerea* fr. in a raspberry plantation. Trans Br Mycol Soc 45:549–559

Käfer E (1965) Origins of translocations in *Aspergillus nidulans*. Genetics 52:217–232

Kars I, Krooshof GH, Wagemakers L, Joosten R, Benen JA, van Kan JA (2005) Necrotizing activity of five *Botrytis cinerea* endopolygalacturonases produced in *Pichia pastoris*. Plant J 43:213–225

Klimpel A, Schulze Gronover C, Williamson B, Stewart JA, Tudzynski B (2002) The adenylate cyclase (BAC) in *Botrytis cinerea* is required for full pathogenicity. Mol Plant Pathol 3:439–450

Kokkelink L, Minz A, Al-Masri M, Giesbert S, Barakat R, Sharon A, Tudzynski P (in press) The small GTPase BcCdc42 affects nuclear division, germination and virulence of the gray mold fungus Botrytis cinerea. Fungal Genet Biol (2011) doi:10.1016/j.fgb.2011.07.007.

Kritzman G, Gilan D, Chet I (1980) Germination-inhibitor in *Botrytis allii* spores. Phytoparasitica 8:73–76

Levis C, Dutertre M, Fortini D, Brygoo Y (1997) Telomeric DNA of *Botrytis cinerea*: a useful tool for strain identification. FEMS Microbiol Lett 157:267–272

Michielse CB, Becker M, Heller J, Moraga J, Collado IG, Tudzynski P (2011) The Botrytis cinerea Reg1 protein, a putative transcriptional regulator, is required for pathogenicity, conidiogenesis, and the production of secondary metabolites. Mol Plant Microbe Interact 24:1074–1085

Noda J, Brito N, González C (2010) The *Botrytis cinerea* xylanase Xyn11A contributes to virulence with its necrotizing activity, not with its catalytic activity. BMC Plant Biol 10:38

Pinedo C, Wang CM, Pradier JM, Dalmais B, Choquer M, Le Pêcheur P, Morgant G, Collado IG, Cane DE, Viaud M (2008) Sesquiterpene synthase from the botrydial biosynthetic gene cluster of the phytopathogen *Botrytis cinerea*. ACS Chem Biol 3:791–801

Rahmann MZ, Honda Y, Islam SZ, Arase S (2002) Effect of metabolic inhibitors on red light-induced resistance of broad bean (*Vicia faba* L.) against *Botrytis cinerea*. J Phytopathol 150:463–468

Reino JL, Hernández-Galán R, Durán-Patrón R, Collado IG (2004) Virulence-toxin production relationship in isolates of the plant pathogenic fungus *Botrytis cinerea*. J Phytopathol 152:563–566

Rodriguez-Romero J, Hedtke M, Kastner C, Müller S, Fischer R (2010) Fungi, hidden in soil or up in the air: light makes a difference. Annu Rev Microbiol 64:585–610

Rolke Y, Liu S, Quidde T, Williamson B, Schouten A, Weltring KM, Siewers V, Tenberge KB, Tudzynski B, Tudzynski P (2004) Functional analysis of $H_2O_2$-generating systems in *Botrytis cinerea*: the major Cu-Zn-superoxide dismutase (BcSOD1) contributes to virulence on French bean, whereas a glucose oxidase (BcGOD1) is dispensable. Mol Plant Pathol 5:17–27

Rui O, Hahn M (2007) The Slt2-type MAP kinase Bmp3 of *Botrytis cinerea* is required for normal saprotrophic growth, conidiation, plant surface sensing and host tissue colonization. Mol Plant Pathol 8:173–184

Schamber A, Leroch M, Diwo J, Mendgen K, Hahn M (2010) The role of mitogen-activated protein (MAP) kinase signalling components and the Ste12 transcription factor in germination and pathogenicity of *Botrytis cinerea*. Mol Plant Pathol 11:105–119

Schulze Gronover C, Kasulke D, Tudzynski P, Tudzynski B (2001) The role of G protein alpha subunits in the infection process of the gray mold fungus *Botrytis cinerea*. Mol Plant Microbe Interact 14:1293–1302

Schulze Gronover C, Schorn C, Tudzynski B (2004) Identification of *Botrytis cinerea* genes up-regulated during infection and controlled by the Galpha subunit BCG1 using suppression subtractive hybridization (SSH). Mol Plant Microbe Interact 17:537–546

Schumacher J, de Larrinoa IF, Tudzynski B (2008a) Calcineurin-responsive zinc finger transcription factor CRZ1 of *Botrytis cinerea* is required for growth, development, and full virulence on bean plants. Eukaryot Cell 7:584–601

Schumacher J, Kokkelink L, Huesmann C, Jimenez-Teja D, Collado I, Barakat R, Tudzynski P, Tudzynski B (2008b) The cAMP-dependent signalling pathway and its role in conidial germination, growth and virulence of the grey mould fungus *Botrytis cinerea*. Mol Plant Microbe Interact 21:1443–1459

Schumacher J, Viaud M, Simon A, Tudzynski B (2008c) The Galpha subunit BCG1, the phospholipase C (BcPLC1) and the calcineurin phosphatase co-ordinately regulate gene expression in the grey mould fungus *Botrytis cinerea*. Mol Microbiol 67:1027–1250

Segmüller N, Ellendorf U, Tudzynski B, Tudzynski P (2007) BcSAK1, a stress-activated mitogen-activated protein kinase, is involved in vegetative differentiation and pathogenicity in *Botrytis cinerea*. Eukaryot Cell 6:211–221

Segmüller N, Kokkelink L, Giesbert S, Odinius D, van Kan J, Tudzynski P (2008) NADPH oxidases are involved in differentiation and pathogenicity in *Botrytis cinerea*. Mol Plant Microbe Interact 21:808–819

Siewers V, Viaud M, Jimenez-Teja D, Collado IG, Schulze Gronover C, Pradier JM, Tudzynski B, Tudzynski P (2005) Functional analysis of the cytochrome P450 monooxygenase gene *bcbot1* of *Botrytis cinerea* indicates that botrydial is a strain-specific virulence factor. Mol Plant Microbe Interact 18:602–612

Suzuki Y, Oda Y (1979) Inhibitory loci of both blue and near ultraviolet lights on lateral-type sclerotial development in *Botrytis cinerea*. Ann Phytopath Soc Japan 45:54–61

Suzuki Y, Kumagai T, Oda Y (1977) Locus of blue and near ultraviolet reversible photoreaction in the stages of conidial development in *Botrytis cinerea*. J Gen Microbiol 98:199–204

Tan KK (1974) Blue-light inhibition of sporulation in *Botrytis cinerea*. J Gen Microbiol 82:191–200

Tan KK (1975) Recovery from the blue-light inhibition of sporulation in *Botrytis cinerea*. Trans Br Mycol Soc 64:223–228

Tan KK, Epton HAS (1973) Effect of light on the growth and sporulation of *Botrytis cinerea*. Trans Br Mycol Soc 61:147–157

Tan KK, Epton HAS (1974) Further studies on light and sporulation in *Botrytis cinerea*. Trans Br Mycol Soc 62:105–112

Tan KK, Epton HAS (1975) Interaction of near-ultraviolet, blue, red, and far-red light in sporulation of *Botrytis cinerea*. Trans Br Mycol Soc 64:215–222

Tani H, Koshino H, Sakuno E, Nakajima H (2005) Botcinins A, B, C, and D, metabolites produced by *Botrytis cinerea*, and their antifungal activity against *Magnaporthe grisea*, a pathogen of rice blast disease. J Nat Prod 68:1768–1772

Tani H, Koshino H, Sakuno E, Cutler HG, Nakajima H (2006) Botcinins E and F and Botcinolide from *Botrytis cinerea* and structural revision of botcinolides. J Nat Prod 69:722–725

Temme N, Tudzynski P (2009) Does *Botrytis cinerea* ignore $H_2O_2$-induced oxidative stress during infection? Characterization of *Botrytis* activator protein 1. Mol Plant Microbe Interact 22:987–998

Tenberge KB, Beckedorf M, Hoppe B, Schouten A, Solf M, von den Driesch M (2002) In situ localization of AOS in host–pathogen interactions. Microsc Microanal 8(Suppl 2):250–251

Thaning C, Nilsson HE (2000) A narrow range of wavelengths active in regulating apothecial development in *Sclerotinia sclerotiorum*. J Phytopathol 148:627–631

Townsend BB (1957) Nutritional factors influencing the production of sclerotia by certain fungi. Ann Bot 21:153–166

Tudzynski P, Kokkelink L (2009) *Botrytis cinerea*: molecular aspects of a necrotrophic life style. In: Deising H (ed) The mycota V, plant relationships, 2nd edn. Springer, Berlin

Van Kan JAL (2006) Licensed to kill: the lifestyle of a necrotrophic plant pathogen. Trends Plant Sci 11:247–253

Viaud M, Brunet-Simon A, Brygoo Y, Pradier JM, Levis C (2003) Cyclophilin A and calcineurin functions investigated by gene inactivation, cyclosporin A inhibition and cDNA arrays approaches in the phytopathogenic fungus *Botrytis cinerea*. Mol Microbiol 50:1451–1465

Willetts HJ (1972) The morphogenesis and possible evolutionary origins of fungal sclerotia. Biol Rev 47:515–536

Williamson B, Duncan GH, Harrison JG, Harding LA, Elad Y, Zimand G (1995) Effect of humidity on infection of rose petals by dry-inoculated conidia of *Botrytis cinerea*. Mycol Res 99:1303–1310

Williamson B, Tudzynski B, Tudzynski P, van Kan JAL (2007) Pathogen profile – *Botrytis cinerea*: the cause of grey mould disease. Mol Plant Pathol 8:561–580

Zheng L, Campbell M, Murphy J, Lam S, Xu JR (2000) The *BMP1* gene is essential for pathogenicity in the gray mold fungus *Botrytis cinerea*. Mol Plant Microbe Interact 13:724–732

# Chapter 12
# Morphogenesis, Growth, and Development of the Grass Symbiont *Epichlöe festucae*

**Barry Scott, Yvonne Becker, Matthias Becker, and Gemma Cartwright**

**Abstract** *Epichloë festucae* and its asexual derivative *Neotyphodium lolii* are mutualistic symbionts that confer on their *Festuca* and *Lolium* grass hosts, protection from various biotic and abiotic stresses. The genetic tractability of *E. festucae* has led to its adoption as a model experimental system to study fungal–grass symbiotic interactions. Growth of *E. festucae* in *Lolium perenne* is both epiphytic and endophytic. Endophytic growth is characterized by hyphal tip growth and branching in the meristematic tissues but in the leaves hyphae divide and extend by intercalary growth in synchrony with the same pattern of growth of the leaves. Forward and reverse genetics approaches have shown that the NADPH oxidase (Nox) complex and MAP kinase signaling pathways are crucial for maintaining this restrictive pattern of hyphal growth in the leaves. Disruption of genes that encode components of these signaling complexes leads to proliferative (pathogenic) growth in the host and a breakdown in the symbiosis. This chapter provides an overview of morphogenesis, growth, and development of *E. festucae* in culture and in planta and an oversight of what is currently known about the fungal signaling mechanisms required for maintaining a balanced symbiosis.

## 12.1 Introduction

*Epichloë festucae* (Ascomycota, Clavicipitaceae) and its asexual derivative *Neotyphodium lolii* form symbiotic associations (symbiota) with temperate grasses of the *Festuca* and *Lolium* genera within the subfamily Pooideae (Leuchtmann et al. 1994; Schardl et al. 1994). Natural hosts for *E. festucae* include the fine (*F. rubra*) and hard (*F. longifolia*) fescues but some strains also form stable mutualistic

B. Scott (✉) • Y. Becker • M. Becker • G. Cartwright
Institute of Molecular Biosciences, Massey University, Private Bag 11 222, Palmerston North 4442, New Zealand
e-mail: d.b.scott@massey.ac.nz

J. Pérez-Martín and A. Di Pietro (eds.), *Morphogenesis and Pathogenicity in Fungi*, Topics in Current Genetics 22, DOI 10.1007/978-3-642-22916-9_12,

associations with perennial ryegrass, *Lolium perenne*, a grass that is relatively easy to inoculate and grow, making it an ideal host to study these endophyte–grass associations (Christensen et al. 1997; Scott et al. 2007).

*E. festucae* and *N. lolii* systemically colonize the vegetative and reproductive aerial tissues of grasses but not the roots. The growth of these biotrophic fungi within leaves is tightly regulated with usually just a single hypha found between adjacent plant cells and a low overall biomass. However, during the reproductive phase of grass growth, *E. festucae* has the potential to switch from restricted to proliferative growth on some tillers to form external sexual reproductive structures (stromata) that prevent emergence of the host inflorescence, a disease known as "choke" (Fig. 12.1).

Numerous studies have established that *E. festucae*, *N. lolii*, and related *Neotyphodium* hybrid species form predominantly mutualistic symbiotic associations with their hosts (Moon et al. 2004; Schardl et al. 2009). The major benefits to the fungal symbiont are access to nutrients from the host apoplast and a means of dissemination through the host seed. Benefits to the host include increased tolerance to both biotic (e.g., insect and mammalian herbivory) and abiotic (e.g., drought) stresses (Schardl et al. 2009).

While there are at least ten different *Epichloë* species, the focus of this review will be on *E. festucae*, which is proving to be an ideal experimental system to study the molecular and cellular mechanisms that underlie fungal symbiotic interactions with grasses (Schardl 2001; Scott et al. 2007). The recent release of genome sequences for *E. festucae* strains E2368 and E984 (Fl1) has provided further impetus to studies with this species (http://csbio-l.csr.uky.edu/endophyte/).

**Fig. 12.1** Stromata of *E. festucae* on fine fescue. Image courtesy of James F. White Jr and Mónica S. Torres, Rutgers University

## 12.2 *E. festucae/N. lolii* Lifecycle

The symbiotic interaction between *E. festucae* and its grass host has been described as pleitropic because the nature of the interaction varies depending on the physiological state of the host and stage of development (Michalakis et al. 1992). When the host is in the vegetative state, hyphae proliferate within the aerial tissues but there are no visible pathogenic symptoms. When the host undergoes reproductive development there are two developmental pathways the endosymbiont can follow; it can maintain restrictive growth within the host to colonize the seed and be vertically transmitted (asexual life cycle) or alternatively, on some tillers, switch to proliferative growth and enter the sexual cycle, to be horizontally transmitted (Scott and Schardl 1993). Vertical transmission is the sole mechanism for dissemination of *N. lolii* because it has lost the ability to enter the sexual cycle. Interestingly, there are no reports to date of "choke" in associations between *E. festucae* and perennial ryegrass. In contrast, symbiota between the broad host range *E. typhina* and perennial ryegrass readily form stromata (Chung and Schardl 1997).

Vertical transmission of endophyte through the seeds is a very efficient process for dissemination of the endophyte. Our understanding of *E. festucae/N. lolii* colonization of seeds and embryos comes principally from microscopy studies conducted on associations between *E. festucae* and red fescue (Sampson 1933), *N. lolii* and perennial ryegrass (Philipson and Christey 1986), and *E. festucae* and perennial ryegrass (May et al. 2008). During reproductive development of the plant, hyphae grow from the vegetative apex into the inflorescence primordium and floral apices, to infect the ovary and ovules. Immediately after fertilization, hyphae gain entry to the embryo sac. During early embryogenesis, hyphae can be found on the surface of the embryo. As the embryo matures, hyphae become widespread throughout the embryo and surrounding tissues including the plumule apex, embryo axis, the aleurone layer, and between the scutellum and the endosperm. During seed germination, hyphae colonize the developing shoot apex. Further colonization of the shoot apical meristem (SAM), leaf primordia, sheaths, and blades of leaves results in systemic infection of aerial tissues.

Entry of *E. festucae* into the sexual cycle follows a switch from restrictive endophytic growth within the mesophyll tissue to proliferative epiphytic growth on the surface of the leaf that surrounds the emerging inflorescence. A distinct band of epiphytic growth gives rise to a stroma bearing the female and male (spermatia) reproductive structures. The mating system is heterothallic (Schardl et al. 1997; Bultman and Leuchtmann 2009), with transfer of spermatia from one stroma to another by female anthomyiid flies of the genus *Botanophila* (Bultman et al. 1998; Bultman and Leuchtmann 2009). Deposition of spermatia is associated with complex oviposition behavior, analogous to insect pollination of angiosperms. Release of specific volatiles from the fungal stroma trigger fly visitations and spermatia transfer (Schiestl et al. 2006; Steinebrunner et al. 2008b, c). These insect-attracting volatiles also have antimicrobial activity suggesting an original role in microbial defense (Steinebrunner et al. 2008a). Interestingly, lower levels of bioprotective alkaloids are found in stroma-bearing grasses than asymptomatic ones (Leuchtmann et al. 2000). Ascospores released from perithecia may be transferred to stigmata of florets on other grass panicles. Germinating hyphae penetrate the stigmatic surface to colonize the transmitting tissue and eventually the ovule (Chung and Schardl 1997). This stage of the growth cycle is very similar to infection of rye plants by the closely related, ergot-producing fungus, *Claviceps purpurea* (Tudzynski and Scheffer 2004).

## 12.3 *E. festucae* as a Model Experimental Organism to Study Fungal Endophyte–Grass Symbiotic Interactions

While *N. lolii* is the most common naturally occurring endophytic symbiont of perennial ryegrass it is quite intractable to genetic analysis, principally because of its very slow growth rate in culture. By comparison, *E. festucae* is relatively fast growing, forming colonies of approximately 1 cm in diameter after 1 week of growth at 22°C on potato dextrose agar (PDA) medium. Furthermore, the colony morphology of *E. festucae* is stable in axenic culture. This is in distinct contrast to many *N. lolii* strains, which tend to develop sectors of altered morphology (Christensen and Latch 1991). This phenotypic plasticity is commonly observed for fungi isolated from the natural environment then grown on synthetic media under laboratory conditions. However, the molecular basis for these morphological changes is poorly understood (Slepecky and Starmer 2009).

*E. festucae* is readily transformed enabling both homologous and nonhomologous recombination of various constructs into the genome to generate reporter-tagged strains or targeted gene deletions (Tanaka et al. 2005; Young et al. 2005). The development of *E. festucae* strains tagged with the reporters GusA or GFP have been crucial for monitoring endophyte morphology, growth, and development within the host plant. These reporter strains have provided crucial insights into the process of host colonization throughout the life cycle of the grass (Takemoto et al. 2006, 2011; Tanaka et al. 2006, 2008; Christensen et al. 2008; May et al. 2008). Gene replacements can be generated in *E. festucae* at variable frequencies, depending on the locus targeted and the size of flanking sequences used to prepare the construct (Scott et al. 2007). Given cultures of *E. festucae* do not produce large numbers of conidia, homokaryons are generated by subculturing the edge of a colony on to selective media several times. This method is effective because the hyphal apical tip cells contain just a single nucleus, and therefore generate homokaryotic sectors within the mycelium (Trinci 1978; Schmid et al. 2000). Other methods that have been used to generate *E. festucae* replacement mutants include the MultiSite Gateway and Cre/Lox recombination systems (Fleetwood et al. 2007; Florea et al. 2009). Insertional mutagenesis methodology has been developed for *E. festucae* using both *Agrobacterium tumefaciens* T-DNA-mediated transformation and restriction enzyme-mediated plasmid integration (REMI)

(Tanaka et al. 2006, 2007). Both methods have been used successfully to isolate mutants with a disrupted host interaction phenotype (Tanaka et al. 2006; Brasell 2010).

*E. festucae* is a haploid organism with a genome size of approximately 34 Mb and is outcrossing (Leuchtmann et al. 1994; Kuldau et al. 1999). This makes it possible to do sexual crosses using strains of *E. festucae* of different mating types but a major limitation to sexual genetic analysis is the inability to do crosses under laboratory conditions. However, the development of both forward and reverse genetics methodologies (described above), combined with the availability of genome sequences for two strains of *E. festucae* and the ability to use next-generation sequencing technologies for other applications, such as RNAseq and comparative genomics, now makes it possible to do functional studies at a level comparable to the classical experimental fungal systems of *Neurospora crassa* and *Aspergillus nidulans* (Eaton et al. 2010) (http://csbio-l.csr.uky.edu/endophyte/).

A further advantage of working with *E. festucae* rather than *N. lolii* is the relative ease with which mutualistic symbiotic associations can be established. Because these fungi do not naturally penetrate leaves of their grass host, as do most phytopathogenic fungi, synthetic associations are established by inserting mycelia in a small incision made across the SAM (Latch and Christensen 1985). Although this technique does damage the primary tiller, the subsequent tillers which grow from the crown are undamaged and systemically infected with hyphae. Frequencies of perennial ryegrass infection with wild-type *E. festucae* are in the range of 80–95%. Using this method it is now possible to readily infect endophyte-free perennial ryegrass seedlings to study endophyte–host interactions (Christensen et al. 2002). Although the growth of *E. festucae* in these novel associations is slightly more vigorous than *N. lolii*, the hyphae still grow parallel to the leaf axis, are infrequently branched, and the pattern of growth is synchronized with that of the host throughout the life cycle of the grass (Takemoto et al. 2006; Tanaka et al. 2006).

## 12.4 *E. festucae* Growth in Culture

### *12.4.1 Vegetative Hyphal Growth*

*E. festucae* cultivated on PDA grows as a white, fluffy mycelium with a growth rate of between 1 and 3 mm per day (Fig. 12.2a) and can utilize a range of different carbon sources including sucrose, glucose, and mannitol. Nitrogen sources that support growth include many amino acids, nitrate, and ammonia. Cultures often start to senesce 2–3 weeks after cultivation when nutrients become limiting. Mutants such as Δ*sakA* (stress-activated mitogen-activated protein kinase) show an enhanced tendency to senesce, especially under stress conditions (Eaton et al. 2008).

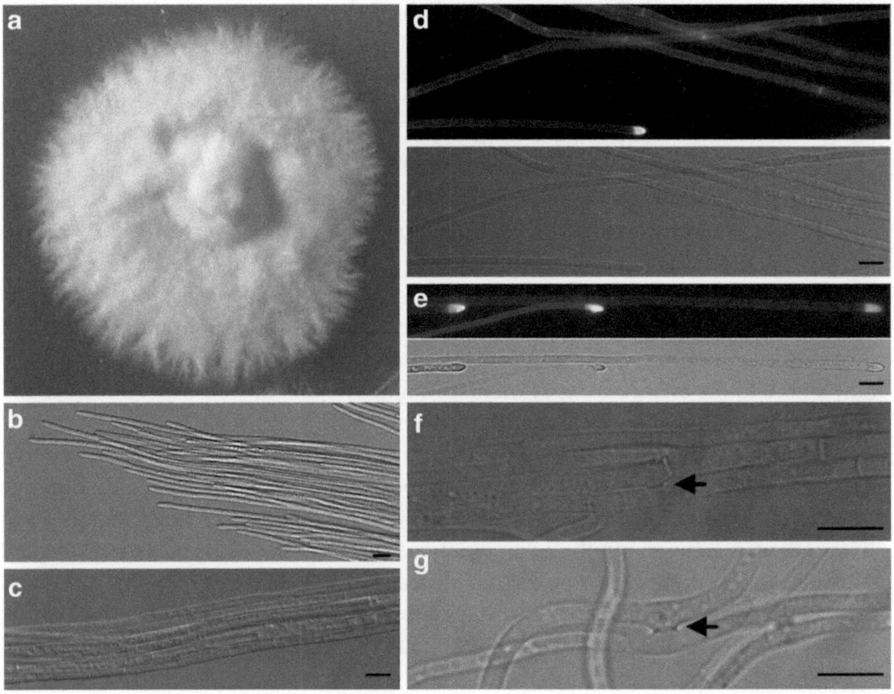


**Fig. 12.2** Culture phenotype of *E. festucae*. (**a**) Colony growing on PDA. (**b** and **c**) Light micrographs of hyphae from the colony edge. (**d** and **e**) Bright field and DIC images of hyphae stained with Calcofluor white. (**f** and **g**) Light micrographs showing lateral fusion of adjacent hyphae. *Bar* = 10 μm

The colony edges of *E. festucae* and *N. lolii* cultures are characterized by the presence of aggregates of long unbranched hyphae with uninucleate hyphal tips (Fig. 12.2b, c) (Schmid et al. 2000). The tips and septa fluoresce brightly when stained with Calcoflour white, indicative of active chitin synthesis close to the growing tip (Fig. 12.2d, e). Hyphae at the colony edge are aligned parallel to one another and appear to be stuck together by an adhesive of unknown composition (Fig. 12.2b, c).

In the middle of the colony, lateral branching is more frequent and is almost always observed at the cell end proximal to the growing tip (Takemoto et al. 2006). Here the mycelium forms two distinct layers of hyphae (Fig. 12.3). The first layer is closely attached to the agar, providing colony access to nutrients. The second layer is comprised of several hyphal bundles, usually >10, growing on top of one another, together with a highly branched aerial hyphal network that is characterized by the presence of coiled or beehive-like structures of closely attached hyphae that often form conidiophores (Fig. 12.3d–f).

In the central, older part of the colony, hyphae frequently fuse resulting in formation of an interlinked hyphal network (Fig. 12.2f, g). The mechanisms responsible for these fusion events remain to be studied but may be promoted by the close attachment of hyphae to one another.

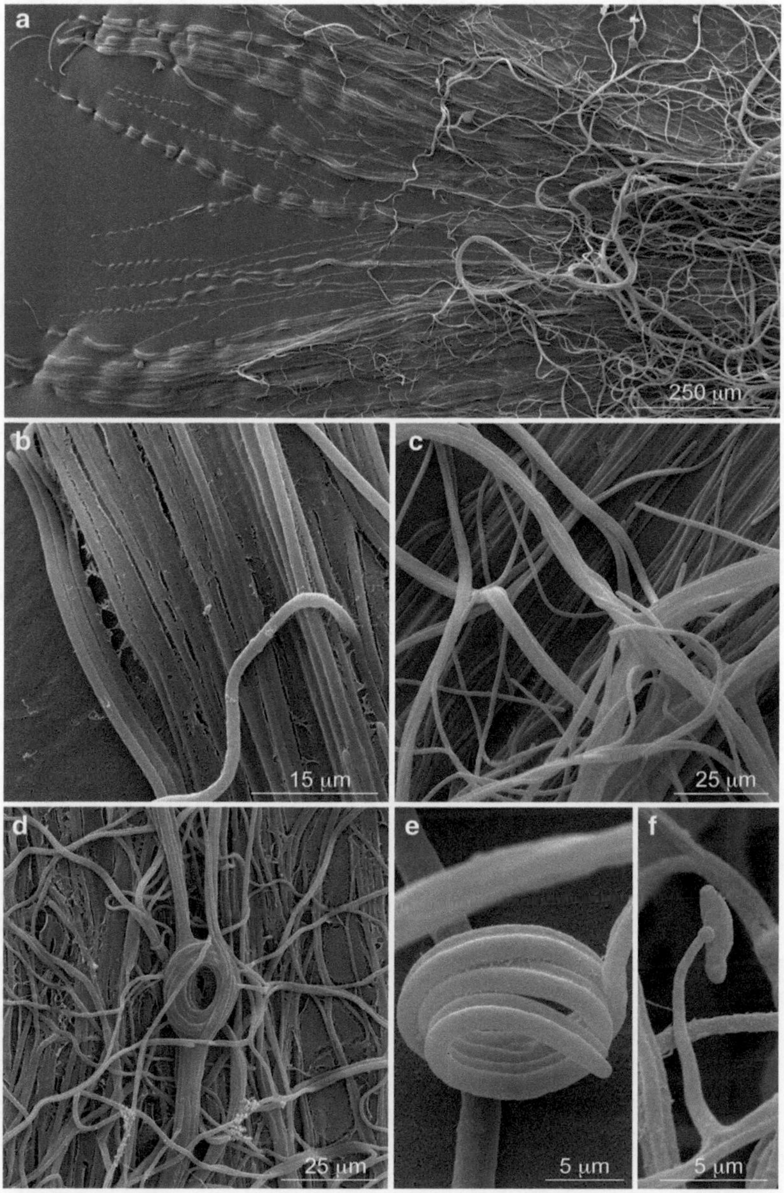


**Fig. 12.3** SEM images of *E. festucae* colony structure on PDA. (**a**) Layers of hyphae at colony edge. (**b**) Straight hyphae growing in parallel and attached to agar at colony edge. (**c**) Hyphal bundles forming second layer of hyphae in colony. (**d** and **e**) Coil-like structures. (**f**) Conidiophore and conidium

## *12.4.2 Conidiogenesis and Conidiospore Germination in Culture*

The ability of *E. festucae* to undergo asexual reproduction in culture is quite variable, but most strains seem to sporulate rather sparsely. While nutrient limitation seems to increase sporulation, the growth rate under these conditions is much reduced. On PDA, sporulation is mainly observed at an early stage of growth on cultures that have a fluffy appearance and have lots of aerial hyphae. Mature conidiophores are formed either at the tips of long, thin, unbranched hyphae or from branches of hyphal coils (Fig. 12.4). The mature conidiophore is a simple unicellular phialide that is 10–20 μm in length. Spores are ejected from the top of the phialide and often remain stuck to the phialide after release (Fig. 12.4b, c). The production of a new conidium or mechanical shearing results in release of the older conidium from the tip of the phialide, a mechanism of release similar to that observed for microcycle conidiation in *E. typhina* following ascospore germination (Bacon and Hinton 1988). Conidia of *E. festucae* are oval shaped (3 × 5 μm) and uninucleate. Germination of spores readily occurs in water or PDA media. Two different patterns of germination were observed in *E. festucae* Fl1. Initially a single germ tube emerges from one end of the conidium then, either a branch is formed or a second germ tube emerges from the opposite end. The first germ tube then forms a branch followed by branching in both germ tubes (Fig. 12.4d, e). After 12–24 h a

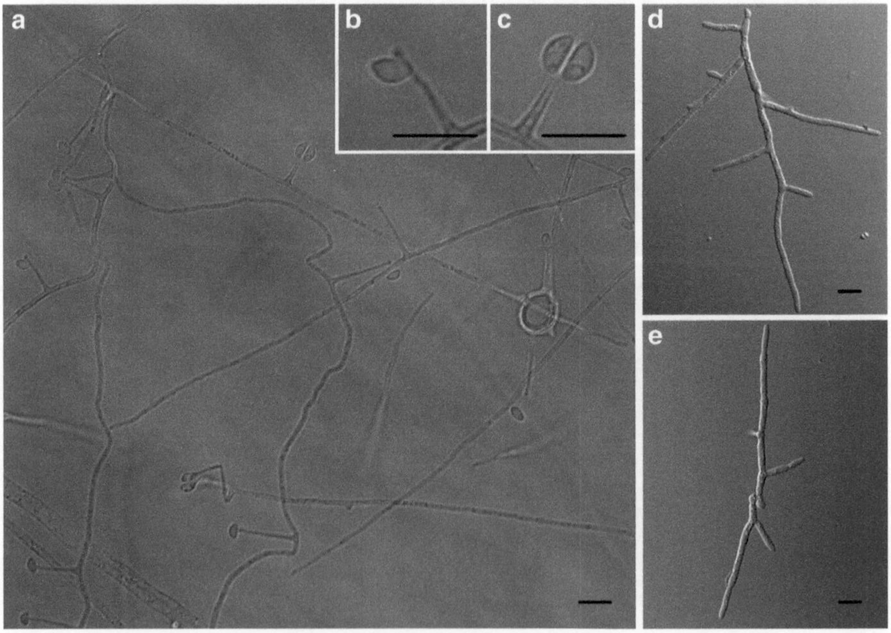


**Fig. 12.4** Conidia formation and germination in *E. festucae*. (**a**) Hyphae growing on PDA showing conidiophores and conidia. (**b** and **c**) Enlargements of sectors of (**a**). (**d** and **e**) Germination of *E. festucae* conidia on PDA

range of germ tube lengths are observed. At 72 h a dense mycelial net is formed that expands in three dimensions in both liquid and on solid medium. In *A. nidulans*, preferential production of ROS at the hyphal tip during spore germination appears to be associated with enforced apical dominance (Semighini and Harris 2008). It would be interesting to determine if ROS also accumulates at the hyphal tip during germination of *E. festucae* conidia.

Two cytosolic subunits of the *E. festucae* NADPH oxidase (Nox) complex, the small GTPase RacA and BemA (Tanaka et al. 2008; Takemoto et al. 2011), are important for formation of conidia; spore production is completely absent in the Δ*racA* mutant and significantly reduced conidiation rates were observed for the Δ*bemA* mutant (Becker, unpublished results). This effect seems to be unrelated to the function of these proteins in Nox complex regulation, given that deletion of either *noxA* or *noxB* alone has no effect on sporulation. In contrast, deletion of *nox1* in *C. purpurea* significantly lowered the germination rate of conidia (Giesbert et al. 2008).

## 12.5 Endophytic and Epiphytic Growth of *E. festucae*

*E. festucae* has no specialized feeding structures such as haustoria or arbuscules, so relies on direct uptake of nutrients from either the apoplastic space or by way of specialized transport processes between the hypha and attached plant cells (Hinton and Bacon 1985; Christensen et al. 2002). Unlike the obligate biotrophic ascomycetes *Blumeria graminis*, *Erysiphe pisi*, and *Golovinomyces orontii* which lack a number of genes involved in primary metabolism, including genes for nitrate and sulfate assimilation, and thiamine biosynthesis (Spanu et al. 2010), a full complement of these gene sets appear to be present in the genome of *E. festucae* (http://csbio-l.csr.uky.edu/endophyte/).

Colonization appears to be principally by mechanical force rather than breakdown of the plant cell wall, but degradation of the cuticle and possibly the cell walls would be required to breach the outer layers of the plant leaf during stroma formation (Christensen et al. 2002). *E. festucae*/*N. lolii* colonization of vascular bundles is rarely observed in natural associations but does occur in some novel associations (Christensen et al. 1997, 2001). Where colonization of the vascular bundles occurs, hyphae are found to be concentrated in the phloem and the protophloem, and the plant hosts frequently became stunted. While *E. festucae* hyphae are found in the root apical meristem (RAM) cells and the embryo radical, tissues derived directly from the true stem, hyphae are unable to colonize the cells laid down behind the RAM (Christensen and Voissey 2007). Hyphae are never found in lateral roots since these are derived by differentiation of pericycle cells, which are never colonized by the fungus.

In addition to endophytic growth, *Epichloë*/*Neotyphodium* spp. have been observed to grow epiphytically on several grass species (Christensen et al. 1997; Moy et al. 2000). PCR amplification and sequencing of the rDNA ITS region was used to confirm that the species growing on the surface of *Poa ampla* was *E. typhina*. Production of conidia and conidiogenous cells on the leaf surfaces of some grass species suggests that horizontal spread of *E. typhina* may also occur during the asexual life cycle. The presence of these extensive epiphyllous hyphal nets may increase the resistance of the host to fungal pathogens such as *Alternaria* and *Rhizoctonia* through "niche exclusion" or by fungal synthesis of bioprotective metabolites (Moy et al. 2000). We have also observed epiphyllous nets of *E. festucae* on the leaves of *L. perenne* plants grown axenically under controlled environmental conditions (Fig. 12.5). Scanning electron microscopy (SEM) analysis of these mycelial networks on the leaf sheath and blade of *L. perenne*, revealed a high density of hyphae at the base of the blade (Fig. 12.5a). Endophytic hyphae appear to give rise to epiphytic hyphae by breaching the leaf surface layers (Fig. 12.5b). Epiphyllous hyphae were mostly attached to the leaf surface, growing along leaf surface depressions, but frequent bridging of furrows and loss of surface contact was also observed (Fig. 12.5c). Coil-like structures, identical to those observed on agar, frequent hyphal fusions, and the formation of conidiophores and conidiospores were also observed (Fig. 12.5d–f). Whether the conidia are able to germinate and penetrate the leaf surface remains to be tested.

The fact that extension and branching of endophytic hyphae ceases when the plant leaf matures, suggests that the plant controls fungal growth. The average number of hyphal strands in a given section of a leaf remains constant once plant cell division has stopped (Tan et al. 2001; Christensen et al. 2002). However, secondary metabolite production remains high in mature leaves (Tan et al. 2001). This change in fungal metabolism correlates with a change in hyphal ultrastructure (Christensen et al. 2008). In contrast to hyphae in young leaves, hyphae in mature leaves have increased numbers of lipid droplets and crystalloid bodies, and thicker cell walls. Production of secondary metabolites by these nongrowing hyphae in planta is analogous to the general observation that fungi synthesize secondary metabolites in culture when the cells enter the stationary phase of growth. Consistent with this hypothesis, genes for secondary metabolite production are down regulated in a MAP kinase mutant of *E. festucae* that undergoes proliferative growth in the host plant (Eaton et al. 2010).

The special mode of grass growth raises the question of how the endophyte is able to grow and expand in synchrony with the plant cells. A grass plant is made up of growth units called tillers (Fig. 12.6), each comprised of a root system, a very short true stem and a pseudostem. The latter is comprised of leaf sheaths wrapped around the emerging leaf blades. In the reproductive mode of growth, tillers produce an elongated stem and flowerhead. During plant development, axillary buds arise from the SAM and expand to become new shoot apices that generate a new tiller (Soper and Mitchell 1956; Veit 2006). Leaves develop from the subapical meristematic region of the shoot apex (Fig. 12.6f, g). Groups of new cells in the apical meristem form growth centers and develop into leaf primordia. Within each leaf primordium there are two zones of cell division, one that gives rise to the blade and the other the sheath. Intercalary division of cells in both zones combined with leaf expansion push both leaf zones upwards (Fig. 12.6).

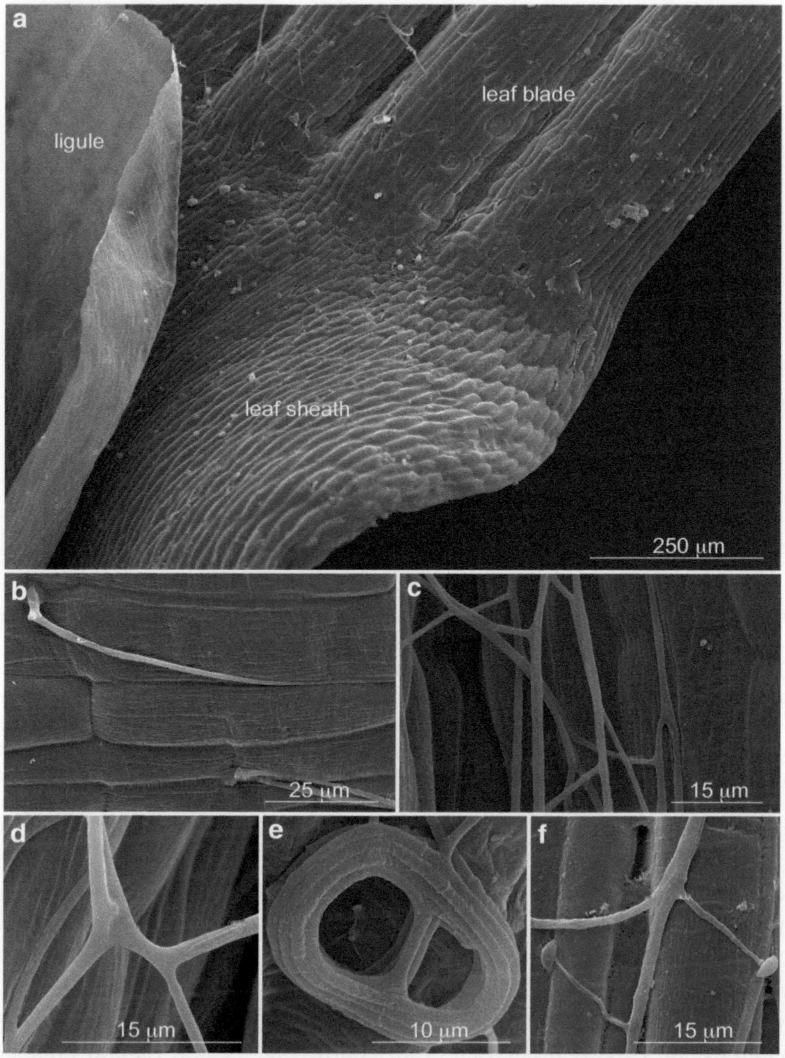


**Fig. 12.5** SEM images of epiphyllous growth of *E. festucae* on perennial ryegrass leaves. (**a**) Sheath–blade transition zone. (**b**) Hyphae emerging from plant leaf. (**c**) Hyphae spanning ridges of blade. (**d**) Hyphal fusion. (**e**) Hyphal coil formation. (**f**) Conidiophore and conidia development

Endophytes show two very distinctive growth patterns during plant colonization. Both seed-borne and artificially introduced endophytes form a proliferative network of hyphae among the cells below the SAM (Christensen et al. 2008). Hyphae in this zone colonize leaf primordia and axillary buds to form a highly branched mycelial network amongst the dividing plant cells. Colonization of leaves and new tillers takes place from three plant cell division zones: the SAM, the blade intercalary division zone, and the sheath intercalary division zone (Christensen et al. 2008).

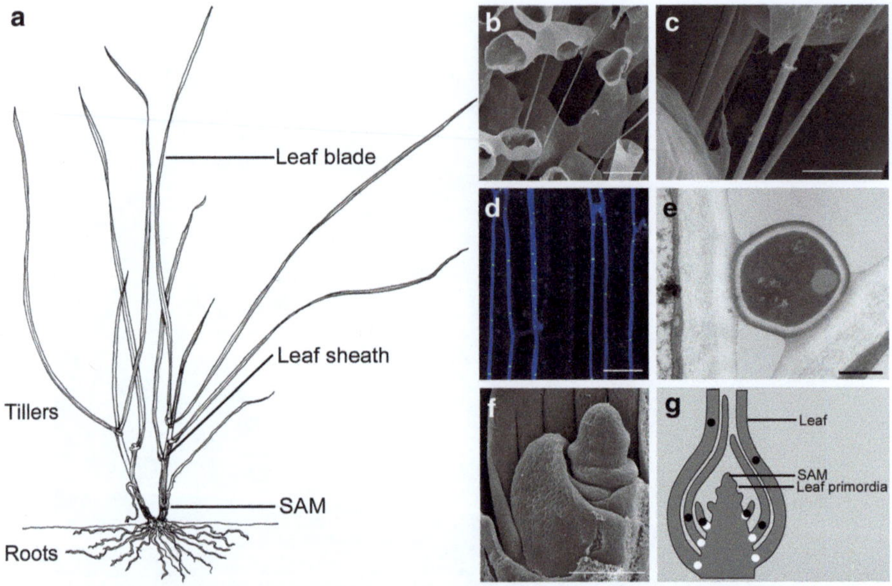


**Fig. 12.6** Growth of *E. festucae* in planta. (**a**) Sketch showing vegetative morphology of perennial ryegrass. (**b** and **c**) SEM images of stretched hyphae in intercellular spaces of blade tissue. (**d**) Confocal depth series image (1 μm) of longitudinal sections through perennial ryegrass (*Lolium perenne*) leaves showing hyphae stained with Alexafluor (WGA-AF488) and *aniline blue*. The image shows hyphae (*fluorescent blue*) growing in close association with plant cells. Strongly illuminated *green points* indicate hyphal septa. (**e**) Electron transmission image of *E. festucae* hyphae attached to plant cell walls in leaf sheath. (**f**) SEM image of perennial ryegrass SAM and young leaves. (**g**) Schematic of perennial ryegrass shoot apex showing leaf primordia and emerging leaves and positions of sheath base (*white dot*) and blade base (*black dot*)

In contrast to the proliferative growth pattern within the meristematic tissues, hyphae colonizing the sheath and blades have a more restricted pattern of growth; typically single hyphae, aligned parallel to the longitudinal leaf axis, are observed between plant cells (Fig. 12.6d). These cells appear to be firmly attached to plant cell walls in the expansion zone of leaves and are probably not free to slide between expanding plant cells (Christensen et al. 2008) (Fig. 12.6e). Lateral branches are infrequently formed and only rarely are two hyphae found in any intercellular space. Frequent hyphal fusion events are seen in *L. perenne* meristem, sheath and blade tissue colonized by *E. festucae*, resulting in an interlinked hyphal network, that would promote cell–cell signaling and nutrient exchange (Christensen et al. 2001; Tanaka et al. 2006).

How this switch in the pattern of growth occurs during leaf colonization has been an outstanding question, particularly if it is assumed fungal hyphae grow exclusively by tip extension. Such a growth pattern would result in shearing of the attached hyphae, particularly the lateral hyphal branches, during cell expansion of the leaf sheath, and blade tissue. The other incongruity with this model is the rapid growth rate that is required for hyphal extension to match the rates of leaf cell extension. Leaf tissue can extend at a rate exceeding 1 cm per day, whereas the growth rate of *E. festucae* on PDA is just 1–3 mm per day. Synchronized growth of the hyphae with the developing leaves requires a growth rate that matches that of the expanding leaf. Furthermore, hyphae within a distinct developmental zone of the leaf have a similar age and physiological state, which argues against a mechanism involving fungal colonization of the blade and sheath solely by tip extension. If this was the case, hyphae in any given cross-section would reflect a variety of developmental stages (Christensen et al. 2008; Voisey 2010).

An alternative model of growth in which hyphae extend by intercalary division and extension instead of tip growth within the expansion zone has recently been proposed (Christensen et al. 2008; Voisey 2010). While this model challenges the generally accepted dogma of fungal growth and raises the question of how intercalary growth is regulated, it does explain how growth of endosymbiont and host cells can be synchronized. A key component of this model is the requirement for physical stretching of the fungal cell wall to activate the fungal intercalary growth machinery (Fig. 12.6b, c). Such a mechanism would explain why growth of the fungus ceases when the leaf stops growing. Given the key role that $Ca^{2+}$ signaling plays in hyphal growth and branching (Jackson and Heath 1993; Torralba and Heath 2001), it would be worthwhile to test whether $Ca^{2+}$ transporters have a role in intercalary growth. A model of growth that requires cell wall stretching would require dynamic biochemical changes in the cell wall structure. Signaling for those changes is likely to occur through the cell wall integrity (CWI) MAP kinase pathway (Eaton et al. 2011a). A further requirement for intercalary growth would be a major reorganization of the fungal cytoskeleton; including reorganization of the microtubules and actin, and redirection of vesicle transport. Elucidating the cell and molecular mechanisms that control intercalary growth will be a challenging area for future research, especially as the process can only be studied within the plant.

## 12.6 Dissecting the *E. festucae–L. perenne* Symbiosis by Forward and Reverse Genetics Together with Omics

### *12.6.1 The NADPH Oxidase Complex*

Host stunting, characterized by loss of apical dominance (increased tillering), premature senescence, and frequent death, is a commonly observed phenotype for symbiotic mutants of *E. festucae* (Eaton et al. 2011b). The first report of an *E. festucae* mutant that induced this host phenotype came from a REMI mutagenesis screen that generated a plasmid insertion in *noxA*, encoding an NADPH oxidase (Tanaka et al. 2006). Instead of the restricted in-planta hyphal growth observed for wild-type, hyphae of the Δ*noxA* mutant were hyperbranched, resulting in proliferative growth throughout the meristematic and mature leaf tissue, and a dramatic increase in overall hyphal biomass. Furthermore, extensive colonization of the vascular bundles, including both phloem and xylem tissue was observed. The Δ*noxA* mutant appears to be deficient in the ability to switch from proliferative growth in the meristematic tissue to synchronized, restricted growth in the leaves. The increase in hyphal biomass in the older leaves is indicative of a loss of synchronization between fungal and plant growth. These results have led to the hypothesis that ROS signaling, mediated by either superoxide ($O_2^-$) itself, the direct product of NoxA, or the dismutated product, $H_2O_2$, is required for the switch from proliferative, polarized tip growth in the meristematic tissue, to intercalary extension in the expanding leaf (Scott and Eaton 2008; Eaton et al. 2011b).

In contrast to *E. festucae* colonization of leaves, *C. purpurea* colonization of ovaries initially involves restricted polarized tip growth through the transmitting tissue followed by a switch to proliferative growth and formation of the sclerotium once it taps into the nutrients of the phloem at the base of the ovule (Haarmann et al. 2009). In both cases NoxA/Nox1 appears to be required for the growth transition, suggesting that ROS play a key role in regulating hyphal growth and branching in the host. If this is the case, tight regulation of both ROS production and scavenging mechanisms would be required for efficient signaling between host and symbiont to maintain either a mutualistic (*E. festucae*) or a pathogenic (*C. purpurea*) interaction (Nathues et al. 2004; Scheffer et al. 2005; Tanaka et al. 2006). Activation of NoxA requires recruitment of cytoplasmic components NoxR, a homolog of the mammalian p67phox, and the small GTPase RacA (Scott and Eaton 2008). Targeted deletion of *E. festucae noxR* or *racA* resulted in a stunted host phenotype similar to that observed for Δ*noxA* (Takemoto et al. 2006; Tanaka et al. 2008). The pathogenicity phenotype of Δ*noxR* mutants of *B. cinerea* and *M. oryzae* is the same as that of the respective Δ*noxA*/Δ*noxB* double mutant, suggesting NoxR regulates both Nox enzymes (Egan et al. 2007; Segmüller et al. 2008). The Δ*rac1* mutant of *C. purpurea* has a severe defect in pathogenicity being unable to penetrate the surface of the stigmatic hairs (Rolke and Tudzynski 2008).

While no direct homologues of the mammalian Nox accessory proteins, p40phox and p47phox, have been identified in fungal genomes, two proteins with a similar functional role have recently been identified in *E. festucae*. Using yeast two hybrid and coimmunoprecipitation assays NoxR was shown to interact with homologs of the yeast polarity proteins, Bem1 and Cdc24, and the PB1 (*P*hox and *B*em*1*) domains found in these proteins were essential for these interactions (Takemoto et al. 2011). GFP-labeled fusions of these proteins preferentially localized to actively growing hyphal tips and septa both in culture and in planta. An *E. festucae* Δ*bemA* mutant was symbiotically defective but the host phenotype was much less severe than the phenotype observed for the Δ*noxA*, Δ*noxR*, and Δ*racA* mutants. The inability to isolate an *E. festucae* Δ*cdc24* mutant suggests that this gene is essential. Based on the protein interaction assays, Cdc24 is proposed to be the guanine-nucleotide exchange (GEF) factor for activation of RacA at the plasma membrane (Takemoto et al. 2011).

While the key components of the fungal Nox complex appear to have now been identified, very little is still known about how this complex is regulated. By analogy with what is known for mammalian systems, two potential targets for regulation are

NoxR and RacA. In mammalian systems p21-activated kinases (Paks) play a key role in regulating the activity of the Nox complex. Paks interact with the small GTPases Rac and Cdc42 and control actin dynamics and phosphorylation of the mammalian NADPH oxidase components p47phox and p22phox (Martyn et al. 2005). In *S. cerevisiae* two paks, Ste20 and Cla4, have been shown to have important roles in polarized growth and actin organization. Whether Paks have a role in regulating the Nox complex in filamentous fungi remains to be determined but a Δ*cla4* mutant of *C. purpurea* has a severe reduction in host pathogenicity and a strong interaction was seen between Cla4 and Rac1 in a yeast two-hybrid interaction assay (Rolke and Tudzynski 2008). Another key protein is the Rho GDP-dissociation inhibitor RhoGDI, which is required to sequester GTPases in the cytoplasm and assist with the delivery and removal of GTPases to and from the plasma membrane (DerMardirossian et al. 2004). Whether, RhoGDI is involved in regulating Nox function in filamentous fungi remains to be determined.

### 12.6.2 Other Pathways

In addition to components of the Nox complex, mutations in the high osmolarity growth (Hog1) and cell integrity (Slt2) MAP kinase pathways in *E. festucae* severely disrupt the symbiosis, giving rise to plants that are severely stunted (Becker and Scott unpublished; Eaton et al. 2010). As with mutants in the Nox complex there is a switch from restricted to proliferative hyphal growth in the host plant tissue. RNAseq analysis of wild-type and *sakA* (stress-activated MAP kinase) mutant associations has provided important insights into the changes in both plant and fungal gene expression that accompany the switch from a mutualistic to pathogenic host interaction (Eaton et al. 2011b). In the mutant association there is a dramatic upregulation of fungal hydrolases, transporters, and genes involved in translation. In contrast, there is a dramatic decrease in expression of genes involved in the production of bioprotective secondary metabolites. These results highlight the power of deep mRNA sequencing to define a minimal set of candidate fungal and plant genes required for maintaining a mutualistic symbiotic interaction (Cox et al. 2010; Eaton et al. 2010, 2011b).

Given endophytes exclusively colonize the apoplast during in-planta growth, it will be crucial to analyze the metabolite (e.g., amino acid and carbohydrate) and protein composition of the apoplast to gain a better understanding of the nutritional requirements of the endophyte and mechanisms of cell–cell communication. Comparisons of the proteomes of both E− (endophyte negative) and E+ plant material (Zhang et al. 2011), and cytoplasmic versus secreted fractions of free-living cultures (Bassett et al. 2009), have been carried out. Despite the very low biomass of fungal symbiont to plant host (<1/100), a fungal Cu/Zn superoxide dismutase was detected by 2-D PAGE analysis (Zhang et al. 2011). The abundance of this protein may provide protection for the endophyte from oxidative stress. The higher levels of a host pathogenesis-related class 10 (PR10) protein found in

E+ compared to E− plants suggests there is some host defense response mounted against the colonizing endophyte.

Metabolic profiling using direct-infusion ion trap mass spectroscopy is a powerful tool to gain further insight into metabolic interactions between the endophyte and its host plant. An unbiased screen of E− and E+ plant material confirmed the presence of known metabolites such as peramine and mannitol in E+ tissue but also identified a number of novel metabolites including a group of cyclic oligopeptides (Cao et al. 2008). In an independent study, levels of nitrate and several amino acids decreased in endophyte-infected perennial ryegrass, whereas water-soluble carbohydrates, some organic acids, and chlorogenic acid increased (Rasmussen et al. 2008). The same group has shown that the metabolite profile and endophyte biomass of E+ plants can change dramatically depending on the host genotype and the nitrogen supply (Rasmussen et al. 2007, 2008). This approach will be a powerful tool for better understanding changes in the apoplast metabolome during endophyte colonization.

## 12.7 ROS Signaling

The severe growth phenotype of Δ*noxA*-infected plants suggests fungal-produced ROS may be involved in a signaling pathway to control hyphal growth and morphology in planta, thereby maintaining the symbiosis. Indeed, depleting flavoprotein-generated ROS from *E. festucae* cultures using DPI causes hyperbranching of hyphae, strongly supporting the concept that endogenous levels of ROS play a key role in controlling hyphal growth of the endophyte (Takemoto et al. 2006). Given the highly reactive and potentially damaging nature of ROS, cells must employ efficient systems for sensing and responding to changing levels of ROS during these cell signaling events. In contrast to the global changes in cellular redox status that occur under oxidative stress conditions, ROS signaling is likely to be mediated by a localized and transient increase of ROS which are then rapidly detoxified to maintain redox homeostasis. An increase in ROS is observed at the tips of growing *E. festucae* hyphae using NBT (Tanaka et al. 2006). Basi-petal gradients of ROS, as detected by NBT staining, have also been observed at the tips of emerging conidial germ tubes of *A. nidulans* (Semighini and Harris 2008). In the presence of DPI conidia fail to germinate and hyphal morphogenesis is disrupted, strongly suggesting that flavoprotein-induced synthesis of ROS is required for establishment and maintenance of hyphal polarity. These observations suggest that ROS has a key signaling role, both spatially and temporally, in polarized hyphal growth (Semighini and Harris 2008).

How ROS signal within multicellular organisms to trigger cell differentiation and development processes remains a key question, but there is consensus that oxidation and reduction of thiol-containing proteins is the main mechanism for oxidant-triggered signal transduction. While there is a huge range of thiol-containing proteins in the cell, including transcription factors, metabolic enzymes (e.g., GAPDH), cell cycle proteins (e.g., Cdc25), signaling regulators (e.g., protein tyrosine phosphatases), and antioxidant defense proteins (e.g., thioredoxin), whether a particular protein is used for signaling will depend on the relative reactivity of that protein for a specific oxidant and its abundance in the cell. There are two general mechanisms proposed for redox regulation (Winterbourn and Hampton 2008). The first is a thermodynamic model where changes in redox buffers such as glutathione result in oxidation of thiol proteins, the target of which depends on the redox potential of the reactive cysteines. However, direct thiol disulphide reactions are slow and there is increasing evidence that cells are unable to maintain thermodynamic equilibrium during increases in oxidant levels during signaling processes. The second is a kinetic model involving a more specific cellular response by a few oxidant-sensitive protein targets. Transmission of the signal would involve transient oxidation of the target in response to the oxidant followed by enzymatic reduction to its basal oxidation state. A good example of this mechanism is the bacterial OxyR transcriptional activator, which undergoes a conformational change upon oxidation to convert it from an inactive to active form. A variation of this second model is a two-step process initiated by a very reactive thiol sensor protein, which once oxidized, facilitates the oxidation of other target proteins by specific protein–protein interactions that facilitate thiol exchange (Delaunay et al. 2002). The paradigm of this mechanism is the Gpx3/Yap1 oxidative stress signal transduction pathway in *S. cerevisiae*. Simulation of the ability of various thiol proteins to be oxidized by increasing concentrations of hydrogen peroxide has identified the peroxiredoxins as the most likely cellular protein targets for hydrogen peroxide-mediated signaling because of their high reactivity and concentration in the cell (Winterbourn and Hampton 2008).

While Yap1 has been demonstrated to be crucial in conferring cellular protection to oxidative stress in culture and during plant host colonization for a range of filamentous fungi (Molina and Kahmann 2007; Lin et al. 2009; Guo et al. 2011), whether this pathway is also involved in cell signaling associated with oxidant-driven cellular differentiation remains to be determined. The other important signaling pathway for cellular protection to oxidative stress is the high osmolarity-regulated MAP kinase pathway. However, the *E. festucae* Δ*sakA* mutant is no more sensitive to hydrogen peroxide-induced oxidative stress than is the wild-type, implying redundancy in the system (Eaton et al. 2008).

One of the major challenges in studying ROS signaling, particularly for plant-associated fungi, is having the available tools to monitor the chemical nature of the oxidant species, the concentrations, and cellular localization. Several methods are routinely used to detect ROS in planta, including the use of cerium chloride ($CeCl_3$) 3, 3′-diaminobenzidine (DAB) and the fluorescent probe 2′,7′-dichorodihydrofluorescein diacetate ($H_2$DCFDA) (Tanaka et al. 2006; Egan et al. 2007). ROS can also be monitored using luminol or lucigenin-derived chemiluminescence (Bai et al. 2001; Belden et al. 2007). While cell-specific changes in ROS levels can be visualized with some of these methods, in general they provide a more global picture of ROS production. A relatively new and more dynamic methodology involves fluorescent protein-based redox probes. These include redox-sensitive versions of YFP (rxYFP) and GFP (roGFP) (Ostergaard et al. 2001, 2004; Dooley et al. 2004; Hanson et al. 2004; Meyer and Dick 2010; Mishina et al. 2011). Probes have been developed that are specific for redox species, GSH, or hydrogen peroxide, and for particular cellular organelles. The rapid and reversible changes in fluorescence in response to the redox status of a cell or organelle, provides a powerful new approach for studying oxidant signaling. The use of these probes should help discriminate where the ROS is being produced, which is a significant issue in studying fungal–plant interactions. Insights into the contribution of ROS by each of the partners will provide a better understanding to the role of ROS in controlling or modulating both pathogenic and symbiotic interactions.

## 12.8 The Fine Balance Between Symbiosis and Pathogenesis

Genetic analysis of the symbiotic interaction between *E. festucae* and *L. perenne* highlights how finely balanced the interaction is between grass host and fungal symbiont. Single gene mutations in key signaling pathways lead to a switch from restrictive to proliferative growth of the fungus in the host and a breakdown in the mutualistic interaction. Identifying the chemical nature of the metabolite signals exchanged between fungus and host, how those signals are sensed and transduced, to maintain a beneficial interaction, will be key areas of future study on this unique and ecologically interesting symbiosis.

**Acknowledgments** This research was supported by grants from the Tertiary Education Commission (TEC) to the Bio-Protection Research Centre, the Royal Society of New Zealand Marsden Fund (MAU0701), Massey University, and by a Top Achiever Doctoral Scholarship to GC from TEC. We thank Dimitry Sokolov and Doug Hopcroft (Manawatu Microscopy and Imaging Centre, IMBS) for technical assistance with microscopy. We also thank Carla Eaton for comments on the manuscript.

## References

Bacon CW, Hinton DM (1988) Ascosporic iterative germination in *Epichloë typhina*. Trans Br Mycol Soc 90:563–569

Bai Z, Harvey LM, McNeil B (2001) Use of the chemiluminescent probe lucigenin to monitor the production of the superoxide anion radical in a recombinant *Aspergillus niger* (B1-D). Biotechnol Bioeng 75:204–211

Bassett SA, Bond JJ, Kwan FY, McCulloch AF, Haynes PA, Johnson RD, Bryan GT, Jordan TW (2009) Proteomic analysis of a filamentous fungal endophyte using EST datasets. Proteomics 9:2295–2300

Belden WJ, Larrondo LF, Froehlich AC, Shi M, Chen CH, Loros JJ, Dunlap JC (2007) The band mutation in *Neurospora crassa* is a dominant allele of *ras-1* implicating RAS signaling in circadian output. Genes Dev 21:1494–1505

Brasell E (2010) Identification of genes regulating the plant-specific expression of the *ltmM* gene in *Epichloë festucae*. M.Sc. thesis. Massey University, Palmerston North, New Zealand

Bultman TL, Leuchtmann A (2009) Biology of the *Epichloë-Botanophila* interaction: an intriguing association between fungi and insects. Fungal Biol Rev 22:131–138

Bultman TL, White JF Jr, Bowdish TI, Welch AM (1998) A new kind of mutualism between fungi and insects. Mycol Res 102:235–238

Cao M, Koulman A, Johnson LJ, Lane GA, Rasmussen S (2008) Advanced data-mining strategies for the analysis of direct-infusion ion trap mass spectrometry data from the association of perennial ryegrass with its endophytic fungus, *Neotyphodium lolii*. Plant Physiol 146:1501–1514

Christensen MJ, Latch GCM (1991) Variation among isolates of *Acremonium* endophytes (*A. coenophialum* and possibly *A. typhinum*) from tall fescue (*Festuca arundinacea*). Mycol Res 95:1123–1126

Christensen MJ, Voissey CR (2007) The biology of the endophyte/grass partnership. In: Popay AJ, Thom ER (eds) Proceedings of the 6th international symposium on fungal endophytes of grasses grasslands research and practice series no, 13. New Zealand Grassland Association, Christchurch, pp 123–133

Christensen MJ, Ball OJ-P, Bennett RJ, Schardl CL (1997) Fungal and host genotype effects on compatibility and vascular colonization by *Epichloë festucae*. Mycol Res 101:493–501

Christensen MJ, Bennett RJ, Schmid J (2001) Vascular bundle colonisation by *Neotyphodium* endophytes in natural and novel associations with grasses. Mycol Res 105:1239–1245

Christensen MJ, Bennett RJ, Schmid J (2002) Growth of *Epichloë/Neotyphodium* and p-endophytes in leaves of *Lolium* and *Festuca* grasses. Mycol Res 106:93–106

Christensen MJ, Bennett RJ, Ansari HA, Koga H, Johnson RD, Bryan GT, Simpson WR, Koolaard JP, Nickless EM, Voisey CR (2008) *Epichloë* endophytes grow by intercalary hyphal extension in elongating grass leaves. Fungal Genet Biol 45:84–93

Chung K-R, Schardl CL (1997) Sexual cycle and horizontal transmission of the grass symbiont, *Epichloë typina*. Mycol Res 101:295–301

Cox MP, Eaton CJ, Scott B (2010) Exploring molecular signaling in plant-fungal symbioses using high throughput RNA sequencing. Plant Signal Behav 5:1353–1358

Delaunay A, Pflieger D, Barrault MB, Vinh J, Toledano MB (2002) A thiol peroxidase is an $H_2O_2$ receptor and redox-transducer in gene activation. Cell 111:471–481

DerMardirossian C, Schnelzer A, Bokoch GM (2004) Phosphorylation of RhoGDI by Pak1 mediates dissociation of Rac GTPase. Mol Cell 15:117–127

Dooley CT, Dore TM, Hanson GT, Jackson WC, Remington SJ, Tsien RY (2004) Imaging dynamic redox changes in mammalian cells with green fluorescent protein indicators. J Biol Chem 279:22284–22293

Eaton CJ, Jourdain I, Foster SJ, Hyams JS, Scott B (2008) Functional analysis of a fungal endophyte stress-activated MAP kinase. Curr Genet 53:163–174

Eaton CJ, Cox MP, Ambrose B, Becker M, Hesse U, Schardl CL, Scott B (2010) Disruption of signaling in a fungal-grass symbiosis leads to pathogenesis. Plant Physiol 153:1780–1794

Eaton C, Mitic M, Scott B (2011a) Signalling in the *Epichloë festucae*-perennial ryegrass mutualistic symbiotic interaction. In: Perotto S, Baluska F (eds) Signalling and communication in plant symbiosis (communication and signaling in plants). Springer, Heidelberg

Eaton CJ, Cox MP, Scott B (2011b) What triggers grass endophytes to switch from mutualism to pathogenism? Plant Sci 180:190–195

Egan M, Wang Z-Y, Jones MA, Smirnoff N, Talbot NJ (2007) Generation of reactive oxygen species by fungal NADPH oxidases is required for rice blast disease. Proc Natl Acad Sci USA 104:11772–11777

Fleetwood DJ, Scott B, Lane GA, Tanaka A, Johnson RD (2007) A complex ergovaline gene cluster in epichloë endophytes of grasses. Appl Environ Microbiol 73:2571–2579

Florea S, Andreeva K, Machado C, Mirabito PM, Schardl CL (2009) Elimination of marker genes from transformed filamentous fungi by unselected transient transfection with a Cre-expressing plasmid. Fungal Genet Biol 46:721–730

Giesbert S, Schurg T, Scheele S, Tudzynski P (2008) The NADPH oxidase Cpnox1 is required for full pathogenicity of the ergot fungus *Claviceps purpurea*. Mol Plant Pathol 9:317–327

Guo M, Chen Y, Du Y, Dong Y, Guo W, Zhai S, Zhang H, Dong S, Zhang Z, Wang Y, Wang P, Zheng X (2011) The bZIP transcription factor MoAP1 mediates the oxidative stress response and is critical for pathogenicity of the rice blast fungus *Magnaporthe oryzae*. PLoS Pathog 7: e1001302

Haarmann T, Rolke Y, Giesbert S, Tudzynski P (2009) Ergot: from witchcraft to biotechnology. Mol Plant Pathol 10:563–577

Hanson GT, Aggeler R, Oglesbee D, Cannon M, Capaldi RA, Tsien RY, Remington SJ (2004) Investigating mitochondrial redox potential with redox-sensitive green fluorescent protein indicators. J Biol Chem 279:13044–13053

Hinton DM, Bacon CW (1985) The distribution and ultrastructure of the endophyte of toxic tall fescue. Can J Microbiol 63:36–42

Jackson SL, Heath IB (1993) Roles of calcium ions in hyphal tip growth. Microbiol Rev 57:367–382

Kuldau GA, Tsai H-F, Schardl CL (1999) Genome sizes of *Epichloë* species and anamorphic hybrids. Mycologia 91:776–782

Latch GCM, Christensen MJ (1985) Artificial infection of grasses with endophytes. Ann Appl Biol 107:17–24

Leuchtmann A, Schardl CL, Siegel MR (1994) Sexual compatibility and taxonomy of a new species of Epichloë symbiotic with fine fescue grasses. Mycologia 86:802–812

Leuchtmann A, Schmidt D, Bush LP (2000) Different levels of protective alkaloids in grasses with stroma-forming and seed-transmitted *Epichloë/Neotyphodium* endophytes. J Chem Ecol 26:1025–1036

Lin CH, Yang SL, Chung KR (2009) The YAP1 homolog-mediated oxidative stress tolerance is crucial for pathogenicity of the necrotrophic fungus *Alternaria alternata* in citrus. Mol Plant Microbe Interact 22:942–952

Martyn KD, Kim M-J, Quinn MT, Dinauer MC, Knaus UG (2005) p21-activated kinase (Pak) regulates NADPH oxidase activation in human neutrophils. Blood 106:3962–3969

May KJ, Bryant MK, Zhang X, Ambrose B, Scott B (2008) Patterns of expression of a lolitrem biosynthetic gene in the *Epichloë festucae*-perennial ryegrass symbiosis. Mol Plant Microbe Interact 21:188–197

Meyer AJ, Dick TP (2010) Fluorescent protein-based redox probes. Antioxid Redox Signal 13:1–30

Michalakis Y, Olivieri I, Renaud F, Raymond M (1992) Pleiotropic action of parasites: how to be good for the host. Trends Ecol Evol 7:59–62

Mishina NM, Tyurin-Kuzmin PA, Markvicheva KN, Vorotnikov AV, Tkachuk VA, Laketa V, Schultz C, Lukyanov S, Belousov VV (2011) Does cellular hydrogen peroxide diffuse or act locally? Antioxid Redox Signal 14:1–7

Molina L, Kahmann R (2007) An *Ustilago maydis* gene involved in $H_2O_2$ detoxification is required for virulence. Plant Cell 19:2293–2309

Moon CD, Craven KD, Leuchtmann A, Clement SL, Schardl CL (2004) Prevalence of interspecific hybrids amongst asexual fungal endophytes of grasses. Mol Ecol 13:1455–1467

Moy M, Belanger F, Duncan R, Freehoff A, Leary C, Meyer W, Sullivan R, WJ JF (2000) Identification of epiphyllous mycelial nets on leaves of grasses infected by clavicipitaceous endophytes. Symbiosis 28:291–302

Nathues E, Joshi S, Tenberge KB, von den Driesch M, Oeser B, Baumer N, Mihlan M, Tudzynski P (2004) CPTF1, a CREB-like transcription factor, is involved in the oxidative stress response in the phytopathogen *Claviceps purpurea* and modulates ROS level in its host *Secale cereale*. Mol Plant Microbe Interact 17:383–393

Ostergaard H, Henriksen A, Hansen FG, Winther JR (2001) Shedding light on disulfide bond formation: engineering a redox switch in green fluorescent protein. EMBO J 20:5853–5862

Ostergaard H, Tachibana C, Winther JR (2004) Monitoring disulfide bond formation in the eukaryotic cytosol. J Cell Biol 166:337–345

Philipson MN, Christey MC (1986) The relationship of host and endophyte during flowering, seed formation, and germination of *Lolium perenne*. New Zeal J Bot 24:125–134

Rasmussen S, Parsons AJ, Bassett S, Christensen MJ, Hume DE, Johnson LJ, Johnson RD, Simpson WR, Stacke C, Voisey CR, Xue H, Newman JA (2007) High nitrogen supply and carbohydrate content reduce fungal endophyte and alkaloid concentration in *Lolium perenne*. New Phytol 173:787–797

Rasmussen S, Parsons AJ, Fraser K, Xue H, Newman JA (2008) Metabolic profiles of *Lolium perenne* are differentially affected by nitrogen supply, carbohydrate content, and fungal endophyte infection. Plant Physiol 146:1440–1453

Rolke Y, Tudzynski P (2008) The small GTPase Rac and the p21-activated kinase Cla4 in *Claviceps purpurea*: interaction and impact on polarity, development and pathogenicity. Mol Microbiol 68:405–423

Sampson K (1933) The systemic infection of grasses by *Epichloë typhina* (Pers.) Tul. Trans Br Mycol Soc 18:30–47

Schardl CL (2001) *Epichloë festucae* and related mutualistic symbionts of grasses. Fungal Genet Biol 33:69–82

Schardl CL, Leuchtmann A, Tsai H-F, Collett MA, Watt DM, Scott DB (1994) Origin of a fungal symbiont of perennial ryegrass by interspecific hybridization of a mutualist with the ryegrass choke pathogen, *Epichloë typhina*. Genetics 136:1307–1317

Schardl CL, Leuchtmann A, Chung K-R, Penny D, Siegel MR (1997) Coevolution by common descent of fungal symbionts (*Epichloë* spp.) and grass hosts. Mol Biol Evol 14:133–143

Schardl CL, Scott B, Florea S, Zhang D (2009) Epichloë endophytes: clavicipitaceous symbionts of grasses. In: Deising HB (ed) The mycota volume V – plant relationships. Springer, Berlin, pp 275–306

Scheffer J, Chen C, Heidrich P, Dickman MB, Tudzynski P (2005) A CDC42 homologue in *Claviceps purpurea* is involved in vegetative differentiation and is essential for pathogenicity. Eukaryot Cell 4:1228–1238

Schiestl FP, Steinebrunner F, Schulz C, Von Reub S, Francke W, Weymuth C, Leuchtmann A (2006) Evolution of 'pollinator'-attracting signals in fungi. Biol Lett 2:401–404

Schmid J, Spiering MJ, Christensen MJ (2000) Metabolic activity, distribution, and propagation of grass endophytes in planta: investigations using the GUS reporter gene system. In: Bacon CW, White JF Jr (eds) Microbial endophytes. Dekker, New York, pp 295–322

Scott B, Eaton CJ (2008) Role of reactive oxygen species in fungal cellular differentiations. Curr Opin Microbiol 11:488–493

Scott B, Schardl C (1993) Fungal symbionts of grasses: evolutionary insights and agricultural potential. Trends Microbiol 1:196–200

Scott B, Takemoto D, Tanaka A, Young CA, Bryant MK, May KJ (2007) Functional analysis of the *Epichloë festucae*-perennial ryegrass symbiosis. In: Popay AJ, Thom ER (eds) Proceedings of the 6th international symposium on fungal endophytes of grasses grasslands research and practice series no. 13. New Zealand Grassland Association, Christchurch, pp 433–441

Segmüller N, Kokkelink L, Giesbert S, Odinius D, van Kan J, Tudzynski P (2008) NADPH oxidases are involved in differentiation and pathogenicity in *Botrytis cinerea*. Mol Plant Microbe Interact 21:808–819

Semighini CP, Harris SD (2008) Regulation of apical dominance in *Aspergillus nidulans* hyphae by reactive oxygen species. Genetics 179:1919–1932

Slepecky RA, Starmer WT (2009) Phenotypic plasticity in fungi: a review with observations on *Aureobasidium pullulans*. Mycologia 101:823–832

Soper K, Mitchell KJ (1956) The developmental anatomy of perennial ryegrass (*Lolium perenne* L.). N Z J Sci Technol 37:484–505

Spanu PD et al (2010) Genome expansion and gene loss in powdery mildew fungi reveal tradeoffs in extreme parasitism. Science 330:1543–1546

Steinebrunner F, Schiestl FP, Leuchtmann A (2008a) Ecological role of volatiles produced by *Epichloë*: differences in antifungal toxicity. FEMS Microbiol Ecol 64:307–316

Steinebrunner F, Schiestl FP, Leuchtmann A (2008b) Variation of insect attracting odor in endophytic *Epichloë* fungi: phylogenetic constrains versus host influence. J Chem Ecol 34:772–782

Steinebrunner F, Twele R, Francke W, Leuchtmann A, Schiestl FP (2008c) Role of odour compounds in the attraction of gamete vectors in endophytic *Epichloë* fungi. New Phytol 178:401–411

Takemoto D, Tanaka A, Scott B (2006) A $p67^{Phox}$-like regulator is recruited to control hyphal branching in a fungal-grass mutualistic symbiosis. Plant Cell 18:2807–2821

Takemoto D, Kamakura S, Saikia S, Becker Y, Wrenn R, Tanaka A, Sumimoto H, Scott B (2011) Polarity proteins Bem1 and Cdc24 are components of the filamentous fungal NADPH oxidase complex. Proc Natl Acad Sci USA 108:2861–2866

Tan YY, Spiering MJ, Scott V, Lane GA, Christensen MJ, Schmid J (2001) In planta regulation of extension of an endophytic fungus and maintenance of high metabolic rates in its mycelium in the absence of apical extension. Appl Environ Microbiol 67:5377–5383

Tanaka A, Tapper BA, Popay A, Parker EJ, Scott B (2005) A symbiosis expressed non-ribosomal peptide synthetase from a mutualistic fungal endophyte of perennial ryegrass confers protection to the symbiotum from insect herbivory. Mol Microbiol 57:1036–1050

Tanaka A, Christensen MJ, Takemoto D, Park P, Scott B (2006) Reactive oxygen species play a role in regulating a fungus-perennial ryegrass mutualistic association. Plant Cell 18:1052–1066

Tanaka A, Wrenn RE, Takemoto D, Scott B (2007) *Agrobacterium tumefaciens* T-DNA mediated transformation of *Epichloë festucae*. In: Popay AJ, Thon ER (eds) Proceedings of the 6th international symposium on fungal endophytes of grasses grassland research and practice series no. 13. New Zealand Grassland Association, Christchurch, pp 469–472

Tanaka A, Takemoto D, Hyon GS, Park P, Scott B (2008) NoxA activation by the small GTPase RacA is required to maintain a mutualistic symbiotic association between *Epichloë festucae* and perennial ryegrass. Mol Microbiol 68:1165–1178

Torralba S, Heath IB (2001) Cytoskeletal and $Ca^{2+}$ regulation of hyphal tip growth and initiation. Curr Top Dev Biol 51:135–187

Trinci APJ (1978) The duplication cycle and branching in fungi. In: Burnett JH, Trinci APJ (eds) Symposium of the British Mycological Society. Cambridge University Press, Queen Elizabeth College, London, pp 319–357

Tudzynski P, Scheffer J (2004) *Claviceps purpurea*: molecular aspects of a unique pathogenic lifestyle. Mol Plant Pathol 5:377–388

Veit B (2006) Stem cell signalling networks in plants. Plant Mol Biol 60:793–810

Voisey CR (2010) Intercalary growth in hyphae of filamentous fungi. Fungal Biol Rev 24:123–131

Winterbourn CC, Hampton MB (2008) Thiol chemistry and specificity in redox signaling. Free Radic Biol Med 45:549–561

Young CA, Bryant MK, Christensen MJ, Tapper BA, Bryan GT, Scott B (2005) Molecular cloning and genetic analysis of a symbiosis-expressed gene cluster for lolitrem biosynthesis from a mutualistic endophyte of perennial ryegrass. Mol Genet Genomics 274:13–29

Zhang N-X, Zhang S, Borchert S, Richardson K, Schmid J (2011) High levels of a fungal superoxide dismutase and increased concentration of a PR-10 plant protein in associations between the endophytic fungus *Neotyphodium lolii* and ryegrass. Mol Plant Microbe Interact 24(8):984–992

# Chapter 13
# Cryptococcus–Neutrophil Interaction

**Asfia Qureshi and Maurizio Del Poeta**

**Abstract** The involvement of neutrophils in the fight against the human pathogen *Cryptococcus neoformans* has predominantly focused on the molecules released during the oxidative burst and those formed in the granular subsets of neutrophils. Very little is known about which and how host signaling pathways regulate the battle. Recent discoveries highlight that the host sphingolipid pathway is important for controlling cryptococcal killing by neutrophils in vitro and, potentially, also in the lung environment. This chapter is timely in describing new developments and findings in this field.

## List of Abbreviations

| | |
|---|---|
| CF | Cystic fibrosis |
| CFTR | Cystic fibrosis transmembrane conductance regulator |
| DAG | Diacylglycerol |
| ESI-MS | Electrospray ionization-mass spectrometry |

A. Qureshi
Department of Biochemistry and Molecular Biology, Medical University of South Carolina, Charleston, SC 29425, USA

M. Del Poeta (✉)
Department of Biochemistry and Molecular Biology, Medical University of South Carolina, Charleston, SC 29425, USA

Department of Microbiology and Immunology, Medical University of South Carolina, Charleston, SC 29425, USA

Department of Craniofacial Biology, Medical University of South Carolina, Charleston, SC 29425, USA

Division of Infectious Diseases, Medical University of South Carolina, Charleston, SC 29425, USA
e-mail: delpoeta@musc.edu

J. Pérez-Martín and A. Di Pietro (eds.), *Morphogenesis and Pathogenicity in Fungi*, Topics in Current Genetics 22, DOI 10.1007/978-3-642-22916-9_13,

| | |
|---|---|
| GCS | Glucosylceramide synthase |
| IPC | Inositol phosphoryl ceramide |
| Ipc1 | Inositol phosphoryl ceramide synthase |
| Isc1 | Inositol phosphospingolipid phospholipase C1 |
| MALDI-MSI | Matrix-assisted laser desorption ionization-mass spectrometric imaging |
| MPO | Myeloperoxidase |
| PC | Phosphotidyl choline |
| Pkc1 | Protein kinase C1 |
| ROS | Reactive oxygen species |
| SM | Sphingomyelin |
| SMS | Sphingomyelin synthase |

## 13.1 *Cryptococcus* spp.

The genus *Cryptococcus* comprises of simple saprophytic yeasts phylogenetically related to higher fungi such as mushrooms and puffballs. Of the more than 50 species known, only *C. neoformans* and *C. gattii* are considered to be pathogenic to humans (Kwon-Chung et al. 2011). *C. neoformans* was first isolated from fermented peach juice by Sanfelice (1894). That same year, Busse, a pathologist, described the yeast he had isolated from the tibial lesions of a 31-year-old female (Busse 1894). The following year, a surgeon named Buschke reported the same isolate from the same patient, thus identifying a new yeast with pathogenic potential (Buschke 1895). Today, more than 117 years later, the total containment of a cryptococcal infection by the host remains elusive.

The four serotypes of *Cryptococcus* responsible for most human and animal infections include *C. gattii* serotypes B and C, which are the causative agents of cryptococcosis in immunocompetent individuals and are associated with eucalyptus trees; *C. neoformans* var. *neoformans* serotype D and *C. neoformans* var. *grubii* serotype A, both prevalent in immunocompromised hosts globally and most often found in soil contaminated with bird faeces. *Cryptococcus* holds the notable distinction as being the only eukaryotic pathogen that produces a polysaccharide capsule, which consists of glucuronoxylomannan (GXM) and galactoxylomannan (GalXM), as well as mannoprotein (Bose et al. 2003). These components all occupy spatially separate and discrete regions in the capsule of *C. neoformans* (Jesus et al. 2010). This antiphagocytic polysaccharide capsule is the major virulence factor of the fungus, with melanin being the second-most important virulence component, and the ability to grow at 37°C a third virulence factor. The regulation of these virulence factors has been reviewed recently in several excellent works (Li and Mody 2010; Alspaugh et al. 2011; Nielsen and Kwon-Chung 2011; Idnurm et al. 2011; Pfaller et al. 2011; Fox et al. 2011).

Mechanisms of host invasion and host immune response are both active areas of research. Upon inhalation into the lung, *C. neoformans* encounters the first line of defense, namely alveolar macrophages. This interaction is important in the process of dissemination (Shao et al. 2005; Mansour and Levitz 2002) in the immunocompromised host, because if *C. neoformans* is not contained in this organ, it disseminates, leading to the development of the most dangerous form of *C. neoformans* infection: cryptococcal meningitis. Studies using either primary or macrophage cell line have found that these host cells may function as a double-edge sword, as the microbe can survive and grow within macrophages (Lee et al. 1995; Feldmesser et al. 2000, 2001). In addition, internalized fungal cells can eventually be extruded by macrophages without killing the host cells (Tucker and Casadevall 2002; Alvarez and Casadevall 2006; Ma et al. 2006). As a consequence, if macrophages are not able to efficiently kill the fungus, both of these processes will increase the likelihood of a disseminated cryptococcosis. Indeed, it has been shown that depletion of alveolar macrophages may actually delay the development of cryptococcal meningoencephalitis and, thus, increase mice survival (Kechichian et al. 2007; Alvarez and Casadevall 2006; McQuiston et al. 2010). This phenomenon may be explained by hypothesizing that if not activated, such as in the condition of immunodeficiency, macrophages may serve as a safe "haven" for *C. neoformans*, which will use them to grow within and/or to be transported to different sites of the body. Thus, macrophages need to be activated in order to control *C. neoformans* infection, with the consequent production of anti-inflammatory cytokines and the recruitment of neutrophils, which are essential for granuloma formation (Goldman et al. 1994; Huffnagle et al. 1991a, b). Although several studies have elucidated the role and mechanisms by which macrophages control *C. neoformans* infection (Feldmesser et al. 2001; Fan et al. 2005; Shao et al. 2005; Alvarez and Casadevall 2006; Ma et al. 2006; Garcia et al. 2008; Stano et al. 2009; Rittershaus et al. 2006; Luberto et al. 2003; McQuiston and Del Poeta 2011), very little is known on the mechanisms by which neutrophils neutralize the fungus during the lung infection.

## 13.2 Neutrophils

Neutrophils play important roles in host defense against all classes of infectious agents and represent 90–95% of granulocytes. They constitute the second line of defense against pathogens after alveolar macrophages because, once an inflammatory response is initiated, neutrophils are the first cells to be recruited to the site of infection (Schleimer et al. 1989). Their targets include fungi, bacteria, viruses, and cancer cells (Ratcliffe et al. 1988). The microbicidal arsenal of neutrophils include the formation of reactive oxygen and nitrogen species, hydrolytic enzymes, and antimicrobial peptides, all of which target microbes (Smith 1994). Studies have shown that the production of these weapons of microbial destruction are coordinated by several mediators that include cytokines, neuroendocrine factors and, very recently, by bioactive lipids. Hence, this chapter will mainly focus on the review of this latter class of compounds.

Neutrophils are terminally differentiated cells that contain cytoplasmic granules rich in receptors, enzymes, and antimicrobial proteins (Borregaard et al. 1993). There are three specialized granule subsets involved in these processes: (1) the azurophilic granules (also known as primary granules), (2) the specific granules (also called secondary granules), and (3) the gelatinase granules (also named tertiary granules) reviewed in (Faurschou and Borregaard 2003). Among other peptides, the azurophilic granules contain defensins, which have been shown to dramatically increase during differentiation of HL-60 cells toward the granulocytic lineage (Herwig et al. 1998; Wang et al. 2004; Tsutsumi-Ishii et al. 2000) and to be extremely cytotoxic toward *C. neoformans* (Alcouloumre et al. 1993; Wang et al. 2009). The azurophilic granules also contain myeloperoxidase (MPO), an additional factor with antifungal activity, as mice lacking MPO are hyper-susceptible to *C. neoformans* (Aratani et al. 2002a, 2006) and other fungal infections (Aratani et al. 1999, 2002a, b).

Neutrophils are produced and mature in human bone marrow at a rate of $10^{11}$ cells per day (Cannistra and Griffin 1988). They are released into the circulation where they spend less than 10 h before migrating to tissue where they survive no longer than 2 days (Itoh et al. 1998). These short-lived neutrophils are thought to undergo apoptosis prior to being removed by macrophages (Maggini et al. 2009).

The importance of neutrophils in host defense has been well illustrated in several excellent articles (Aratani 2006; Urban et al. 2006; Ellerbroek et al. 2004; Brummer 1998; Smith 1994). Interestingly, neutropenia induced by chemotherapy or rare inherited defects in neutrophil function are major risk factors for developing potentially fatal bacterial and fungal infections (Malech and Gallin 1987). Gram-negative bacilli and Staphylococcus infections are most common in neutropenic patients, followed by pathogenic infections due to *Candida albicans* and *Pseudomonas aeroginosa* (Schaffner et al. 1986; Smith 1994). Surprisingly though, neutropenic patients are not at particular risk for developing cryptococcal infections. However, patients receiving immunosuppressive therapy and AIDS patients are more susceptible to cryptococcosis due to low CD4+ T cell counts, and often have phagocytic and chemotactic defects of their neutrophils and monocytes (Bender et al. 1988; Diamond and Erickson 1982). It has been shown that neutrophils from persons with late-stage HIV infection had significantly impaired IL-8 production, which correlated with impaired complement receptor function (Monari et al. 1999).

Notably, neutrophils are more effective killers of fungi than macrophages, and they possess the innate ability to kill microbes without a cell-mediated activation. Interestingly, transfusion of human neutrophils significantly improved the survival of mice challenged with fungal organisms (Spellberg et al. 2005, 2007), suggesting that these cells are able to control the infection through their antifungal activity.

Due to the extremely short lifetime of neutrophils, any laboratory research on these cells leads by necessity to the use of model systems. The well-known model for neutrophils is the human promyelocytic leukemia cell line, HL-60, derived from the peripheral blood of a patient with acute promyelocytic leukemia (Collins et al. 1977; Gallagher et al. 1979). The majority of these cells are promyelocytic in their morphology and histochemistry, but up to 15% of them show morphological characteristics of myelocytes, metamyelocytes, and polymorphonuclear leukocytes (PMNs) (Newburger et al. 1979). However, this latter population is unable to wield any significant microbicidal effects (Harris and Ralph 1985). HL-60 cells can be induced to terminally differentiate to mature granulocytes by incubation with a wide variety of compounds, including dimethylsulfoxide, retinoic acid, dimethylformamide, and other polar solvents (Breitman et al. 1980; Collins et al. 1978; Collins 1987; Harris and Ralph 1985; Mukherjee et al. 1985). These mature HL-60 cells have many of the functional characteristics of normal primary peripheral blood granulocytes, including the ability to phagocytize, the surface expression of complement receptors, chemotaxis capability, and the ability to reduce nitroblue tetrazolium (Collins et al. 1978, 1979; Newburger et al. 1979). The functional changes that have been observed to correlate with differentiation include a dramatic increase of superoxide production, and degranulation as measured by release of b-glucoronidase, lysozyme, and peroxidase during several days (3–6) of in vitro incubation (Newburger et al. 1979), rendering the cells particularly suitable for biochemical and genetic studies. This cell line therefore provides a unique system for studying human myeloid differentiation in vitro.

However, this is a model system and, as such, it is therefore important to understand the capacity and limitations of this system. This was particularly illustrated by Itoh and coworkers who monitored the expression profile of these cells and compared it with that of peripheral primary granulocytes, which represent the activities of neutrophils (Itoh et al. 1998). Although gene activities are not necessarily reflected by the abundance of mRNA, gene expression profiling leads to a best approximation due to other methods not being available. While a number of genes were in common, about seven genes were found in human peripheral granulocytes only, including b 2-microglobulin, granulocyte colony stimulating factor receptor, and tumor necrosis factor receptor (Itoh et al. 1998). A reduced response to IL-8 compared to neutrophils was also observed by measuring motile properties of differentiated HL-60 cells and this is likely due to the low expression of IL-8 receptors in the latter (Hauert et al. 2002). Nevertheless, HL-60 cells remain a valid model system to study the molecular mechanisms of neutrophil killing abilities.

## 13.3 Cryptococcal–Neutrophil Interaction

To date, 14 reviews have been published on the cryptococcal–neutrophil interaction, although the majority do not focus solely on *Cryptococcus*, but instead cover a range of pathogens. They all highlight a key role for MPO produced by neutrophils in host defense against fungal infection. Indeed, MPO-deficient mice have reduced cytotoxicity to several microorganisms including *C. neoformans*, demonstrating that the MPO-dependent oxidative system is important for host defense against fungi (Aratani 2006; Aratani et al. 2006). From the cryptococcal point of view, the polysaccharide capsule prevents neutrophil migration and exerts a negative effect on T cell activation at the site of inflammation (Ellerbroek et al. 2004; Vecchiarelli 2000; Kozel et al. 1988). Neutrophils have been noted to rapidly ingest limited amounts of the capsular component GXM, and then expel or degrade it after 1 h of incubation, whereas macrophages continuously uptake GXM for up to 1 week of incubation (Monari et al. 2003). Accumulation of GXM by neutrophils is accompanied by reduced anti-cryptococcal activity (Monari et al. 2003).

Recent studies have investigated whether the presence of common antifungal drugs, such as amphotericin B, fluconazole, and voriconazole, may enhance the killing activity of neutrophils (Chiller et al. 2002). Neutrophils alone resulted in 61% inhibition of *C. neoformans* after 24 h of incubation. Voriconazole alone at 0.01 and 0.05 mg/ml resulted in 48% inhibition and 19% killing. Interestingly, the addition of 0.01 and 0.05 mg/ml of voriconazole to neutrophils enhanced killing of *C. neoformans* by 51% and 71%, respectively. If granulocyte colony-stimulating factor (G-CSF) was also added to the mix, then the ability of neutrophils to kill the fungus was further enhanced to over 92%. These data suggest that the use of voriconazole in combination with G-CSF in a clinical setting would have good efficacy in the treatment of *C. neoformans* infections especially in environments in which the fungus is surrounded by neutrophils.

While it is appreciated that neutrophils kill *C. neoformans* via generation of fungal oxidants, use of inhibitors and scavengers of the respiratory burst with neutrophils only partially reversed the anti-cryptococcal activity, suggesting that both oxidative and nonoxidative mechanisms are operative (Mambula et al. 2000). To define the mediators of nonoxidative anti-cryptococcal activity, neutrophils were fractionated into their granular fractions. Antifungal activity of the primary granules was significantly reduced by pronase treatment, boiling, high ionic strength and magnesium, but not calcium treatment. Fractionation of the primary granules by HPLC identified primarily defensins, but displayed several overlapping multiple compounds, confirming that multiple proteins, including defensins, are responsible for the activity of primary granules.

A plethora of articles exist on the well-studied methods that neutrophils employ in their role to overcome a *C. neoformans* attack as the second line of defense, but how the secretion of antimicrobial factors of neutrophils is regulated is still not fully understood. Recent discoveries however highlight that the neutrophil sphingolipid pathway may be involved in the regulation of such secretion.

## 13.4 Bioactive Lipids

Sphingolipids not only serve as integral components of eukaryotic cell membranes (Dickson and Lester 1999; Hannun and Luberto 2000), but also act as signaling molecules in many cellular functions, playing critical roles in regulating patho-biological processes in several diseases including inflammation and infectious diseases.

In microbial cells, sphingolipids have been implicated in heat stress response (Jenkins et al. 1997; Patton et al. 1992), endocytosis (Munn and Riezman 1994; Zanolari et al. 2000), signal transduction (Obeid et al. 2002), apoptosis (Cheng et al. 2003), and fungal pathogenesis (Cheng et al. 2003; Luberto et al. 2001). In *C. albicans*, for example, a systematic screen of a homozygous deletion library revealed that the glycolipid glucosylceramide is required specifically for virulence (Noble et al. 2010). In *C. neoformans*, a novel regulatory pathway initiated by the sphingolipid enzyme inositol phosphoryl ceramide (IPC) synthase (Ipc1) has been identified (Heung et al. 2004, 2005). The effects of Ipc1 on melanin production was found to be potentially propagated through the production of diacylglycerol (DAG) and subsequent activation of protein kinase C1 (Pkc1) through its C1 domain. This putative Ipc1–DAG–Pkc1 pathway was the first evidence that the sphingolipid pathway is critical to the regulation of fungal virulence and established a key role for DAG generated from sphingolipid metabolism (Heung et al. 2005). In addition to Ipc1, other enzymes of the sphingolipid pathway have also been identified as essential regulators of *C. neoformans* growth in the intracellular environment. The inositol phosphosphingolipid-phospholipase C1 (Isc1), an enzyme that metabolizes fungal inositol sphingolipids into phytoceramide, is essential for intracellular growth, as it provides protection from acidic, oxidative, and nitrosative stresses of the phagolysosome microenvironment through Pma1-dependent mechanism(s) (Garcia et al. 2008; Shea et al. 2006). While Isc1 is essential for *C. neoformans* survival within the host phagolysosome, glycosylceramide synthase (GCS) is essential for extracellular growth (Kechichian et al. 2007; Rittershaus et al. 2006). GCS is involved in controlling cell-cycle progression and fungal growth in environments characterized by a neutral/alkaline pH and physiological concentrations of $CO_2$, such as the lung alveolar spaces and bloodstream. The *ΔGCS* mutant strain is contained within lung granulomas and therefore does not disseminate to other organs.

In protozoa, the sphingolipid pathway is dependent on the organism and is not conserved among all members of the subkingdom. For example, the etiologic agents of malaria, *Plasmodium* species, produce mammalian-like sphingolipids, while kinetoplastid protozoa such as *Giardia*, *Leishmania*, and *Trypanosoma* species contain fungal-like sphingolipids. With the sequencing of the *Plasmodium falciparum* clone 3D7 genome and the *Leishmania major* (Friedlin strain) genome completed, the fields of genomics and proteomics can be used to help elucidate the sphingolipid-metabolizing enzymes of these protozoa (Gardner et al. 2002; Ivens et al. 2005). Glucosylceramide plays a vital role in regulating cell cycle progression, membrane trafficking, and stage differentiation in *G. lamblia* (Stefanic et al. 2010). Indeed, current data on sphingolipids in pathogenic protozoans illustrates the roles these lipids play in multiple aspects of cellular life including differentiation, replication, trafficking, and the synthesis of virulence factors (Stefanic et al. 2010; Richmond et al. 2010; Zhang et al. 2010). In the intracellular protozoan *Plasmodium falciparum*, which resides in red blood cells and is the most virulent of the four *Plasmodium* species infecting humans, both sphinogomyelin synthase (SMS) (Haldar et al. 1991; Elmendorf and Haldar 1993, 1994; Ansorge et al. 1995) and

GCS (Couto et al. 2004; Gerold and Schwarz 2001) activities have been observed, suggesting that *Plasmodium* has a conserved mammalian-like sphingolipid biosynthetic pathway. Inhibition of these two enzymes leads to increased intracellular concentrations of ceramide and results in growth inhibition of the protozoan, suggesting that both these enzymes are crucial to the survival and virulence of *Plasmodium* (Pankova-Kholmyansky and Flescher 2006). This is important because exogenously added ceramide and sphingomyelinase cause a dose- and time-dependent inhibition of *P. falciparum* growth (Pankova-Kholmyansky et al. 2003), suggesting that this approach could be exploited as potential alternative therapeutic strategies.

The kinetoplastid protozoan *Leishmania* live extracellularly within the sand fly midgut, but within an acidified, fusogenic phagolysosome of the macrophage within the mammalian host (Zhang et al. 2010). The predominant sphingolipids reported in *L. donovani*, the species associated with fatal visceral infections (and eventually among all the *Leishmania* species), are the IPCs, more typical to those found in fungi than in mammalian hosts (Hsu et al. 2007; Wassef et al. 1985; Kaneshiro et al. 1986). Among the protozoan surface membrane glycolipids, which include lipophosphoglycan and glycosylinositolphospholipids, IPCs have been found to be the most abundant class (Denny et al. 2001; Ralton et al. 2002). However, no IPC synthase has been identified in this parasite so far.

In host cells, the acid sphingomyelinase-ceramide system plays an important role in controlling the infection by bacterial pathogens such as *Neisseria gonorrhoeae*, *Escherichia coli*, *Staphylococcus aureus*, *Listeria monocytogenes*, *Salmonella typhimurium*, and *Pseudomonas aeruginosa* (Grassme et al. 2008). In the case of *P. aeruginosa*, formation of plasma membrane ceramide-enriched platforms (lipid rafts) enables internalization of the bacterium, which triggers apoptosis of the host cell and ultimately kills the pathogen. Infection by *P. aeruginosa* is often associated with cystic fibrosis, with mutations in the cystic fibrosis transmembrane conductance regulator (CFTR) protein. Localization of CFTR protein to lipid rafts is necessary for uptake of *P. aeruginosa* by host cells. In CF patients, the mutated form of CFTR protein does not localize to rafts, altering the microbe internalization and increasing the extracellular proliferation and severity of the disease (Grassme et al. 2008; Kowalski and Pier 2004).

Another intriguing sphingolipid metabolizing activity in host cells is carried out by SMS which is encoded by two genes: *SMS1* and *SMS2* (Huitema et al. 2004; Tafesse et al. 2006, 2007). SMS transfers a choline phosphate moiety from phosphatidylcholine (PC) to ceramide, thereby producing sphingomyelin (SM) and DAG (Garcia et al. 2008; Tafesse et al. 2007; Villani et al. 2008). This enzyme is particularly important because not only does it produce SM, a key component of cellular membranes, but also because it regulates the level of two bioactive lipid molecules such as ceramide and DAG. Since (a) ceramide can regulate transcription factors, such as NF-kB involved in cytokine production (Miskolci et al. 2003), (b) DAG controls antifungal activity by neutrophils through reactive oxygen species (ROS) production (Graham et al. 2007), and (c) SM has been implicated in controlling the host immune response in phagocytic cells (Gutierrez et al. 2009), the role of SMS in neutrophils against *C. neoformans* is important to understand.

Recently it has been identified that the host sphingolipid pathway in neutrophils is required to exert their killing activity on *C. neoformans* (Qureshi et al. 2010). In particular, the inhibition of SMS activity profoundly impaired the killing ability of neutrophils by preventing the extracellular release of an anticryptococcal factor(s). Indeed, inhibition of protein kinase D1 (Pkd1), which controls vesicular sorting and secretion, and which is regulated by DAG produced by SMS, also totally blocked the extracellular killing activity. The expression of *SMS* genes, SMS activity, and the levels of the lipids regulated by SMS (namely SM and DAG) as measured by electrospray ionization-mass spectrometry (ESI-MS), were up-regulated during neutrophil differentiation (using the HL-60 cell model system) (Qureshi et al. 2010). Therefore it is hypothesized that secretion of anti-cryptococcal factors, such as MPO, defensins, and others, may be under the control of the SMS-DAG-Pkd1 pathway. The use of matrix-assisted laser desorption-ionization mass spectrometric imaging (MALDI-MSI) of CBA/J mouse lung tissue infected with *C. neoformans* revealed that specific SM species were associated with infiltration of neutrophils at the infection site. This study established a key role for SMS in the regulation of the killing activity of neutrophils against *C. neoformans* through a DAG-PKD1-dependent mechanism, and provided new insights into the protective role of host sphingolipids against a cryptococcal infection.

These studies open up a variety of exciting avenues to pursue. An even better understanding of the cryptococcal–neutrophil interaction will delineate how the anti-cryptococcal factor(s) interact with *C. neoformans*. Furthermore, given the antifungal activity of SMS in neutrophils, an up-regulation of SMS activity (e.g., by lentiviral expression) in the HL-60 cell model system may enhance their killing activity toward *Cryptococcus* and other microorganisms that cause infection. This strategy could be employed to replenish the killing activity of these cells, especially in conditions of immunodeficiency in which the antimicrobial activity of resident neutrophils could be impaired. A cell-based immunotherapy that can recapitulate neutrophil functions in neutropenic individuals afflicted with a microbial infection is under development by stably transfecting HL-60 cells with a suicide trap to enable purging of the cells when desired, as well as a bioluminescence marker in order to track cells in vivo in mice (Lin et al. 2010). The results of these studies lay the groundwork for continued translational development of these promising new technologies for the treatment of those infections that are resistant to current medications in the neutropenic host.

In conclusion, this is only the beginning of our understanding of neutrophil–*C. neoformans* interactions in terms of lipid involvement. The availability of *C. neoformans* and human genome sequences with the progressive applications of classical and new innovative molecular, biochemical, and metabolic tools will enable researchers to provide new insights into this battle and potentially reveal new means to enhance the killing activity of neutrophils under conditions of immunodeficiency. For instance, microarray studies have identified new and validated previously discovered factors involved in the regulation of intracellular growth (Fan et al. 2005). Studies using gene deletion libraries have identified new *C. neoformans* genes important to lung infectivity, suggesting that there are additional players in the *C. neoformans* host lung interaction that need to be studied (Price et al. 2008; Moyrand et al. 2007, 2008; Liu et al. 2006, 2008). Finally, the use of host and fungal sphingolipid arrays may add a new dimension of the fungus–neutrophil interaction that will help to integrate the understanding of such a complex and intriguing relationship.

**Acknowledgments** This work was supported in part by NIH Grant R01-AI56168 and R01-AI71142 (to M.D.P) and was conducted in a facility constructed with support from the National Institutes of Health, Grant Number C06 RR015455 from the Extramural Research Facilities Program of the National Center for Research Resources. Maurizio Del Poeta is a Burroughs Welcome New Investigator in Pathogenesis of Infectious Diseases.

## References

Alcouloumre MS, Ghannoum MA, Ibrahim AS, Selsted ME, Edwards JE Jr (1993) Fungicidal properties of defensin NP-1 and activity against *Cryptococcus neoformans* in vitro. Antimicrob Agents Chemother 37:2628–2632

Alspaugh JA, Nichols CB, Xue C, Shen WC, Wang P (2011) G-Protein signaling pathways: regulating morphogenesis and virulence of *Cryptococcus*. In: Heitman J, Kozel TR, Kwon-Chung KJ, Perfect JR, Casadevall A (eds) Cryptococcus: from human pathogen to model yeast. ASM, Washington, DC

Alvarez M, Casadevall A (2006) Phagosome extrusion and host-cell survival after *Cryptococcus neoformans* phagocytosis by macrophages. Curr Biol 16:2161–2165

Ansorge I, Jeckel D, Wieland F, Lingelbach K (1995) *Plasmodium falciparum*-infected erythrocytes utilize a synthetic truncated ceramide precursor for synthesis and secretion of truncated sphingomyelin. Biochem J 308:335–341

Aratani Y (2006) Role of myeloperoxidase in the host defense against fungal infection. Nihon Ishinkin Gakkai zasshi 47:195–199

Aratani Y, Koyama H, Nyui S, Suzuki K, Kura F, Maeda N (1999) Severe impairment in early host defense against *Candida albicans* in mice deficient in myeloperoxidase. Infect Immun 67:1828–1836

Aratani Y, Kura F, Watanabe H, Akagawa H, Takano Y, Suzuki K, Dinauer MC, Maeda N, Koyama H (2002a) Critical role of myeloperoxidase and nicotinamide adenine dinucleotide phosphate-oxidase in high-burden systemic infection of mice with *Candida albicans*. J Infect Dis 185:1833–1837

Aratani Y, Kura F, Watanabe H, Akagawa H, Takano Y, Suzuki K, Dinauer MC, Maeda N, Koyama H (2002b) Relative contributions of myeloperoxidase and NADPH-oxidase to the early host defense against pulmonary infections with *Candida albicans* and *Aspergillus fumigatus*. Med Mycol 40:557–563

Aratani Y, Kura F, Watanabe H, Akagawa H, Takano Y, Ishida-Okawara A, Suzuki K, Maeda N, Koyama H (2006) Contribution of the myeloperoxidase-dependent oxidative system to host defence against *Cryptococcus neoformans*. J Med Microbiol 55:1291–1299

Bender BS, Davidson BL, Kline R, Brown C, Quinn TC (1988) Role of the mononuclear phagocyte system in the immunopathogenesis of human immunodeficiency virus infection and the acquired immunodeficiency syndrome. Rev Infect Dis 10:1142–1154

Borregaard N, Lollike K, Kjeldsen L, Sengelov H, Bastholm L, Nielsen MH, Bainton DF (1993) Human neutrophil granules and secretory vesicles. Eur J Haematol 51:187–198

Bose I, Reese AJ, Ory JJ, Janbon G, Doering TL (2003) A yeast under cover: the capsule of *Cryptococcus neoformans*. Eukaryot Cell 2:655–663

Breitman TR, Selonick SE, Collins SJ (1980) Induction of differentiation of the human promyelocytic leukemia cell line (HL-60) by retinoic acid. Proc Natl Acad Sci USA 77 (5):2936–2940

Brummer E (1998) Human defenses against *Cryptococcus neoformans*: an update. Mycopathologia 143:121–125

Buschke A (1895) Uber eine durch Coccidien hervorgerufene Krankheit des Menschen. Dtsch Med Wochenschr 21:14

Busse O (1894) Uber parasitare Zelleinschlusse und ihre Zuchtung. Zentralbl Bakteriol 16:175–180

Cannistra SA, Griffin JD (1988) Regulation of the production and function of granulocytes and monocytes. Semin Hematol 25:173–188

Cheng J, Park TS, Chio LC, Fischl AS, Ye XS (2003) Induction of apoptosis by sphingoid long-chain bases in *Aspergillus nidulans*. Mol Cell Biol 23:163–177

Chiller T, Farrokhshad K, Brummer E, Stevens DA (2002) Effect of granulocyte colony-stimulating factor and granulocyte-macrophage colony-stimulating factor on polymorphonuclear neutrophils, monocytes or monocyte-derived macrophages combined with voriconazole against *Cryptococcus neoformans*. Med Mycol 40:21–26

Collins SJ (1987) The HL-60 promyelocytic leukemia cell line: proliferation, differentiation, and cellular oncogene expression. Blood 70:1233–1244

Collins SJ, Gallo RC, Gallagher RE (1977) Continuous growth and differentiation of human myeloid leukaemic cells in suspension culture. Nature 270:347–349

Collins SJ, Ruscetti FW, Gallagher RE, Gallo RC (1978) Terminal differentiation of human promyelocytic leukemia cells induced by dimethyl sulfoxide and other polar compounds. Proc Natl Acad Sci USA 75:2458–2462

Collins SJ, Ruscetti FW, Gallagher RE, Gallo RC (1979) Normal functional characteristics of cultured human promyelocytic leukemia cells (HL-60) after induction of differentiation by dimethylsulfoxide. J Exp Med 149:969–974

Couto AS, Caffaro C, Uhrig ML, Kimura E, Peres VJ, Merino EF, Katzin AM, Nishioka M, Nonami H, Erra-Balsells R (2004) Glycosphingolipids in *Plasmodium falciparum*. Presence of an active glucosylceramide synthase. Eur J Biochem 271:2204–2214

Denny PW, Field MC, Smith DF (2001) GPI-anchored proteins and glycoconjugates segregate into lipid rafts in Kinetoplastida. FEBS Lett 491:148–153

Diamond RD, Erickson NF III (1982) Chemotaxis of human neutrophils and monocytes induced by *Cryptococcus neoformans*. Infect Immun 38:380–382

Dickson RC, Lester RL (1999) Metabolism and selected functions of sphingolipids in the yeast *Saccharomyces cerevisiae*. Biochim Biophys Acta 1438:305–321

Ellerbroek PM, Walenkamp AM, Hoepelman AI, Coenjaerts FE (2004) Effects of the capsular polysaccharides of *Cryptococcus neoformans* on phagocyte migration and inflammatory mediators. Curr Med Chem 11:253–266

Elmendorf HG, Haldar K (1993) Identification and localization of ERD2 in the malaria parasite *Plasmodium falciparum*: separation from sites of sphingomyelin synthesis and implications for organization of the Golgi. EMBO J 12:4763–4773

Elmendorf HG, Haldar K (1994) *Plasmodium falciparum* exports the Golgi marker sphingomyelin synthase into a tubovesicular network in the cytoplasm of mature erythrocytes. J Cell Biol 124:449–462

Fan W, Kraus PR, Boily MJ, Heitman J (2005) *Cryptococcus neoformans* gene expression during murine macrophage infection. Eukaryot Cell 4:1420–1433

Faurschou M, Borregaard N (2003) Neutrophil granules and secretory vesicles in inflammation. Microbes Infect 5:1317–1327

Feldmesser M, Kress Y, Novikoff P, Casadevall A (2000) *Cryptococcus neoformans* is a facultative intracellular pathogen in murine pulmonary infection. Infect Immun 68:4225–4237

Feldmesser M, Tucker S, Casadevall A (2001) Intracellular parasitism of macrophages by *Cryptococcus neoformans*. Trends Microbiol 9:273–278

Fox DS, Djordjevic JT, Sorrell T (2011) Signaling cascades and enzymes as *Cryptococcus* virulence factors. In: Heitman J, Kozel TR, Kwon-Chung KJ, Perfect JR, Casadevall A (eds) Cryptococcus: from human pathogen to model yeast. ASM, Washington, DC

Gallagher R, Collins S, Trujillo J, McCredie K, Ahearn M, Tsai S, Metzgar R, Aulakh G, Ting R, Ruscetti F, Gallo R (1979) Characterization of the continuous, differentiating myeloid cell line (HL-60) from a patient with acute promyelocytic leukemia. Blood 54:713–733

Garcia J, Shea J, Alvarez-Vasquez F, Qureshi A, Luberto C, Voit EO, Del Poeta M (2008) Mathematical modeling of pathogenicity of *Cryptococcus neoformans*. Mol Syst Biol 4:183

Gardner MJ, Hall N, Fung E, White O, Berriman M, Hyman RW, Carlton JM, Pain A, Nelson KE, Bowman S, Paulsen IT, James K, Eisen JA, Rutherford K, Salzberg SL, Craig A, Kyes S, Chan MS, Nene V, Shallom SJ, Suh B, Peterson J, Angiuoli S, Pertea M, Allen J, Selengut J, Haft D, Mather MW, Vaidya AB, Martin DM, Fairlamb AH, Fraunholz MJ, Roos DS, Ralph SA, McFadden GI, Cummings LM, Subramanian GM, Mungall C, Venter JC, Carucci DJ, Hoffman SL, Newbold C, Davis RW, Fraser CM, Barrell B (2002) Genome sequence of the human malaria parasite *Plasmodium falciparum*. Nature 419:498–511

Gerold P, Schwarz RT (2001) Biosynthesis of glycosphingolipids de-novo by the human malaria parasite *Plasmodium falciparum*. Mol Biochem Parasitol 112:29–37

Goldman D, Lee SC, Casadevall A (1994) Pathogenesis of pulmonary *Cryptococcus neoformans* infection in the rat. Infect Immun 62:4755–4761

Graham DB, Robertson CM, Bautista J, Mascarenhas F, Diacovo MJ, Montgrain V, Lam SK, Cremasco V, Dunne WM, Faccio R, Coopersmith CM, Swat W (2007) Neutrophil-mediated oxidative burst and host defense are controlled by a Vav-PLCgamma2 signaling axis in mice. J Clin Invest 117:3445–3452

Grassme H, Becker KA, Zhang Y, Gulbins E (2008) Ceramide in bacterial infections and cystic fibrosis. Biol Chem 389:1371–1379

Gutierrez MG, Gonzalez AP, Anes E, Griffiths G (2009) Role of lipids in killing mycobacteria by macrophages: evidence for NF-kappaB-dependent and -independent killing induced by different lipids. Cell Microbiol 11:406–420

Haldar K, Uyetake L, Ghori N, Elmendorf HG, Li WL (1991) The accumulation and metabolism of a fluorescent ceramide derivative in *Plasmodium falciparum*-infected erythrocytes. Mol Biochem Parasitol 49:143–156

Hannun YA, Luberto C (2000) Ceramide in the eukaryotic stress response. Trends Cell Biol 10:73–80

Harris P, Ralph P (1985) Human leukemic models of myelomonocytic development: a review of the HL-60 and U937 cell lines. J Leukoc Biol 37:407–422

Hauert AB, Martinelli S, Marone C, Niggli V (2002) Differentiated HL-60 cells are a valid model system for the analysis of human neutrophil migration and chemotaxis. Int J Biochem Cell Biol 34:838–854

Herwig S, Su Q, Tempst P (1998) Drug-activated multiple pathways of defensin mRNA regulation in HL-60 cells are defined by reversed roles of participating protein kinases. Leuk Res 22:913–925

Heung LJ, Luberto C, Plowden A, Hannun YA, Del Poeta M (2004) The sphingolipid pathway regulates Pkc1 through the formation of diacylglycerol in *Cryptococcus neoformans*. J Biol Chem 279:21144–21153

Heung LJ, Kaiser AE, Luberto C, Del Poeta M (2005) The role and mechanism of diacylglycerol-protein kinase C1 signaling in melanogenesis by *Cryptococcus neoformans*. J Biol Chem 280:28547–28555

Hsu FF, Turk J, Zhang K, Beverley SM (2007) Characterization of inositol phosphorylceramides from *Leishmania major* by tandem mass spectrometry with electrospray ionization. J Am Soc Mass Spectrom 18:1591–1604

Huffnagle GB, Yates JL, Lipscomb MF (1991a) Immunity to a pulmonary *Cryptococcus neoformans* infection requires both CD4+ and CD8+ T cells. J Exp Med 173:793–800

Huffnagle GB, Yates JL, Lipscomb MF (1991b) T cell-mediated immunity in the lung: a *Cryptococcus neoformans* pulmonary infection model using SCID and athymic nude mice. Infect Immun 59:1423–1433

Huitema K, van den Dikkenberg J, Brouwers JF, Holthuis JC (2004) Identification of a family of animal sphingomyelin synthases. EMBO J 23:33–44

Idnurm A, Bahn YS, Shen WC, Rutherford JC, Muhlschlegel FA (2011) Sensing extracellular signals in *Cryptococcus neoformans*. In: Heitman J, Kozel TR, Kwon-Chung KJ, Perfect JR, Casadevall A (eds) Cryptococcus: from human pathogen to model yeast. ASM, Washington, DC

Itoh K, Okubo K, Utiyama H, Hirano T, Yoshii J, Matsubara K (1998) Expression profile of active genes in granulocytes. Blood 92:1432–1441

Ivens AC, Peacock CS, Worthey EA, Murphy L, Aggarwal G, Berriman M, Sisk E, Rajandream MA, Adlem E, Aert R, Anupama A, Apostolou Z, Attipoe P, Bason N, Bauser C, Beck A, Beverley SM, Bianchettin G, Borzym K, Bothe G, Bruschi CV, Collins M, Cadag E, Ciarloni L, Clayton C, Coulson RM, Cronin A, Cruz AK, Davies RM, De Gaudenzi J, Dobson DE, Duesterhoeft A, Fazelina G, Fosker N, Frasch AC, Fraser A, Fuchs M, Gabel C, Goble A, Goffeau A, Harris D, Hertz-Fowler C, Hilbert H, Horn D, Huang Y, Klages S, Knights A, Kube M, Larke N, Litvin L, Lord A, Louie T, Marra M, Masuy D, Matthews K, Michaeli S, Mottram JC, Muller-Auer S, Munden H, Nelson S, Norbertczak H, Oliver K, O'Neil S, Pentony M, Pohl TM, Price C, Purnelle B, Quail MA, Rabbinowitsch E, Reinhardt R, Rieger M, Rinta J, Robben J, Robertson L, Ruiz JC, Rutter S, Saunders D, Schafer M, Schein J, Schwartz DC, Seeger K, Seyler A, Sharp S, Shin H, Sivam D, Squares R, Squares S, Tosato V, Vogt C, Volckaert G, Wambutt R, Warren T, Wedler H, Woodward J, Zhou S, Zimmermann W, Smith DF, Blackwell JM, Stuart KD, Barrell B, Myler PJ (2005) The genome of the kinetoplastid parasite, *Leishmania major*. Science 309:436–442

Jenkins GM, Richards A, Wahl T, Mao C, Obeid L, Hannun Y (1997) Involvement of yeast sphingolipids in the heat stress response of *Saccharomyces cerevisiae*. J Biol Chem 272:32566–32572

Jesus MD, Nicola AM, Chow SK, Lee IR, Nong S, Specht CA, Levitz SM, Casadevall A (2010) Glucuronoxylomannan, galactoxylomannan, and mannoprotein occupy spatially separate and discrete regions in the capsule of *Cryptococcus neoformans*. Virulence 1:500–508

Kaneshiro ES, Jayasimhulu K, Lester RL (1986) Characterization of inositol lipids from *Leishmania donovani* promastigotes: identification of an inositol sphingophospholipid. J Lipid Res 27:1294–1303

Kechichian TB, Shea J, Del Poeta M (2007) Depletion of alveolar macrophages decreases the dissemination of a glucosylceramide-deficient mutant of *Cryptococcus neoformans* in immunodeficient mice. Infect Immun 75:4792–4798

Kwon-Chung KJ, Boekhout T, Wickes BL, Fell JW (2011) Systematics of the genus Cryptococcus and its type species C. neoformans. In: Heitman J, Kozel TR, Kwon-Chung KJ, Perfect JR, Casadevall A (eds) Cryptococcus: from human pathogen to model yeast. ASM, Washington, DC

Kowalski MP, Pier GB (2004) Localization of cystic fibrosis transmembrane conductance regulator to lipid rafts of epithelial cells is required for *Pseudomonas aeruginosa*-induced cellular activation. J Immunol 172:418–425

Kozel TR, Pfrommer GS, Guerlain AS, Highison BA, Highison GJ (1988) Role of the capsule in phagocytosis of *Cryptococcus neoformans*. Rev Infect Dis 10(Suppl 2):S436–S439

Lee SC, Kress Y, Zhao ML, Dickson DW, Casadevall A (1995) *Cryptococcus neoformans* survive and replicate in human microglia. Lab Invest 73:871–879

Li SS, Mody CH (2010) Cryptococcus. Proc Am Thorac Soc 7:186–196

Lin L, Ibrahim AS, Baquir B, Fu Y, Applebaum D, Schwartz J, Wang A, Avanesian V, Spellberg B (2010) Safety and efficacy of activated transfected killer cells for neutropenic fungal infections. J Infect Dis 201:1708–1717

Liu X, Hu G, Panepinto J, Williamson PR (2006) Role of a VPS41 homologue in starvation response, intracellular survival and virulence of *Cryptococcus neoformans*. Mol Microbiol 61:1132–1146

Liu OW, Chun CD, Chow ED, Chen C, Madhani HD, Noble SM (2008) Systematic genetic analysis of virulence in the human fungal pathogen *Cryptococcus neoformans*. Cell 135:174–188

Luberto C, Toffaletti DL, Wills EA, Tucker SC, Casadevall A, Perfect JR, Hannun YA, Del Poeta M (2001) Roles for inositol-phosphoryl ceramide synthase 1 (IPC1) in pathogenesis of *C. neoformans*. Genes Dev 15:201–212

Luberto C, Martinez-Marino B, Taraskiewicz D, Bolanos B, Chitano P, Toffaletti DL, Cox GM, Perfect JR, Hannun YA, Balish E, Del Poeta M (2003) Identification of App 1 as a regulator of phagocytosis and virulence of *Cryptococcus neoformans*. J Clin Invest 112:1080–1094

Ma H, Croudace JE, Lammas DA, May RC (2006) Expulsion of live pathogenic yeast by macrophages. Curr Biol 16:2156–2160

Maggini J, Raiden S, Salamone G, Trevani A, Geffner J (2009) Regulation of neutrophil apoptosis by cytokines, pathogens and environmental stressors. Front Biosci 14:2372–2385

Malech HL, Gallin JI (1987) Current concepts: immunology. Neutrophils in human diseases. New Engl J Med 317:687–694

Mambula SS, Simons ER, Hastey R, Selsted ME, Levitz SM (2000) Human neutrophil-mediated nonoxidative antifungal activity against *Cryptococcus neoformans*. Infect Immun 68:6257–6264

Mansour MK, Levitz SM (2002) Interactions of fungi with phagocytes. Curr Opin Microbiol 5:359–365

McQuiston T, Del Poeta M (2011) The interaction of *Cryptococcus neoformans* with host macrophages and neutrophils. In: Heitman J, Kozel TR, Kwon-Chung KJ, Perfect JR, Casadevall A (eds) Cryptococcus: from human pathogen to model yeast. ASM, Washington, DC

McQuiston T, Luberto C, Del Poeta M (2010) Role of host sphingosine kinase 1 in the lung response against Cryptococcosis. Infect Immun 78:2342–2352

Miskolci V, Castro-Alcaraz S, Nguyen P, Vancura A, Davidson D, Vancurova I (2003) Okadaic acid induces sustained activation of NFkappaB and degradation of the nuclear IkappaBalpha in human neutrophils. Arch Biochem Biophys 417:44–52

Monari C, Casadevall A, Pietrella D, Bistoni F, Vecchiarelli A (1999) Neutrophils from patients with advanced human immunodeficiency virus infection have impaired complement receptor function and preserved Fcgamma receptor function. J Infect Dis 180:1542–1549

Monari C, Retini C, Casadevall A, Netski D, Bistoni F, Kozel TR, Vecchiarelli A (2003) Differences in outcome of the interaction between *Cryptococcus neoformans* glucuronoxylomannan and human monocytes and neutrophils. Eur J Immunol 33:1041–1051

Moyrand F, Fontaine T, Janbon G (2007) Systematic capsule gene disruption reveals the central role of galactose metabolism on *Cryptococcus neoformans* virulence. Mol Microbiol 64:771–781

Moyrand F, Lafontaine I, Fontaine T, Janbon G (2008) UGE1 and UGE2 regulate the UDP-glucose/UDP-galactose equilibrium in *Cryptococcus neoformans*. Eukaryot Cell 7:2069–2077

Mukherjee AB, Czirbik RJ, Parsa NZ, Testa JR (1985) Induction of terminal differentiation and nuclear appendage(s) formation in a human myeloid leukaemia cell line (HL-60). Cytobios 44:109–118

Munn AL, Riezman H (1994) Endocytosis is required for the growth of vacuolar H(+)-ATPase-defective yeast: identification of six new END genes. J Cell Biol 127:373–386

Newburger PE, Chovaniec ME, Greenberger JS, Cohen HJ (1979) Functional changes in human leukemic cell line HL-60. A model for myeloid differentiation. J Cell Biol 82:315–322

Nielsen K, Kwon-Chung KJ (2011) A role for mating in *Cryptococcal virulence*. In: Heitman J, Kozel TR, Kwon-Chung KJ, Perfect JR, Casadevall A (eds) Cryptococcus: from human pathogen to model yeast. ASM, Washington, DC

Noble SM, French S, Kohn LA, Chen V, Johnson AD (2010) Systematic screens of a *Candida albicans* homozygous deletion library decouple morphogenetic switching and pathogenicity. Nat Genet 42:590–598

Obeid LM, Okamoto Y, Mao C (2002) Yeast sphingolipids: metabolism and biology. Biochim Biophys Acta 1585:163–171

Pankova-Kholmyansky I, Flescher E (2006) Potential new antimalarial chemotherapeutics based on sphingolipid metabolism. Chemotherapy 52:205–209

Pankova-Kholmyansky I, Dagan A, Gold D, Zaslavsky Z, Skutelsky E, Gatt S, Flescher E (2003) Ceramide mediates growth inhibition of the *Plasmodium falciparum* parasite. Cell Mol Life Sci 60:577–587

Patton JL, Srinivasan B, Dickson RC, Lester RL (1992) Phenotypes of sphingolipid-dependent strains of *Saccharomyces cerevisiae*. J Bacteriol 174:7180–7184

Pfaller MA, Lodge JK, Ghannoum MA (2011) Drug resistance in *Cryptococcus*: epidemiology and molecular mechanisms. In: Heitman J, Kozel TR, Kwon-Chung KJ, Perfect JR, Casadevall A (eds) Cryptococcus: from human pathogen to model yeast. ASM, Washington, DC

Price MS, Nichols CB, Alspaugh JA (2008) The *Cryptococcus neoformans* Rho-GDP dissociation inhibitor mediates intracellular survival and virulence. Infect Immun 76(12):5729–5737

Qureshi A, Subathra M, Grey A, Schey K, Del Poeta M, Luberto C (2010) Role of sphingomyelin synthase in controlling the antimicrobial activity of neutrophils against *Cryptococcus neoformans*. PLoS One 5(12):e15587

Ralton JE, Mullin KA, McConville MJ (2002) Intracellular trafficking of glycosylphosphatidylinositol (GPI)-anchored proteins and free GPIs in *Leishmania mexicana*. Biochem J 363:365–375

Ratcliffe DR, Nolin SL, Cramer EB (1988) Neutrophil interaction with influenza-infected epithelial cells. Blood 72:142–149

Richmond GS, Gibellini F, Young SA, Major L, Denton H, Lilley A, Smith TK (2010) Lipidomic analysis of bloodstream and procyclic form *Trypanosoma brucei*. Parasitology 137:1357–1392

Rittershaus PC, Kechichian TB, Allegood JC, Merrill AH Jr, Hennig M, Luberto C, Del Poeta M (2006) Glucosylceramide synthase is an essential regulator of pathogenicity of *Cryptococcus neoformans*. J Clin Invest 116:1651–1659

Sanfelice F (1894) Contributo alla morfologia e biologia dei blastomiceti che si sviluppano nei succhi di alcuni frutti. Ann Igien 4:463–495

Schaffner A, Davis CE, Schaffner T, Markert M, Douglas H, Braude AI (1986) In vitro susceptibility of fungi to killing by neutrophil granulocytes discriminates between primary pathogenicity and opportunism. J Clin Invest 78:511–524

Schleimer RP, Freeland HS, Peters SP, Brown KE, Derse CP (1989) An assessment of the effects of glucocorticoids on degranulation, chemotaxis, binding to vascular endothelium and formation of leukotriene B4 by purified human neutrophils. J Pharmacol Exp Therap 250:598–605

Shao X, Mednick A, Alvarez M, van Rooijen N, Casadevall A, Goldman DL (2005) An innate immune system cell is a major determinant of species-related susceptibility differences to fungal pneumonia. J Immunol 175:3244–3251

Shea JM, Kechichian TB, Luberto C, Del Poeta M (2006) The cryptococcal enzyme inositol phosphosphingolipid-phospholipase C confers resistance to the antifungal effects of macrophages and promotes fungal dissemination to the central nervous system. Infect Immun 74:5977–5988

Smith JA (1994) Neutrophils, host defense, and inflammation: a double-edged sword. J Leukoc Biol 56:672–686

Spellberg BJ, Collins M, French SW, Edwards JE Jr, Fu Y, Ibrahim AS (2005) A phagocytic cell line markedly improves survival of infected neutropenic mice. J Leukoc Biol 78:338–344

Spellberg BJ, Collins M, Avanesian V, Gomez M, Edwards JE Jr, Cogle C, Applebaum D, Fu Y, Ibrahim AS (2007) Optimization of a myeloid cell transfusion strategy for infected neutropenic hosts. J Leukoc Biol 81:632–641

Stano P, Williams V, Villani M, Cymbalyuk ES, Qureshi A, Huang Y, Morace G, Luberto C, Tomlinson S, Del Poeta M (2009) App 1: an antiphagocytic protein that binds to complement receptors 3 and 2. J Immunol 182(1):84–91

Stefanic S, Spycher C, Morf L, Fabrias G, Casas J, Schraner E, Wild P, Hehl AB, Sonda S (2010) Glucosylceramide synthesis inhibition affects cell cycle progression, membrane trafficking, and stage differentiation in *Giardia lamblia*. J Lipid Res 51:2527–2545

Tafesse FG, Ternes P, Holthuis JC (2006) The multigenic sphingomyelin synthase family. J Biol Chem 281:29421–29425

Tafesse FG, Huitema K, Hermansson M, van der Poel S, van den Dikkenberg J, Uphoff A, Somerharju P, Holthuis JC (2007) Both sphingomyelin synthases SMS1 and SMS2 are required for sphingomyelin homeostasis and growth in human HeLa cells. J Biol Chem 282:17537–17547

Tsutsumi-Ishii Y, Hasebe T, Nagaoka I (2000) Role of CCAAT/enhancer-binding protein site in transcription of human neutrophil peptide-1 and -3 defensin genes. J Immunol 164:3264–3273

Tucker SC, Casadevall A (2002) Replication of *Cryptococcus neoformans* in macrophages is accompanied by phagosomal permeabilization and accumulation of vesicles containing polysaccharide in the cytoplasm. Proc Natl Acad Sci USA 99:3165–3170

Urban CF, Lourido S, Zychlinsky A (2006) How do microbes evade neutrophil killing? Cell Microbiol 8:1687–1696

Vecchiarelli A (2000) Immunoregulation by capsular components of *Cryptococcus neoformans*. Med Mycol 38:407–417

Villani M, Subathra M, Im YB, Choi Y, Signorelli P, Del Poeta M, Luberto C (2008) Sphingomyelin synthases regulate production of diacylglycerol at the Golgi. Biochem J 414:31–41

Wang N, Su Q, Boeckh-Herwig S, Yaneva M, Tempst P (2004) Delayed-late activation of a myeloid defensin minimal promoter by retinoids and inflammatory mediators. Leuk Res 28:879–889

Wang Y, Jiang Y, Gong T, Cui X, Li W, Feng Y, Wang B, Jiang Z, Li M (2009) High-level expression and novel antifungal activity of mouse beta defensin-1 mature peptide in *Escherichia coli*. Appl Biochem Biotechnol 160:213–221

Wassef MK, Fioretti TB, Dwyer DM (1985) Lipid analyses of isolated surface membranes of *Leishmania donovani* promastigotes. Lipids 20:108–115

Zanolari B, Friant S, Funato K, Sutterlin C, Stevenson BJ, Riezman H (2000) Sphingoid base synthesis requirement for endocytosis in *Saccharomyces cerevisiae*. EMBO J 19:2824–2833

Zhang K, Bangs JD, Beverley SM (2010) Sphingolipids in parasitic protozoa. Adv Exp Med Biol 688:238–248

# Index




J. Pérez-Martín and A. Di Pietro (eds.), *Morphogenesis and Pathogenicity in Fungi*, Topics in Current Genetics 22, DOI 10.1007/978-3-642-22916-9,

**D**



**E**

## M



## N



## O



## P

MIX
Papier aus verantwortungsvollen Quellen
Paper from responsible sources
FSC® C105338
FSC www.fsc.org

If you have any concerns about our products, you can contact us on
**ProductSafety@springernature.com**

In case Publisher is established outside the EU, the EU authorized representative is:
**Springer Nature Customer Service Center GmbH**
**Europaplatz 3, 69115 Heidelberg, Germany**

Printed by Libri Plureos GmbH
in Hamburg, Germany